DNA Computing Based Genetic Algorithm

Jili Tao · Ridong Zhang · Yong Zhu

DNA Computing Based Genetic Algorithm

Applications in Industrial Process Modeling and Control

Springer

Jili Tao
School of Information Science
and Engineering
NingboTech University
Ningbo, Zhejiang, China

Ridong Zhang
The Belt and Road Information
Research Institute
Hangzhou Dianzi University
Hangzhou, Zhejiang, China

Yong Zhu
School of Information Science
and Engineering
NingboTech University
Ningbo, Zhejiang, China

ISBN 978-981-15-5405-6 ISBN 978-981-15-5403-2 (eBook)
https://doi.org/10.1007/978-981-15-5403-2

This Springer imprint is published by the registered company Springer Nature Singapore Pte Ltd.
The registered company address is: 152 Beach Road, #21-01/04 Gateway East, Singapore 189721, Singapore

Contents

1 Introduction 1
1.1 Standard Genetic Algorithm 1
1.2 State of Art for GA 2
1.2.1 Theoretical Research of GA 2
1.2.2 Encoding Problem of GA 4
1.2.3 Constraint Handling in GA 4
1.2.4 Multi-objective Genetic Algorithm 6
1.2.5 Applications of GA 8
1.3 DNA Computing Based GA 11
1.3.1 DNA Molecular Structure of DNA Computing 11
1.3.2 Biological Operators of DNA Computing 12
1.3.3 DNA Computing Based Genetic Algorithm 13
1.4 The Main Content of This Book 13
References 16

2 DNA Computing Based RNA Genetic Algorithm 25
2.1 Introduction 25
2.2 RNA-GA Based on DNA Computing 26
2.2.1 Digital Encoding of RNA Sequence 26
2.2.2 Operations of RNA Sequence 27
2.2.3 Encoding and Operators in RNA-GA 28
2.2.4 The Procedure of RNA-GA 34
2.3 Global Convergence Analysis of RNA-GA 34
2.4 Performance of the RNA-GA 37
2.4.1 Test Functions 37
2.4.2 Adaptability of the Parameters 39
2.4.3 Comparisons Between RNA-GA and SGA 41
2.5 Summary 52
Appendix 54
References 54

3 DNA Double-Helix and SQP Hybrid Genetic Algorithm 57
3.1 Introduction 57
3.2 Problem Description and Constraint Handling 58
3.3 DNA Double-Helix Hybrid Genetic Algorithm (DNA-DHGA) 59
3.3.1 DNA Double-Helix Encoding 59
3.3.2 DNA Computing Based Operators 61
3.3.3 Hybrid Genetic Algorithm with SQP 64
3.3.4 Convergence Rate Analysis of DNA-DHGA 66
3.4 Numeric Simulation 70
3.4.1 Test Functions 70
3.4.2 Simulation Analysis 70
3.5 Summary 75
Appendix 76
References 78

4 DNA Computing Based Multi-objective Genetic Algorithm 81
4.1 Introduction 81
4.2 Multi-objective Optimization Problems 83
4.3 DNA Computing Based MOGA (DNA-MOGA) 84
4.3.1 RNA Encoding 84
4.3.2 Pareto Sorting and Density Information 85
4.3.3 Elitist Archiving and Maintaining Scheme 86
4.3.4 DNA Computing Based Crossover and Mutation Operators 89
4.3.5 The Procedure of DNA-MOGA 90
4.3.6 Convergence Analysis of DNA-MOGA 90
4.4 Simulations on Test Functions by DNA-MOGA 92
4.4.1 Test Functions and Performance Metrics 92
4.4.2 Calculation Results 93
4.5 Summary 96
Appendix 96
References 99

5 Parameter Identification and Optimization of Chemical Processes 101
5.1 Introduction 101
5.2 Problem Description of System Identification 103
5.2.1 Lumping Models for a Heavy Oil Thermal Cracking Process 104
5.2.2 Parameter Estimation of FCC Unit Main Fractionator 105

5.3 Gasoline Blending Recipe Optimization 111
5.3.1 Formulation of Gasoline Blending Scheduling 111
5.3.2 Optimization Results for Gasoline Blending Scheduling 113
5.4 Summary 116
Appendix 116
References 117

6 GA-Based RBF Neural Network for Nonlinear SISO System 119
6.1 Introduction 119
6.2 The Coke Unit 122
6.3 RBF Neural Network 123
6.4 RNA-GA Based RBFNN for Temperature Modeling 126
6.4.1 Encoding and Decoding 126
6.4.2 Fitness Function 127
6.4.3 Operators of RBFNN Optimization 128
6.4.4 Procedure of the Algorithm 130
6.4.5 Temperature Modeling in a Coke Furnace 131
6.5 Improved MOEA Based RBF Neural Network for Chamber Pressure 135
6.5.1 Encoding of IMOEA 138
6.5.2 Optimization Objectives of RBFNN Model 143
6.5.3 Operators of IMOEA for RBFNN 143
6.5.4 The Procedure of IMOEA 145
6.5.5 The Chamber Pressure Modeling in a Coke Furnace 146
6.6 PCA and INSGA-II Based RBFNN Disturbance Modeling of Chamber Pressure 154
6.6.1 RV Criterion in PCA Variable Selection 154
6.6.2 Encoding of RBFNN 156
6.6.3 Operators of INSGA-II 156
6.6.4 The Procedure of Improved NSGA-II 158
6.6.5 Main Disturbance Modeling of Chamber Pressure 158
6.7 Summary 164
References 164

7 GA Based Fuzzy Neural Network Modeling for Nonlinear SISO System 167
7.1 Introduction 167
7.2 T-S Fuzzy Model 169
7.2.1 T-S Fuzzy ARX Model 169
7.2.2 T-S Fuzzy Plus Tah Function Model 171

7.3 Improved GA based T-S Fuzzy ARX Model Optimization 172
7.3.1 Hybrid Encoding Method 173
7.3.2 Objectives of T-S Fuzzy Modeling 174
7.3.3 Operators of IGA for T-S Fuzzy Model 175
7.3.4 Optimization Procedure 177
7.3.5 Computing Complexity Analysis 177
7.3.6 Oxygen Content Modeling by Fuzzy ARX Model 178
7.4 IGA Based Fuzzy ARX Plus Tanh Function Model 182
7.4.1 Encoding of IGA for Fuzzy Neural Network 182
7.4.2 Operators of IGA for New Fuzzy Model 184
7.4.3 Liquid Level Modeling by Nonlinear Fuzzy Neural Network 185
7.5 Summary 186
References 190

8 PCA and GA Based ARX Plus RBF Modeling for Nonlinear DPS 193
8.1 Introduction 193
8.2 DPS Modeling Issue 194
8.2.1 Time/Space Separation via PCA 195
8.2.2 Decoupled ARX Model Identification 198
8.2.3 RBF Neural Network Modeling 199
8.2.4 Structure and Parameter Optimization by GA 201
8.2.5 Encoding Method 202
8.3 Simulation Results 204
8.3.1 Catalytic Rod 205
8.3.2 Heat Conduction Equation 215
8.4 Summary 216
References 219

9 GA-Based Controller Optimization Design 221
9.1 Introduction 221
9.2 Non-minimal State-Space Predictive Function PID Controller 223
9.2.1 Process Model Formulation 223
9.2.2 PID Controller Design 225
9.2.3 GA-Based Weighting Matrix Tuning 226
9.2.4 The Chamber Pressure Control by PFC-PID 228
9.3 RNA-GA-Based Fuzzy Neuron Hybrid Controller 236
9.3.1 Neuron Controller 236
9.3.2 Simple Fuzzy PI Control 237
9.3.3 Fuzzy Neuron Hybrid Control (FNHC) 239
9.3.4 Parameters Optimization of RNA-GA 240

9.3.5 Continuous Steel Casting Description 241
9.3.6 FNHC Controller Performance Analysis 243
9.4 Stabilization Subspaces Based MOGA for PID Controller Optimization 249
9.4.1 Generalized Hermite-Biehler Theorem 249
9.4.2 Hermite-Biehler Theorem Based PID Controller Stabilizing 250
9.4.3 Optimizing PID Controller Parameters Based on Stabilization Subspaces 253
9.4.4 Simulation for Optimization of PID Controllers 254
9.5 Summary 257
References 258

10 Further Idea on Optimal Q-Learning Fuzzy Energy Controller for FC/SC HEV 261
10.1 Introduction 261
10.2 FC/SC HEV System Description 263
10.3 Q-Learning Based Fuzzy Energy Management Controller 263
10.3.1 Fuzzy Energy Management Controller 264
10.3.2 Q-Learning in HEV Energy Control 267
10.3.3 GA Optimal Q-Learning Algorithm 269
10.3.4 Initial Value Optimization of Q-Table 269
10.3.5 Procedure of Improved Q-Learning Fuzzy EMS 271
10.3.6 Real-Time Energy Management 272
10.4 Summary 273
References 273

Chapter 1
Introduction

This chapter first reviews the research status of genetic algorithm theory, encoding problem, constrained optimization, and multi-objective optimization. Secondly, it briefly introduces the biological basis, the problems of DNA biocomputing, and the significance of the combination of genetic algorithm and DNA computing. Finally, the main work and organizational structure of this book are introduced.

Genetic algorithms (GAs) are one of the evolutionary computing techniques, which have been widely used to solve complex optimization problems that are known to be difficult for traditional calculus-based optimization techniques [1–3]. These traditional optimization methods generally require the problem to possess certain mathematical properties, such as continuity, differentiability, and convexity. [4], which cannot be satisfied in many practical problems [5, 6]. As such, GA, that does not require these requirements, has been considered and often adopted as an efficient optimization tool in many applications.

1.1 Standard Genetic Algorithm

Since professor Holland of Michigan University proposed GA in 1975 [7], it has achieved numerous achievements in solving various complex optimization problems [8–10]. In 1985, the first international conference on genetic algorithm was held in Carnegie Mellon University. In 1989, Goldberg and Michalewicz published an important monograph: Genetic Algorithms in Search, Optimization, and Machine Learning [11]. Since then, genetic algorithms entered a period of vigorous development, the applications of GAs were more extensive, and the theoretical research was more in-depth. After that, IEEE Transactions on Evolutionary Computation started its publication in 1997. Michalewicz described clearly some real number genetic operators and their typical applications in his book: Genetic Algorithms + Data Structures = Evolution Programs [12]. There are a lot of articles in Elsevier, IEEE, and other kinds of databases [13–17]. However, GAs are not always superior for

J. Tao et al., *DNA Computing Based Genetic Algorithm*,
https://doi.org/10.1007/978-981-15-5403-2_1

```
Initialize the parameters and generate the initial population
Calculate the value of the fitness function of each individual
While ( Stop condition is not satisfied)
    Calculate the fitness values of all individuals in the population
    Compute the selection probability of each individual
    Select N individual as the parents of crossover and mutation operators
    for i=0; i<N/2; i++
        Select two parents in terms of selection rate
        Randomly generate r between 0 and 1
        if r>Pc
            Save two parent unchanged to the next population
        else
            Execute crossover operation and produce two children
            Save them to the next population
        end
        Randomly generate r between 0 and 1
        If r<Pm
          Execute the mutation operation and produce a new individual
          Save it to the next population
        end
    end
Output the optimal solution
```

Fig. 1.1 The basic framework of SGA

specific problems in a certain field. A lot of basic research and theoretical innovation is still needed to be further developed. Among all of the various GAs, standard GA (SGA) is a basic one. In SGA, each variable of the problem to be optimized can be described by a binary chromosome with length l, that is, the range of the variable is $[0, 2^l - 1]$. If there are n variables, the length of the binary chromosome becomes $L = n \times l$. By executing the selection, crossover, and mutation operators, the chromosome population will evolve and converge to the optimal ones.

The framework of SGA is shown in Fig. 1.1, where there are four basic parameters, i.e., crossover rate P_c, mutation rate P_m, population size N, and encoding length l. l is set according to the precision requirements for the problem being solved, and the other parameters recommended by Schaffer are that: $N = 20 - 200, P_c = 0.5 - 1.0$, and $P_m = 0 - 0.05$.

1.2 State of Art for GA

1.2.1 Theoretical Research of GA

Though the simplicity of SGA computing is witnessed, its operating mechanism is complicated. With its applications in complex optimization problems and industrial processes, the theoretical basis of GA has obtained more and more attention.

(1) Schema theory [7]: Suppose that the probabilities of crossover and mutation are P_c and P_m, the distance of Schemata S is $\delta(S)$, the length of S is l, the order is $o(S)$, the number of instances of S at generation t is $N(S,t)$, the expected number of instances of a schema S at $t+1$ generation is $E[N(S,t+1)]$, the following inequality will be satisfied: $E[|S \cap p(t+1)|] \geq N(S,t)r(S,t)\left[1 - p_c\frac{\delta(S)}{l-1} - o(S)p_m\right]$, where fitness ratio $r(S,t) = \frac{f(S,t)}{\bar{f}(t)}$ is an important parameter expressing the fitness of a schema $f(S,t)$ relative to the average fitness of the population $\bar{f}(t)$. Readers can refer to [18] to gain the detailed derivation process. The theorem states that the number of schemata with low order, short defined distance, and excess average fitness will increase exponentially with the number of iteration. Poli et al. utilized schema theory and its hidden parallel principle to explain the evolution mechanism of GA, which can be used to analyze the different strategies for encoding and genetic operators [19–21]. However, schema theory was found insufficient and not rigorous [22].

(2) Building block Hypothesis [23]: By implementing selection, crossover, and mutation operators, the gene of an individual can be spliced together, and a new chromosome with higher fitness is generated. The hypothesis explains the basic idea of solving various problems by genetic algorithm, i.e., by splicing between gene blocks, a better solution can be obtained [24, 25]. However, the building block hypothesis has never been proved.

(3) New models of GA

To understand the mechanism of genetic algorithm better, a variety of new models have been constructed, such as the Markov chain model [26], axiomatic model [29, 30], integral operator model [31, 32], and paradox [33, 34], where the most typical one is the Markov chain model. There are mainly three Markov chain models of genetic algorithm: population model [27, 35], Vose model [35, 36], and Cerf perturbed model [37]. The population Markov chain model treats the population iterative sequence as a finite state Markov chain and uses some general properties of the model transition probability matrix to analyze the extreme behavior of the genetic algorithm. Under the assumption of infinite population, Vose and Liepins modeled genetic algorithm selection and recombination as the interleaving of linear and quadratic operators. Moreover, spectral analysis of the underlying matrices drew preliminary conclusions about fixed points and their stability, an explicit formula was also obtained relating the nonuniform Walsh transform to the dynamics of genetic search [38]. Cerf considered the genetic algorithm as a special form of simulated annealing model and used the stochastic perturbation theory to study the extreme behavior and convergence speed of the genetic algorithm, showing how a delicate interaction between the perturbations and the selection pressure forced the convergence toward the global maxima of the fitness function [39].

In summary, the existing research results have been achieved under a certain or specific condition. When using the Markov chain model to analyze the convergence of genetic algorithms, it is difficult to obtain the transition probability, which hinders the study of the finite-time behavior of genetic algorithms. However, it is helpful to grasp the characteristics of the genetic algorithm and improve the efficiency and accuracy of the solution from the perspective of the stochastic process and mathematical statistics. Till now, most of the algorithm analysis is still based on the Markov chain model [28, 40, 41].

1.2.2 Encoding Problem of GA

Binary encoding is used widely in GA; however, there exists a Hamming cliff problem, i.e., when 15(01111) is changed to 16(10000), all bits in the chromosome need to be inversed. There are various encodings to improve the performance of GA.

(1) Gray encoding [42]: It is a variant of the binary encoding. Only one bit is different between the Gray code corresponding to two consecutive integers, and the difference between any two integers is the Hamming distance of Gray coding, which overcomes the Hamming cliff problem of the binary coding and improves the local search capability of the genetic algorithm.
(2) Real encoding [43]: For some multi-dimension, high-precision continuous optimization problems, binary coding is not convenient to reflect the specific knowledge of the problem and brings the mapping error when the continuous variables are discretized. Real coding of GA is to overcome these shortcomings, that is, each gene is represented by a real number to decrease the computational complexity and improve the computational efficiency.
(3) Symbol encoding [44]: The gene in the chromosome is taken from a symbol set without numerical meaning. These symbols can be the characters or numbers. The advantages of symbolic coding lie in its easier utilization of the specific knowledge and algorithms associated with the problem.

In addition, there are various other encoding forms, such as multi-value encoding [45], dynamic parameter encoding [46], delta encoding [47], hybrid encoding [48], and DNA encoding [49, 50]. When solving the optimal problems, the encoding form should be selected carefully in terms of their characteristics and advantages.

1.2.3 Constraint Handling in GA

Many optimization problems involve inequality and/or equality constraints, which are regarded as constrained optimization problems. In order to solve constrained optimization problems using GAs, various constraint handling methods have been developed.

(1) Penalty function method

Penalty function method is a popular approach to handle constraints, especially for inequality constraints. It was first proposed by Courant [51], and various improved forms were proposed afterward [52–55]. The penalty parameters keep constant in the whole evolution processes when using the static penalty function method. The main disadvantage is that the penalty parameters are determined based on specific issues and lack versatility [56, 57]. Though the dynamic penalty function method [58] has better optimization performances than static ones, it is also difficult to design the dynamic penalty parameters. Adaptive penalty function method [59, 60] can change the penalty parameters according to the feedback information during the search processes; however, the ranges of the parameters need to be defined beforehand, and their variation ranges should be defined to avoid the abrupt change of the penalty parameters.

(2) Specific encoding and operators

A specific encoding strategy is proposed for the specific constrained optimization problem. The specific operators are designed to be suitable for the specific encoding. The feasibility of the solution can then be guaranteed and the constraint problem is solved [61]. However, this kind of method cannot be popularized easily and mainly applied to the problem that is difficult to obtain feasible solutions.

(3) Repair method

Some repair programs were utilized to repair the infeasible solutions from the constraint set during the evolution processes [62]. The advantage of the repair method is that there are not too many additional requirements for individual encoding, genetic operators, etc., and it is expected that the optimal solution can be eventually approached from both sides of the feasible and infeasible solutions. Its disadvantage is the dependency on the problem itself at the expense of expanding the search space. In some problems, the repair process was even more complex than the original problem [44].

(4) Separation of objectives and constraints

Co-evolution method, multi-objective optimization method, etc., can be used to implement the separation. In the co-evolution method, the constraints and the fitness function were evolved simultaneously using two populations, which was similar to the adaptive penalty method and capable of obtaining satisfactory results [63]. However, using historical records to calculate an individual's fitness value may cause the evolution to stagnate. The heuristic rule was introduced to separate the feasible solution and the infeasible solution and made the fitness value of the feasible solution better than that of the infeasible solution [64, 65]. The separation method has defects in the maintenance of population diversity, even may be invalid, especially when the feasible domain is too small in the entire search space. Multi-objective optimization method transformed the constrained optimization problems into multi-objective optimization problems by redefining each objective function [66], which was not sensitive to parameter settings. However, the optimization results are not superior to those obtained using the penalty function.

(5) Hybrid method

Since the structure of the genetic algorithm is open, it is easy to combine with other algorithms, such as steepest descent method [67], quasi-Newton method [68], simplex method [69], augmented Lagrange method [70], and sequence quadratic programming [71]. The hybrid search algorithm is mainly experimental, and no theory has yet been researched. Moreover, other parameters will be introduced, and the main limitation of the traditional penalty function method is still not solved. If combined with the random evolution algorithm, it actually becomes another repair method [72]. However, if there are no feasible solutions in the initial population, it is not clear how the algorithm performs and whether it is suitable for nonbinary encoding. Applying a hybrid approach may sacrifice the versatility of GA, but there are some methods that take advantage of both specific knowledge and versatility, which is an effective way to improve the performance of basic genetic algorithms [60].

The constraint handling in Chaps. 3 and 9 is designed based on the idea that feasible solutions are always better than infeasible solutions by combining the penalty function method, separation method, and hybrid method.

1.2.4 Multi-objective Genetic Algorithm

For multi-objective optimization problems, genetic algorithm either expresses multiple objective functions as a single target or finds the Pareto optimal set in the evolution processes. The most popular multi-objective optimization methods based on the genetic algorithm are as follows:

(1) Weighted-sum method

Weighted sums of multiple objectives is the most common method [73] that converts multiple objective functions to one fitness function. GA can then be applied to solve the multi-objective problem without any modification. However, the optimization results are affected directly by the values of weights. Lacking enough information about the problem, the weights are difficult to be determined. Moreover, it cannot be applied to solve the optimization problem of the concave Pareto frontier. Adaptive weighted sum method was then proposed to adapt the concave Pareto front [74, 75].

(2) Vector evaluated genetic algorithm (VEGA)

Schaffer proposed VEGA to mainly improve the selection operator when solving the multi-objective optimization problem [76]. In the evolution processes, subpopulation is generated according to each objective function. For a problem with k objective functions, GA with a population size N will produce k subpopulations with size N/k

at each generation. Then, the crossover and mutation operations are performed to produce the next generation population. Notice that each subpopulation is generated based on a single-objective function regardless of other objective functions. Hence, the solutions obtained by VEGA are locally non-inferior, but not necessarily globally inferior, that is, individuals in different subpopulations are optimized only for single target within the subpopulation. This may make individuals with intermediate performance eliminated in the selection operator because they are not optimal in single target evaluation [77].

(3) Niched Pareto genetic algorithm (NPGA)

In NPGA, the selection operator of SGA is improved using the new selection strategy based on Pareto's superiority and inferiority [78]. Unlike the traditional genetic algorithm that compares only two individuals, NPGA allows competition between multiple individuals. When two competitors are non-inferior or inferior, the competition result is determined according to the sharing factor of the fitness function in the target domain. Since the Pareto rank is only applied to a part of the population, it has a faster calculation speed, but its population size is larger. Moreover, in addition to the sharing factor, an additional parameter is required to indicate the number of individuals participating in the competition [79].

(4) Multi-objective genetic algorithm (MOGA)

MOGA adopts a rank-based fitness assignment strategy [80]. The rank of each individual in the population is determined by the number of superior individuals. All non-inferior individuals have a rank of 1, while other individuals have a rank of 1 plus the number of dominated individuals in the current generation. By using sharing function [81] and species formation [82], the population can be evenly distributed in the Pareto frontier to a certain degree, and avoid premature convergence of the population. MOGA is widely used for solving multi-objective optimization problems, especially in the field of control system design because of its excellent performance. However, its performance relies too much on the appropriate choice of sharing factors [83, 84].

(5) Non-dominated sorting genetic algorithm (NSGA)

In NSGA, the hierarchical selection is used to guarantee the good individuals, and the niche method is adopted to maintain the stability of the excellent subpopulations. Its overall performance and computational complexity are inferior to MOGA, and it is more sensitive to the sharing factor. Deb et al. presented an improved NSGA method, i.e., NSGA-II [85], which utilized the elitist strategy and crowded distance evaluation strategy to overcome the above shortcomings. Furthermore, the computational complexity is reduced from the original $O(\mathrm{MN}^3)$ to $O(\mathrm{MN}^2)$, where M delegates the number of objectives, and N represents the population size. Experiment results showed that NSGA-II was better than several other MOEA algorithms [86], and some successful applications have been achieved by solving the multi-objective optimal problems [87, 88]. However, further efforts are required to maintain the population diversity and solve more complicated multi-objective problems. Recently, NSGA-III has been proposed to solve many objectives and big data optimization problems [89–92].

(6) Pareto archived evolution strategy (PAES)

In PAES, local search is introduced into the multi-objective optimization, and the computational complexity is greatly reduced [93]. Only mutation operator is used in the evolution processes, and the generated individual is limited to a local range; however, some advantages based on the population are lost. Hence, Knowles et al. proposed an improved PAES, i.e., M-PAES, which used a crossover operator and combined PAES with the usage of a population and recombination strategy [94]. Moreover, PAES-II was proposed using region-based selection where the unit of selection was a hyper box in objective space [95]. In addition, the rigorous running time was analyzed for PAES using a simple mutation operator [96] as a complement of insufficient theoretical analysis of multi-objective evolutionary algorithms (MOEAs).

(7) Strength Pareto evolutionary algorithm (SPEA)

SPEA was developed based on the synthesis of multiple MOEAs described above, where an external non-inferior set was used to save the non-inferior solutions found in each evolution process [97]. In the external non-inferior set, the strength of the ith individual is a real number between [0, 1], and the individual fitness value is equal to its strength value. As for the individual in the current population, the fitness value is equal to the sum of all the strength values superior to the individual. At the same time, individuals in the external non-inferior set can also participate in competition, and Pareto-based niche methods do not require the distance parameters [98]. Zitzler et al. proposed SPEA2, which incorporated in contrast to its predecessor a fine-grained fitness assignment strategy, a density estimation technique, and an enhanced archive truncation method [99]. Many good results have been obtained through applications of SPEA and its modified formats [100, 101].

The content of Chap. 4 is based on the research of SPEA and NSGA.

1.2.5 Applications of GA

Since there exist many complex optimization problems in various fields, the application research on GAs has penetrated into all disciplines, such as artificial intelligence [102], robotics [103], social science [104], bioengineering [105], electronics [106], automatic control [107], and so on. In this book, we will focus on the optimization problems in the field of automatic control.

(1) System parameter identification

The selection of model structure and model parameter identification are two main problems in system identification. There are quite a lot of advantages using genetic algorithm for system identification, such as wide adaptability, stable calculation, high identification accuracy, and simultaneous determination of the model structure and parameters. For a linear system, system order, time lag, and its parameters can be constituted into a chromosome. The modeling error is converted into the value of fitness function. Thus, the system identification problem becomes an optimal problem that can be solved by GA. As for the nonlinear system, since there are not uniform expression of nonlinear system, it is necessary to determine which type of structure is used in advance [108], and various nonlinear modeling tools such as neural networks [109, 110], fuzzy neural networks [111, 112], and support vector machine [113, 114] can also be utilized.

Due to the universal approximation capability of neural networks, it has been successfully applied in the nonlinear system modeling. However, how to determine the neural network structure still lacks theoretical basis. Genetic algorithm combined with artificial neural network is another active research field, which has been successfully applied in structure optimization of several types of neural networks, such as back-propagation (BP) neural network [115, 116], radial basis function (RBF) network [117, 118], recurrent neural network [119, 120], and fuzzy neural network [111, 112]. Esposito et al. optimized the output layer weights of RBF neural networks by GA [121]. Vesin et al. proposed a simplex reproduction scheme to solve the center selection problem of an RBF network. This algorithm was easy to implement, but the centers of the radial basis functions were selected among the input samples, which was difficult to reflect the input–output relationship in many cases [122]. Moreover, GA was utilized to simultaneously optimize the structure and centers of RBF network by minimizing model output error, but data morbidity can occur in the evolution process [118]. There are still many literature of neural network optimization using GA, and most of them focused on the optimization of connection weights, structures, and neural network parameters [115, 123, 124]. We cannot list them here due to space limitation and vast literature.

In the first part of Chap. 5 in this book, system parameter identification is carried out assuming that the model structure is known. The nonlinear system of Chaps. 6 and 7 is based on the nonlinear modeling tools, and the genetic algorithm is applied to solve the nonlinear system modeling and optimization problem using RBF neural network and T-S fuzzy neural network.

(2) Controller optimization

When designing the control system, almost all controllers need to tune parameters in advance, such as PID controller parameter tuning [125–128], selection of poles in pole placement controller [129–131], membership function determination of fuzzy controller [132, 133], and weights setting of neural network controller [134, 135]. The parameters can be calculated either in terms of mathematical model, or by trial and error, or by optimization methods. The parameters of the controller will directly affect the performances of the whole control system. Genetic algorithms are appropriate to off-line optimization of various types of controller parameters, such as controller structure optimization, controller parameter optimization, and controller structure and parameter optimization simultaneously [125, 129, 134, 136].

Except for a single-objective genetic algorithm, there are a large number of multi-objective genetic algorithms (MOGAs) for controller design. Fonesca et al. used MOGA to solve the optimization problem of the low-pressure spool speed governor of a Pegasus turbine engine [137]. Chen et al. introduced a fuzzy controller multi-objective optimal design method to guarantee H_2 and $H\infty$ reference tracking performance simultaneously [138]. In non-dominated Pareto frontier based multi-objective optimization methods [80, 139, 140], all possible non-inferior solutions can be obtained, and users can choose the final results according to control system requirement. PID controller and its improved format designed in Ch9 are another useful attempt in this context based on Pareto sorting multi-objective genetic algorithm and SGA, respectively. The research of GA for controller optimization design will continue to be studied in-depth, and readers can refer to a large number of references on controller optimization.

(3) Machine learning optimization

Machine learning is a subfield of Artificial Intelligence that is currently a hot field in computer science. There are many types of machine learning techniques, such as supervised learning [141], unsupervised learning [142], semi-supervised learning [143], reinforcement learning [144], evolutionary learning [145], and deep learning [146]. Humans have very strong skills for sensing the environment, they take actions against what they perceive from the environment [147], which is the basic idea of reinforcement learning (RL). Now, RL is a popular computational approach to learn whereby an agent tries to maximize the total amount of reward it receives, while interacting with a complex, uncertain environment. It is different from supervised learning in the sense that accurate input and output sets are not offered [144]. With strong environment adaptability, RL scheme is receiving considerable attention, and evolution algorithm is also utilized to optimize the RL scheme. In [148], a simple GA was used to search the space of RL policies by designing an appropriate objective function. Such et al. optimized the weights of a deep neural network with a simple, gradient-free GA, and it performed well on hard deep RL problems [149]. Vice versa, RL can be utilized to improve the performance of GA. Liu et al. proposed a reinforcement mutation GA, which used heterogeneous pairing selection instead of random pairing selection and constructed a reinforcement mutation operator by modifying the Q-Learning algorithm [150]. Mariani et al. proposed an improved NSGA-II approach by incorporating a parameter-free self-tuning method using reinforcement learning technique [151].

In Chap. 10, a Q-Learning based fuzzy energy management controller was proposed for real-time hybrid energy vehicle (HEV) energy split; GA is utilized to optimize its initial value of Q-table to satisfy the power requirement and decrease energy consumption and load fluctuation simultaneously. There are a large number of manuscripts about RL and its applications published every year [152], and readers can also follow these references [153–155].

1.3 DNA Computing Based GA

With the optimization problem being more complex, GAs are also continuously improving. DNA computing becomes conducive to improve GA after Dr. Adleman successfully solving the Hamiltonian path problem of 7 vertices with the operation of DNA molecules [156], and opening up a new biomolecular computing, e.g., DNA computing.

1.3.1 DNA Molecular Structure of DNA Computing

DNA computing simulates the double-helix structure and base complementary pairing rules to implement the information encoding. Whether in theory or technology, DNA computing is a challenge for traditional computing methods. In spite of its amazing development in the past decades, the disadvantages of DNA computing, such as the exponential explosion problem and its limitation to solve complex optimization problems in the engineering, have appeared [157].

DNA is the main molecule of DNA computing, which carries genetic information and its basic unit is deoxynucleotide [158]. A polynucleotide or deoxynucleotide is formed by linking 3′ and 5′ phosphodiester bond to four kinds of nucleotide or deoxynucleotide bond in a certain order, which is called a nucleotide sequence or a base sequence. The difference of nucleotides lies in the bases, where adenine and guanine are denoted as A and G and cytosine and thymine are denoted as C and T. The deoxynucleotide or nucleotide linkage has a strict directionality, and 3′-OH of the former nucleotide forms a 3′–5′ phosphodiester bond with the 5′ phosphoric acid of the next nucleotide to form a linear macromolecule without branches, as shown in Fig. 1.2. RNA is another polymer that plays an important role in active cells [159]. Its structure is very close to the DNA's, but the pyrimidine is composed of cytosine and uracil, T of DNA is then replaced by U.

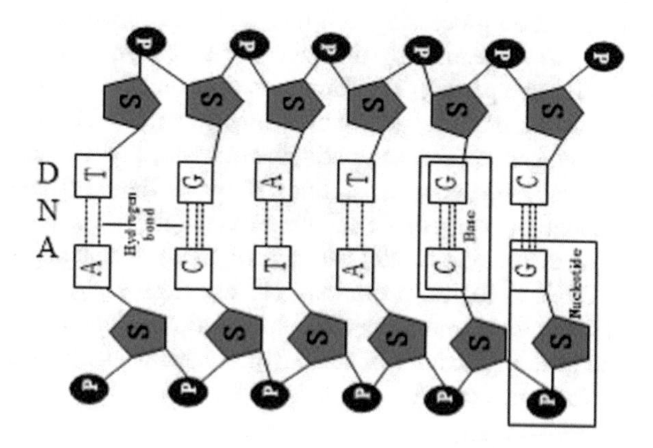

Fig. 1.2 Biological structure of DNA

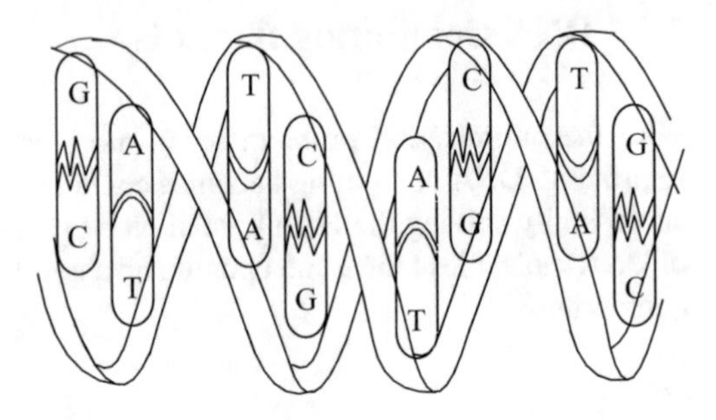

Fig. 1.3 Double-helix structure of DNA

The hydroxyl of one nucleotide can interact with that of another nucleotide to form a weaker hydrogen bond, and the formation of the bond complies with the Watson–Crick complementarity pairing principle, i.e., A pairs T, C pairs G. A single strand of DNA is first formed through a phosphodiester bond. Then, a double-strand DNA molecule can be easily formed using this complementarity principle. Two long nucleotides are joined together by a hydrogen bond between the bases to form a double-strand helical structure, as shown in Fig. 1.3. It has been simplified greatly as two linear chains combined by the Watson–Crick complementarity principle.

1.3.2 Biological Operators of DNA Computing

DNA computing is implemented using various biological operators. Some of them are physical while some are chemical. The physical operators are actually to control the conditions of its biochemical reaction, such as temperature, pH values, and so on. The most important biological operators are DNA denaturation and renaturation, DNA crossover, DNA probe, polymerase chain reaction, DNA gel electrophoresis, DNA reduction, DNA ligation, and DNA extension [160].

In terms of the characteristics of DNA structure and its biological operations, there contains two important units in DNA molecule with computational power: (1) an information processing unit that performs denaturation, replication, and annealing of DNA molecules by the action of an enzyme, i.e., DNA molecules can be cut, copied, and pasted; (2) information store unit through DNA sequences using multiple biological enzymes to simultaneously perform DNA molecule manipulation. The advantages of DNA computing lie in its huge parallelism of information storage and its parallel operation capabilities. However, the experimental implementation of DNA computing is unreliable and the cost is high [161]. At present, most improvement of DNA computing is only theoretically feasible that depends on further improvement of DNA biological operations. Nevertheless, DNA is an important genetic material that carries abundant genetic information. It can promote the genetic mechanism and gene regulation mechanism of organisms, which may be helpful to improve the performance of GA.

1.3.3 DNA Computing Based Genetic Algorithm

Genetic algorithm is similar to DNA computing to some degree, and the new theory of DNA computing, such as DNA encoding, DNA molecular operators, and the creative ideas of DNA computing, may be combined to improve the performance of the genetic algorithm when solving the complicated problems in the fields of automatic control, pattern recognition, decision making, machine learning, etc. Recently, various DNA computing based GAs have been proposed to solve complicated problems. Dio et al. presented a novel GA with DNA double-helix to promote non-metamorphic variation in DNA replication [162]. Yoshikawa et al. combined DNA encoding with pseudo-bacteria GA [163]. Ren et al. proposed a CM-DNAGA based on not only the properties of randomness and the stable tendency of the normal cloud model, but also the idea of GA with the bio-inspired coding method, i.e., DNA [164]. Lin et al. raised an encoding mode based on DNA biological molecular structure to encode the PID controller structure and parameters [165]. Gu et al. proposed various DNA computing based GAs to avoid its shortcoming of premature convergence and solve the Job shop scheduling problem efficiently [166]. Abdullah used DNA and logistic map functions to create the number of initial DNA masks and applied GA to determine the best mask for encryption, and the results were superior to methods available in the literature [167]. The authors presented RNA-based GA, double-helix structure based GA, and other advanced GA to improve the optimization performance and solve the complicated problems in system modeling and control [49, 112].

Recently, various evolution algorithms, such as Ant Colony Algorithm [168–171], Particle swarm optimization [172–174], Cuckoo Search Algorithm [175, 176], Differential Evolution algorithm [177–180], Artificial Fish Swarm algorithm [181, 182], and so on, are emerging. GA was combined with these algorithms to improve the performance [178, 183–185]. The ideas of DNA computing based GA may also be helpful to these algorithms.

1.4 The Main Content of This Book

It can be seen from the above analysis that the research on DNA-GA for complicated optimization problem still needs to be further studied. Accordingly, the specific structure of the book is arranged as shown below (Fig. 1.4):

The book is divided into 10 chapters. The first chapter firstly describes the practical background and research significance of this topic and explains the related characteristics of DNA-GA and the difficulty for system modeling and control technologies. Secondly, the basic genetic algorithm is introduced, and the trend and its application to different optimization problems in research are pointed out. Then, the research topics and related DNA computing and its biological operations are introduced, which paves the way for the subsequent contents of the book.

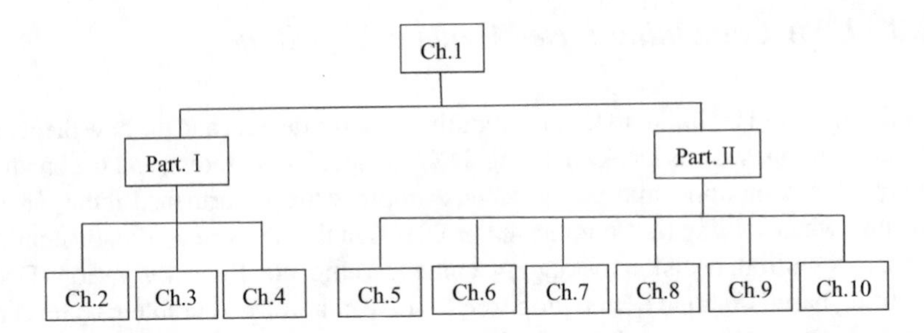

Fig. 1.4 Relation among chapters

Subsequent contents are divided into two parts. Part I mainly studies DNA-GA for unconstrained, constrained, single-objective, and multi-objective function optimization problems, and Part II mainly studies the modeling and control problems in industrial processes. The contents of each chapter of the book are both related to and different from one another. For example, Part I proposes several improved GAs based on DNA and RNA operators, Part II uses improved GA to solve the industrial modeling and control problems; however, some new operators can also be introduced to be appropriate for the different optimization problems. For Chap. 2, RNA-GA is proposed to solve the unconstrained function minimization problem, and Chap. 3 combines SQP algorithm to enhance the constraint handling capability. Chapter 4 presented DNA-MOGA to solve the multi-objective function minimization problems. Chapters 5–10 are based on Chaps. 2–4. Chapter 5 solves the parameter identification problems based on the known system structure. Chapter 6 models the nonlinear system using an optimal RBF neural network considering the network structure simplification and modeling precision. Chapter 7 uses T-S fuzzy neural network to construct the nonlinear system model by utilizing constrained single-objective and multi-objective DNA-GA. Chapter 8 constructs the distributed parameter system (DPS) model using PCA, RBF, and GA. Chapters 9 and 10 solve the controller optimization problems. In Chap. 10, the controller is a Q-Learning based fuzzy controller, and GA is utilized to optimize the parameters of Q-Table. This book proposes some DNA computing based GA and solves the system modeling and controller optimization problems in industrial processes. It can be used as the guiding material for corresponding engineering modeling and control problems, and can also provide some help for engineers and technicians.

Some parts of the material in this book were published in relevant journals by the authors several years ago and included here for relevance and completeness. This includes

Algorithm for the Gasoline Blending Recipe Optimization Problem, 440-451, © 2008 Elsevier Ltd.

Chapter 4 reprinted with permission from Neurocomputing, volume 98, Tao Jili, Fan Qinru, Chen Xiaoming, Zhu Yong, Constraint multi-objective automated synthesis for CMOS operational amplifier, 108–113, © 2012 Elsevier Ltd.

Section 6.4 of Chap. 6 reprinted with permission from Industrial & Engineering Chemistry Research, volume 53, Zhang, Ridong; Tao, Jili; Gao, Furong Temperature Modeling in a Coke Furnace with an Improved RNA-GA Based RBF Network, 3236–3245 © 2014 American Chemical Society.

Section 6.5 of Chap. 6 reprinted with permission from IEEE Transactions on Industrial Electronics, volume 64, Zhang, Ridong; Tao, Jili, Data-Driven Modeling Using Improved Multi-Objective Optimization Based Neural Network for Coke Furnace System, 3147–3155. © 2017 IEEE.

Section 6.6 of Chap. 6 reprinted with permission from Industrial & Engineering Chemistry Research, volume 58, Zhang Ridong, Lv Qiang, Tao, Jili, Gao Furong. Data driven modeling using optimal principle component analysis based neural network and its application to nonlinear coke furnace, 6344–6352. © 2018 American Chemical Society.

Section 7.3 of Chap. 7 reprinted with permission from Industrial & Engineering Chemistry Research, volume 55, Zhang Ridong, Tao Jili, Gao Furong. A new approach of Takagi–Sugeno fuzzy modeling using an improved genetic algorithm optimization for oxygen content in a coke furnace. Industrial & Engineering Chemistry Research, 6465–6474. © 2016 American Chemical Society.

Section 7.4 of Chap. 7 reprinted with permission from IEEE Transactions on Industrial Electronics, volume 65, Zhang Ridong, Tao Jili. A nonlinear fuzzy neural network modeling approach using an improved genetic algorithm, 5882–5892 © 2017 IEEE.

Chapter 8 reprinted with permission from IEEE Trans Neural Netw Learn Syst, volume 29, Zhang Ridong, Tao Jili, Lu Renquan, Jin, Qibing. Decoupled ARX and RBF Neural Network Modeling Using PCA and GA Optimization for Nonlinear Distributed Parameter Systems, 457–469 © 2018 IEEE.

Section 9.2 of Chap. 9 reprinted with permission from Chemometrics and Intelligent Laboratory Systems, volume 137, Tao, Jili; Yu, Zaihe; Zhu, Yong, PFC based PID design using genetic algorithm for chamber pressure in a coke furnace, 155–161 © 2014 Elsevier Ltd.

Section 9.3 of Chap. 9 reprinted from 31st Annual Conference of IEEE Industrial Electronics Society, Tao Jili, Wang Ning. Neuron control based on RBF network and fuzzy scheme for a direct drive robot, 165–170. © 2005 IEEE.

Section 9.4 of Chap. 9 reprinted from Proceedings of the 27th Chinese Control Conference, Jili Tao, Ning Wang, Xiongxiong He, Stabilization subspaces based DPGA for optimizing PID controllers, 38–42. © 2008 IEEE.

References

1. Haupt, R.L., and S.E. Haupt. 2004. *Practical Genetic Algorithms*, 2nd ed. Hoboken, New Jersey: Wiley.
2. Whitley, D. 1994. A genetic algorithm tutorial. *Statistics Computing* 4 (2): 65–85.
3. Zhang, R.D., et al. 2016. A new approach of Takagi-Sugeno fuzzy modeling using improved GA optimization for oxygen content in a coke furnace. *Industrial and Engineering Chemistry Research* 55 (22): 6465–6474.
4. Nocedal, J., and S. Wright. 2006. *Numerical Optimization*. Springer Science & Business Media.
5. Campbell, S.D., et al. 2019. Review of numerical optimization techniques for meta-device design. *Optical Materials Express* 9 (4): 1842–1863.
6. Mohamed, A.W., and A.K. Mohamed. 2019. Adaptive guided differential evolution algorithm with novel mutation for numerical optimization. *International Journal of Machine Learning Cybernetics* 10 (2): 253–277.
7. Holland, J.H. 1975. *Adaptation in Natural and Artificial Systems*. The University of Michigan Press.
8. Dong, H., et al. 2018. A novel hybrid genetic algorithm with granular information for feature selection and optimization. *Applied Soft Computing* 65: S1568494618300048.
9. Edison, E., and T. Shima. 2011. Integrated task assignment and path optimization for cooperating uninhabited aerial vehicles using genetic algorithms. *Computers Operations Research* 38 (1): 340–356.
10. Zhang, R., et al. 2018. Decoupled ARX and RBF neural network modeling using PCA and GA optimization for nonlinear distributed parameter systems. *IEEE Transactions on Neural Networks and Learning Systems* 29 (2): 457–469.
11. Goldberg, D.E. 1989. *Genetics algorithms in search, optimization and machine learning*. Addison-Wesley: MA Publisher.
12. Michalewicz and Zbigniew. 1996. *Genetic Algorithms + Data Structures = Evolution Programs*. Berlin: Springer.
13. Jiang, Y., et al. 2019. Multi-parameter and multi-objective optimisation of articulated monorail vehicle system dynamics using genetic algorithm. *Vehicle System Dynamics* 47: 1–18.
14. Kao, Y.T., and E. Zahara. 2008. A hybrid genetic algorithm and particle swarm optimization for multimodal functions. *Applied Soft Computing* 8 (2): 849–857.
15. Lopez-Garcia, P., et al. 2015. A hybrid method for short-term traffic congestion forecasting using genetic algorithms and cross entropy. *IEEE Transactions on Intelligent Transportation Systems* 17 (2): 557–569.
16. Zhang, R., and J. Tao. 2018. GA based fuzzy energy management system for FC/SC powered HEV considering H2 consumption and load variation. *IEEE Transactions on Fuzzy Systems* 26 (4): 1833–1843.
17. Tavakkoli-Moghaddam, R., J. Safari, and F. Sassani. 2017. Reliability optimization of series-parallel systems with a choice of redundancy strategies using a genetic algorithm. *Reliability Engineering & System Safety* 93 (4): 550–556.
18. Reeves, C., and J.E. Rowe. 2002. *Genetic Algorithms: Principles and Perspectives: A Guide to GA Theory*, vol. 20. Springer Science & Business Media.
19. Poli, R., and N.F. McPhee. 2003. General schema theory for genetic programming with subtree-swapping crossover: Part II. *Evolutionary Computation* 11 (2): 169–206.
20. Poli, R., and W.B. Langdon. 1998. Schema theory for genetic programming with one-point crossover and point mutation. *Evolutionary Computation* 6 (3): 231–252.
21. Poli, R., and N.F. McPhee. 2001. Exact schema theory for GP and variable-length GAs with homologous crossover. In *Proceedings of the 3rd Annual Conference on Genetic and Evolutionary Computation*. Morgan Kaufmann Publishers Inc.
22. Sudholt, D. 2017. How crossover speeds up building block assembly in genetic algorithms. *Evolutionary Computation* 25 (2): 237–274.

23. Forrest, S., and M. Mitchell. 1993. *Relative building-block fitness and the building-block hypothesis*. In *Foundations of Genetic Algorithms*, 109–126. Elsevier.
24. Stephens, C., and H. Waelbroeck. 1999. Schemata evolution and building blocks. *Evolutionary Computation* 7 (2): 109–124.
25. Wu, A.S., and R.K. Lindsay. 1996. A comparison of the fixed and floating building block representation in the genetic algorithm. *Evolutionary Computation* 4 (2): 169–193.
26. Goldberg, D.E., and P. Segrest. 1987. *Finite Markov chain analysis of genetic algorithms*. In *Proceedings of the Second International Conference on Genetic Algorithms*.
27. Suzuki, J. 1995. A Markov chain analysis on simple genetic algorithms. *IEEE Transactions on Systems, Man, Cybernetics* 25 (4): 655–659.
28. Ter Braak, C.J. 2006. A Markov Chain Monte Carlo version of the genetic algorithm differential evolution: easy Bayesian computing for real parameter spaces. *Statistics Computing* 16 (3): 239–249.
29. Leung, K.-S., et al. 2001. A new model of simulated evolutionary computation-convergence analysis and specifications. *IEEE Transactions on Evolutionary Computation* 5 (1): 3–16.
30. Mingzhi, X., Z. Weicai, and J. Licheng. 2003. The convergence of the abstract evolutionary algorithm based on a special selection mechanism. In *Proceedings Fifth International Conference on Computational Intelligence and Multimedia Applications. ICCIMA 2003*. IEEE.
31. Rudolph, G. 1996. *Convergence of evolutionary algorithms in general search spaces*. In *Proceedings of IEEE International Conference on Evolutionary Computation*. IEEE.
32. Hou, J., et al. 2009. Integrating genetic algorithm and support vector machine for polymer flooding production performance prediction. *Journal of Petroleum Science Engineering* 68 (1–2): 29–39.
33. Wang, X.P., and L.M. Cao. 2002. *Genetic Algorithms-Theory, Application and Software Implementation*. Xi' an: Xi' an Jiaotong University Press.
34. Nordin, P., and W. Banzhaf. 1996. *Genetic Reasoning Evolving Proofs with Genetic Search*. Citeseer.
35. Vose, M.D. 1995. Modeling simple genetic algorithms. *Evolutionary Computation* 3 (4): 453–472.
36. Nix, A.E., and M.D. Vose. 1992. Modeling genetic algorithms with Markov chains. *Annals of mathematics artificial intelligence* 5 (1): 79–88.
37. Catoni, O., and R. Cerf. 1997. The exit path of a Markov chain with rare transitions. *Probability Statistics* 1: 95–144.
38. Liepins, G.E., and M.D. Vose. 1991. *Deceptiveness and genetic algorithm dynamics*. In *Foundations of genetic algorithms*, 36–50. Elsevier.
39. Cerf, R. 1998. Asymptotic convergence of genetic algorithms. *Advances in Applied Probability* 30 (2): 521–550.
40. Heo, J.-H., et al. 2011. A reliability-centered approach to an optimal maintenance strategy in transmission systems using a genetic algorithm. *IEEE Transactions on Power Delivery* 26 (4): 2171–2179.
41. Yang, S., et al. 2018. Machine learning approach to decomposing arterial travel time using a hidden Markov model with genetic algorithm. *Journal of Computing in Civil Engineering* 32 (3).
42. Chakraborty, U.K., and C.Z. Janikow. 2003. An analysis of Gray versus binary encoding in genetic search. *Information Sciences* 156 (3–4): 253–269.
43. Ono, I., and S. Kobayashi. 1999. A real-coded genetic algorithm for function optimization using unimodal normal distribution. In *Proceedings of International Conference on Genetic Algorithms*.
44. Jia, H., et al. 2003. A modified genetic algorithm for distributed scheduling problems. *Journal of Intelligent Manufacturing* 14 (3–4): 351–362.
45. Pouryoussefi, S., and Y. Zhang. 2015. Identification of two-phase water–air flow patterns in a vertical pipe using fuzzy logic and genetic algorithm. *Applied Thermal Engineering* 85: 195–206.

46. Schraudolph, N.N., and R.K. Belew. 1992. Dynamic parameter encoding for genetic algorithms. *Machine Learning* 9 (1): 9–21.
47. Mathias, K.E., and L.D. Whitley. 1994. Initial performance comparisons for the delta coding algorithm. In *Proceedings of the First IEEE Conference on Evolutionary Computation. IEEE World Congress on Computational Intelligence*. IEEE.
48. Baskar, S., P. Subbaraj, and M. Rao. 2003. Hybrid real coded genetic algorithm solution to economic dispatch problem. *Computers Electrical Engineering* 29 (3): 407–419.
49. Tao, J., and N. Wang. 2007. DNA computing based RNA genetic algorithm with applications in parameter estimation of chemical engineering processes. *Computers Chemical Engineering* 31 (12): 1602–1618.
50. Ding, Y., and L. Ren. 2000. DNA genetic algorithm for design of the generalized membership-type Takagi-Sugeno fuzzy control system. In *IEEE International Conference on Systems, Man and Cybernetics*. IEEE.
51. Courant, R. 1943. *Variational Methods for the Solution of Problems of Equilibrium and Vibrations*. Verlag nicht ermittelbar.
52. Knypiński, Ł., K. Kowalski, and L. Nowak. 2018. Constrained optimization using penalty function method combined with genetic algorithm. In *ITM Web of Conferences*. EDP Sciences.
53. Li, B., et al. 2011. An exact penalty function method for continuous inequality constrained optimal control problem. *Journal of Optimization Theory Applications* 151 (2): 260.
54. Liu, J., et al. 2016. An exact penalty function-based differential search algorithm for constrained global optimization. *Soft Computing* 20 (4): 1305–1313.
55. Munjiza, A., and K. Andrews. 2000. Penalty function method for combined finite–discrete element systems comprising large number of separate bodies. *International Journal for Numerical Methods in Engineering* 49 (11): 1377–1396.
56. Coello Coello, C.A. 2016. Constraint-handling techniques used with evolutionary algorithms. In *Proceedings of the 2016 on Genetic and Evolutionary Computation Conference Companion*. ACM.
57. Tao, J., X. Chen, and Y. Zhu. 2010. Constraint multi-objective automated synthesis for CMOS operational amplifier. In *Life System Modeling and Intelligent Computing*, 120–127. Springer.
58. Kazarlis, S., and V. Petridis. 1998. Varying fitness functions in genetic algorithms: studying the rate of increase of the dynamic penalty terms. In *International Conference on Parallel Problem Solving from Nature*. Springer.
59. Tessema, B., and G.G. Yen. 2006. A self adaptive penalty function based algorithm for constrained optimization. In *IEEE International Conference on Evolutionary Computation*. IEEE.
60. Tao, J., and N. Wang. 2008. DNA double helix based hybrid GA for the gasoline blending recipe optimization problem. *Chemical Engineering Technology* 31 (3): 440–451.
61. Bakirtzis, A.G., et al. 2002. Optimal power flow by enhanced genetic algorithm. *IEEE Transactions on Power Systems* 17 (2): 229–236.
62. Chootinan, P., and A. Chen. 2006. Constraint handling in genetic algorithms using a gradient-based repair method. *Computers operations research* 33 (8): 2263–2281.
63. Chang, Y.-H. 2010. Adopting co-evolution and constraint-satisfaction concept on genetic algorithms to solve supply chain network design problems. *Expert Systems with Applications* 37 (10): 6919–6930.
64. Powell, D., and M.M. Skolnick. 1993. Using genetic algorithms in engineering design optimization with non-linear constraints. In *Proceedings of the 5th International Conference on Genetic Algorithms*. Morgan Kaufmann Publishers Inc.
65. Dasgupta, D., and Z. Michalewicz. 2013. *Evolutionary Algorithms in Engineering Applications*. Springer Science & Business Media.
66. Cai, Z., and Y. Wang. 2006. *A multiobjective optimization-based evolutionary algorithm for constrained optimization. IEEE Transactions on Evolutionary Computation* 10 (6): 658–675.
67. Lin, Y.-C., F.-S. Wang, and K.-S. Hwang. 1999. A hybrid method of evolutionary algorithms for mixed-integer nonlinear optimization problems. In *Proceedings of the 1999 Congress on Evolutionary Computation-CEC99 (Cat. No. 99TH8406)*. IEEE.

68. Razik, H., C. Defranoux, and A. Rezzoug. 2000. Identification of induction motor using a genetic algorithm and a quasi-Newton algorithm. In *7th IEEE International Power Electronics Congress. Technical Proceedings. CIEP 2000 (Cat. No. 00TH8529)*. IEEE.
69. Wang, Y., et al. 2009. Constrained optimization based on hybrid evolutionary algorithm and adaptive constraint-handling technique. *Structural Multidisciplinary Optimization* 37 (4): 395–413.
70. Costa, L., I.A.E. Santo, and E.M. Fernandes. 2012. A hybrid genetic pattern search augmented Lagrangian method for constrained global optimization. *Applied Mathematics Computation* 218 (18): 9415–9426.
71. Fesanghary, M., et al. 2008. Hybridizing harmony search algorithm with sequential quadratic programming for engineering optimization problems. *Computer Methods in Applied Mechanics Engineering* 197 (33–40): 3080–3091.
72. Belur, S.V. 1997. CORE: Constrained optimization by random evolution. In *Late Breaking Papers at the Genetic Programming 1997 Conference*. Stanford Bookstore.
73. Das, I., and J.E. Dennis. 1997. A closer look at drawbacks of minimizing weighted sums of objectives for Pareto set generation in multicriteria optimization problems. *Structural Optimization* 14 (1): 63–69.
74. Kim, I.Y., and O. De Weck. 2006. Adaptive weighted sum method for multiobjective optimization: a new method for Pareto front generation. *Structural Multidisciplinary Optimization* 31 (2): 105–116.
75. Zhang, R., J. Tao, and F. Gao. 2014. Temperature modeling in a coke furnace with an improved RNA-GA based RBF network. *Industrial Engineering Chemistry Research* 53 (8): 3236–3245.
76. Schaffer, J.D. 1985. Multiple objective optimization with vector evaluated genetic algorithms. In *Proceedings of the First International Conference on Genetic Algorithms and Their Applications, 1985*. Lawrence Erlbaum Associates. Inc., Publishers.
77. Dias, A.H., and J.A. De Vasconcelos. 2002. Multiobjective genetic algorithms applied to solve optimization problems. *IEEE Transactions on Magnetics* 38 (2): 1133–1136.
78. Rey Horn, J., N. Nafpliotis, and D.E. Goldberg. 1994. A niched Pareto genetic algorithm for multiobjective optimization. In *Proceedings of the First IEEE Conference on Evolutionary Computation, IEEE World Congress On Computational Intelligence*. Citeseer.
79. Guria, C., P.K. Bhattacharya, and S.K. Gupta. 2005. Multi-objective optimization of reverse osmosis desalination units using different adaptations of the non-dominated sorting genetic algorithm (NSGA). *Computers Chemical Engineering* 29 (9): 1977–1995.
80. Deb, K. 2001. *Multi-objective Optimization Using Evolutionary Algorithms*, vol. 16. Wiley
81. Goldberg, D.E., and J. Richardson. 1987. Genetic algorithms with sharing for multimodal function optimization. In *Genetic Algorithms and Their Applications: Proceedings of the Second International Conference on Genetic Algorithms*. Hillsdale, NJ: Lawrence Erlbaum.
82. Deb, K. 1989. An investigation of niche and species formation in genetic function optimization. In *Proceedings of the Third International Conference on Genetic algorithms.*
83. Coello, C.C. 1999. An updated survey of evolutionary multiobjective optimization techniques: State of the art and future trends. In *Proceedings of the 1999 Congress on Evolutionary Computation-CEC99 (Cat. No. 99TH8406)*. IEEE.
84. Hu, Z.-H. 2010. A multiobjective immune algorithm based on a multiple-affinity model. *European Journal of Operational Research* 202 (1): 60–72.
85. Deb, K., et al. 2002. A fast and elitist multiobjective genetic algorithm: NSGA-II. *IEEE Transactions on Evolutionary Computation* 6 (2): 182–197.
86. Kannan, S., et al. 2008. Application of NSGA-II algorithm to generation expansion planning. *IEEE Transactions on Power Systems* 24 (1): 454–461.
87. Dhanalakshmi, S., et al. 2011. Application of modified NSGA-II algorithm to combined economic and emission dispatch problem. *International Journal of Electrical Power* 33 (4): 992–1002.
88. Taleizadeh, A.A., P.P. Khaligh, and I. Moon. 2019. Hybrid NSGA-II for an imperfect production system considering product quality and returns under two warranty policies. *Applied Soft Computing* 75: 333–348.

89. Vesikar, Y., K. Deb, and J. Blank. 2018. Reference point based NSGA-III for preferred solutions. In *IEEE Symposium Series on Computational Intelligence (SSCI)*. IEEE.
90. Vesikar, Y., K. Deb, and J. Blank. 2018. Reference point based NSGA-III for preferred solutions. In *2018 IEEE Symposium Series on Computational Intelligence (SSCI)*. IEEE.
91. Li, H., et al. 2019. Comparison between MOEA/D and NSGA-III on a set of novel many and multi-objective benchmark problems with challenging difficulties. *Swarm Evolutionary Computation* 46: 104–117.
92. Yi, J.-H., et al. 2018. An improved NSGA-III Algorithm with adaptive mutation operator for big data optimization problems. *Future Generation Computer Systems* 88: 571–585.
93. Knowles, J., and D. Corne. 1999. The pareto archived evolution strategy: a new baseline algorithm for pareto multiobjective optimisation. In *Congress on Evolutionary Computation (CEC99)*.
94. Knowles, J.D., and D.W. Corne. 2000. M-PAES: A memetic algorithm for multiobjective optimization. In *Proceedings of the 2000 Congress on Evolutionary Computation. CEC00 (Cat. No. 00TH8512)*. IEEE.
95. Corne, D.W., et al. 2001. PESA-II: region-based selection in evolutionary multi-objective optimization. In *Proceedings of the 3rd Annual Conference on Genetic and Evolutionary Computation*. Morgan Kaufmann Publishers Inc.
96. Peng, X., et al. 2018. Running time analysis of the Pareto archived evolution strategy on pseudo-Boolean functions. *Multimedia Tools Applications* 77 (9): 11203–11217.
97. Zitzler, E., and L. Thiele. 1999. Multiobjective evolutionary algorithms: a comparative case study and the strength Pareto approach. *IEEE Transactions on Evolutionary Computation* 3 (4): 257–271.
98. Zitzler, E., K. Deb, and L. Thiele. 2000. Comparison of multiobjective evolutionary algorithms: Empirical results. *Evolutionary Computation* 8 (2): 173–195.
99. Zitzler, E., M. Laumanns, and L. Thiele. 2001. *SPEA2: Improving the Strength Pareto Evolutionary Algorithm*. TIK-report, 103.
100. Rostami, S., and F. Neri. 2016. Covariance matrix adaptation pareto archived evolution strategy with hypervolume-sorted adaptive grid algorithm. *Integrated Computer-Aided Engineering* 23 (4): 313–329.
101. Ding, R., et al. 2019. A novel two-archive strategy for evolutionary many-objective optimization algorithm based on reference points. *Applied Soft Computing* 78: 447–464.
102. Tabassum, M., and K. Mathew. 2014. A genetic algorithm analysis towards optimization solutions. *International Journal of Digital Information Wireless Communications* 4 (1): 124–142.
103. Ayala, H.V.H., and L. dos Santos Coelho. 2012. Tuning of PID controller based on a multiobjective genetic algorithm applied to a robotic manipulator. *Expert Systems with Applications* 39 (10): 8968–8974.
104. Epstein, J.M. 2018. *Nonlinear Dynamics, Mathematical Biology, and Social Science: Wise Use of Alternative Therapies*. CRC Press.
105. Tumuluru, J.S., and R. McCulloch. 2016. Application of hybrid genetic algorithm routine in optimizing food and bioengineering processes. *Foods* 5 (4): 76.
106. Zebulum, R.S., M.A. Pacheco, and M.M.B. Vellasco. 2018. *Evolutionary Electronics: Automatic Design of Electronic Circuits and Systems by Genetic Algorithms*. CRC press.
107. Zou, T., S. Wu, and R. Zhang. 2018. Improved state space model predictive fault-tolerant control for injection molding batch processes with partial actuator faults using GA optimization. *ISA Transactions* 73: 147–153.
108. Naitali, A., and F. Giri. 2016. Wiener-Hammerstein system identification—an evolutionary approach. *International Journal of Systems Science* 47 (1): 45–61.
109. Zhang, R., and J.L. Tao. 2017. Data driven modeling using improved multi-objective optimization based neural network for coke furnace system. *IEEE Transactions on Industrial Electronics* 64 (4): 3147–3155.
110. Kant, G., and K.S. Sangwan. 2015. Predictive modelling and optimization of machining parameters to minimize surface roughness using artificial neural network coupled with genetic algorithm. *Procedia Cirp* 31: 453–458.

111. Ishigami, H., et al. 1995. Structure optimization of fuzzy neural network by genetic algorithm. *Fuzzy Sets Systems* 71 (3): 257–264.
112. Zhang, R., and J. Tao. 2018. A nonlinear fuzzy neural network modeling approach using an improved genetic algorithm. *IEEE Transactions on Industrial Electronics* 65 (7): 5882–5892.
113. Nekoei, M., M. Mohammadhosseini, and E. Pourbasheer. 2015. QSAR study of VEGFR-2 inhibitors by using genetic algorithm-multiple linear regressions (GA-MLR) and genetic algorithm-support vector machine (GA-SVM): a comparative approach. *Medicinal Chemistry Research* 24 (7): 3037–3046.
114. Liu, D., et al. 2014. Short-term wind speed forecasting using wavelet transform and support vector machines optimized by genetic algorithm. *Renewable Energy* 62: 592–597.
115. Lam, H., et al. 2001. Tuning of the structure and parameters of neural network using an improved genetic algorithm. In *IECON'01. 27th Annual Conference of the IEEE Industrial Electronics Society*. IEEE.
116. Yanmin, W., and Y. Pingjing. 2003. Simulation and optimization for thermally coupled distillation using artificial neural network and genetic algorithm. *Chinese Journal of Chemical Engineering* 11 (3): 307–311.
117. Zhang, R., et al. 2018. Data driven modeling using optimal principle component analysis based neural network and its application to nonlinear coke furnace. *Industrial and Engineering Chemistry Research* 57 (18): 6344–6352.
118. Sarimveis, H., et al. 2004. A new algorithm for developing dynamic radial basis function neural network models based on genetic algorithms. *Computers Chemical Engineering* 28 (1–2): 209–217.
119. Angeline, P.J., G.M. Saunders, and J.B. Pollack. 1994. An evolutionary algorithm that constructs recurrent neural networks. *IEEE Transactions on Neural Networks* 5 (1): 54–65.
120. Juang, C.-F. 2004. A hybrid of genetic algorithm and particle swarm optimization for recurrent network design. *IEEE Transactions on Systems, Man, Cybernetics, Part B* 34 (2): 997–1006.
121. Esposito, A., et al. 2000. Approximation of continuous and discontinuous mappings by a growing neural RBF-based algorithm. *Neural Networks* 13 (6): 651–665.
122. Vesin, J.-M., and R. Grüter. 1999. Model selection using a simplex reproduction genetic algorithm. *Signal Processing* 78 (3): 321–327.
123. Lin, C.-T., M. Prasad, and A. Saxena. 2015. An improved polynomial neural network classifier using real-coded genetic algorithm. *IEEE Transactions on Systems, Man, Cybernetics: Systems* 45 (11): 1389–1401.
124. Assunção, F., et al. Evolving the topology of large scale deep neural networks. In *European Conference on Genetic Programming*. 2018. Springer.
125. Thomas, N., and D.P. Poongodi. 2009. Position control of DC motor using genetic algorithm based PID controller. In *Proceedings of the World Congress on Engineering*.
126. Tao, J., Z. Yu, and Y. Zhu. 2014. PFC based PID design using genetic algorithm for chamber pressure in a coke furnace. *Chemometrics Intelligent Laboratory Systems* 137: 155–161.
127. Wang, Y., et al., Predictive fuzzy PID control for temperature model of a heating furnace. In *2017 36th Chinese Control Conference (CCC)*. IEEE.
128. Zhang, R., et al. 2014. New PID controller design using extended nonminimal state space model based predictive functional control structure. *Industrial Engineering Chemistry Research* 53 (8): 3283–3292.
129. Eshtehardiha, S., A. Kiyoumarsi, and M. Ataei. 2007. Optimizing LQR and pole placement to control buck converter by genetic algorithm. In *2007 International Conference on Control, Automation and Systems*. IEEE.
130. Russell, D., A.J. Fleming, and S.S. Aphale. 2015. Simultaneous optimization of damping and tracking controller parameters via selective pole placement for enhanced positioning bandwidth of nanopositioners. *Journal of Dynamic Systems, Measurement, Control* 137 (10).
131. Nevaranta, N., et al. 2020. Adaptive MIMO pole placement control for commissioning of a rotor system with active magnetic bearings. *Mechatronics* 65: 102313.
132. Sanchez, E., T. Shibata, and L.A. Zadeh. 1997. *Genetic Algorithms and Fuzzy Logic Systems: Soft Computing Perspectives*, vol. 7. World Scientific.

133. Lagunes, M.L., et al. 2019. Comparative study of fuzzy controller optimization with dynamic parameter adjustment based on Type 1 and Type 2 fuzzy logic. In *International Fuzzy Systems Association World Congress*. Springer.
134. Zhang, R., et al. 2014. GA based predictive functional control for batch processes under actuator faults. *Chemometrics and Intelligent Laboratory Systems* 137: 67–73.
135. Han, H.-G., et al. 2018. Multi-objective design of fuzzy neural network controller for wastewater treatment process. *Applied Soft Computing* 67: 467–478.
136. Lam, H., et al. 2001. Tuning of the structure and parameters of neural network using an improved genetic algorithm. In *IECON'01. 27th Annual Conference of the IEEE Industrial Electronics Society (Cat. No. 37243)*. IEEE.
137. Fonseca, C.M., and P.J. Fleming. 1995. *Multiobjective Optimization and Multiple Constraint Handling with Evolutionary Algorithms II: Application Example*. The University of Sheffield.
138. Chen, B.S., and S.J. Ho. 2016. Multiobjective tracking control design of T-S fuzzy systems: Fuzzy Pareto optimal approach. *Fuzzy Sets and Systems* 290: 39–55.
139. Gad, S., et al. 2017. Multi-objective genetic algorithm fractional-order PID controller for semi-active magnet orheologically damped seat suspension. *Journal of Vibration* 23 (8): 1248–1266.
140. Mahmoodabadi, M., and H. Jahanshahi. 2016. Multi-objective optimized fuzzy-PID controllers for fourth order nonlinear systems. *International Journal of Engineering Science Technology* 19 (2): 1084–1098.
141. Havlíček, V., et al. 2019. Supervised learning with quantum-enhanced feature spaces. *Nature* 567 (7747): 209–212.
142. Wang, L. 2016. Discovering phase transitions with unsupervised learning. *Physical Review B* 94 (19): 195105.
143. Ashfaq, R.A.R., et al. 2017. Fuzziness based semi-supervised learning approach for intrusion detection system. *Information Sciences* 378: 484–497.
144. Sutton, R.S., and A.G. Barto. 2018. *Reinforcement Learning: An Introduction*. MIT press.
145. Oh, B.K., et al. 2017. Evolutionary learning based sustainable strain sensing model for structural health monitoring of high-rise buildings. *Applied Soft Computing* 58: 576–585.
146. Marsland, S. 2014. *Machine Learning: An Algorithmic Perspective*. Chapman and Hall/CRC.
147. Sharma, P., and M. Kaur. 2013. Classification in pattern recognition: a review. *International Journal of Advanced Research in Computer Science Software Engineering* 3 (4).
148. Moriarty, D.E., A.C. Schultz, and J.J. Grefenstette. 1999. Evolutionary algorithms for reinforcement learning. *Journal of Artificial Intelligence Research* 11: 241–276.
149. Stanley, K.O., et al. 2019. Designing neural networks through neuroevolution. *Nature Machine Intelligence* 1 (1): 24–35.
150. Liu, F., and G. Zeng. 2009. Study of genetic algorithm with reinforcement learning to solve the TSP. *Expert Systems with Applications* 36 (3): 6995–7001.
151. Bora, T.C., V.C. Mariani, and L. dos Santos Coelho. 2019. Multi-objective optimization of the environmental-economic dispatch with reinforcement learning based on non-dominated sorting genetic algorithm. *Applied Thermal Engineering* 146: 688–700.
152. Liu, J., et al. 2020. QMR: Q-learning based multi-objective optimization routing protocol for flying Ad Hoc networks. *Computer Communications* 150: 304–316.
153. Tong, Z., et al. 2020. A scheduling scheme in the cloud computing environment using deep Q-learning. *Information Sciences* 512: 1170–1191.
154. Pi, C.-H., et al. 2020. Low-level autonomous control and tracking of quadrotor using reinforcement learning. *Control Engineering Practice* 95: 104222.
155. Dabney, W., et al. 2020. A distributional code for value in dopamine-based reinforcement learning. *Nature* 1–5.
156. Adleman, L.M. 1994. Molecular computation of solutions to combinatorial problems. *Science* 1021–1024.
157. Ezziane, Z. 2005. DNA computing: applications and challenges. *Nanotechnology* 17 (2): R27.
158. Kari, L., et al. 1998. DNA computing, sticker systems, and universality. *Acta Informatica* 35 (5): 401–420.
159. Shapiro, E., and B. Gil. 2008. RNA computing in a living cell. *Science* 322 (5900): 387–388.

160. Kaiser, C.A., et al. 2007. *Molecular Cell Biology*. WH Freeman.
161. Paun, G., G. Rozenberg, and A. Salomaa. 2005. *DNA Computing: New Computing Paradigms*. Springer Science & Business Media.
162. Doi, H., and M. Furusawa. 1996. Evolution is promoted by asymmetrical mutations in DNA replication-genetic algorithm with double-stranded DNA. *FSTJ* 32 (2): 248–255.
163. Yoshikawa, T., T. Furuhashi, and Y. Uchikawa. 1997. *The* effects of combination of DNA coding method with pseudo-bacterial GA. In *Proceedings of 1997 IEEE International Conference on Evolutionary Computation (ICEC'97)*. IEEE.
164. Zang, W., et al. 2018. A cloud model based DNA genetic algorithm for numerical optimization problems. *Future Generation Computer Systems* 81: 465–477.
165. Jan, H.Y., C.L. Lin, and T.S. Hwang. 2006. Self-organized PID control design using DNA computing approach. *Journal of the Chinese Institute of Engineers* 29 (2): 251–261.
166. Qun, N., and G. Xingsheng. 2004. Flow shop scheduling problems based on DNA evolutionary algorithms. *Journal of Shanghai University(Natural Science)* 10 (B10): 88–92.
167. Enayatifar, R., A.H. Abdullah, and I.F. Isnin. 2014. Chaos-based image encryption using a hybrid genetic algorithm and a DNA sequence. *Optics Lasers in Engineering* 56: 83–93.
168. Dorigo, M., and L.M. Gambardella. 1997. Ant colony system: a cooperative learning approach to the traveling salesman problem. *IEEE Transactions on Evolutionary Computation* 1 (1): 53–66.
169. Rajendran, C., and H. Ziegler. 2004. Ant-colony algorithms for permutation flowshop scheduling to minimize makespan/total flowtime of jobs. *European Journal of Operational Research* 155 (2): 426–438.
170. Dorigo, M., and T. Stützle. 2019. Ant colony optimization: overview and recent advances. In *Handbook of Metaheuristics*, 311–351. Springer.
171. Mirjalili, S., J.S. Dong, and A. Lewis. 2020. Ant Colony optimizer: theory, literature review, and application in AUV path planning. In *Nature-Inspired Optimizers*, 7–21. Springer.
172. Shelokar, P., et al. 2007. Particle swarm and ant colony algorithms hybridized for improved continuous optimization. *Applied Mathematics Computation* 188 (1): 129–142.
173. Kennedy, J., and R. Eberhart. 1995. Particle swarm optimization (PSO). In *Proceeding of IEEE International Conference on Neural Networks, Perth, Australia.*
174. Deng, W., et al. 2019. A novel intelligent diagnosis method using optimal LS-SVM with improved PSO algorithm. *Soft Computing* 23 (7): 2445–2462.
175. Gandomi, A.H., X.-S. Yang, and A.H. Alavi. 2013. Cuckoo search algorithm: a metaheuristic approach to solve structural optimization problems. *Engineering with Computers* 29 (1): 17–35.
176. Zhang, M., et al. 2018. Hybrid multi-objective cuckoo search with dynamical local search. *Memetic Computing* 10 (2): 199–208.
177. Sun, J., Q. Zhang, and E.P. Tsang. 2005. DE/EDA: A new evolutionary algorithm for global optimization. *Information Sciences* 169 (3–4): 249–262.
178. Shi, X., et al. 2005. An improved GA and a novel PSO-GA-based hybrid algorithm. *Information Processing Letters* 93 (5): 255–261.
179. Qin, A.K., V.L. Huang, and P.N. Suganthan. 2008. Differential evolution algorithm with strategy adaptation for global numerical optimization. *IEEE Transactions on Evolutionary Computation* 13 (2): 398–417.
180. Fleetwood, K. 2004. An introduction to differential evolution. In *Proceedings of Mathematics and Statistics of Complex Systems (MASCOS) One Day Symposium, 26th November, Brisbane, Australia.*
181. Neshat, M., et al. 2014. Artificial fish swarm algorithm: a survey of the state-of-the-art, hybridization, combinatorial and indicative applications. *Artificial Intelligence Review* 42 (4): 965–997.
182. Zheng, Z.-X., J.-Q. Li, and P.-Y. Duan. 2019. Optimal chiller loading by improved artificial fish swarm algorithm for energy saving. *Mathematics Computers in Simulation* 155: 227–243.
183. Fang, N., et al. 2014. A hybrid of real coded genetic algorithm and artificial fish swarm algorithm for short-term optimal hydrothermal scheduling. *International Journal of Electrical Power Energy Systems* 62: 617–629.

184. Trivedi, A., et al. 2016. A genetic algorithm–differential evolution based hybrid framework: case study on unit commitment scheduling problem. *Information Sciences* 354: 275–300.
185. Kanagaraj, G., S. Ponnambalam, and N. Jawahar. 2013. A hybrid cuckoo search and genetic algorithm for reliability–redundancy allocation problems. *Computers Industrial Engineering* 66 (4): 1115–1124.

Chapter 2
DNA Computing Based RNA Genetic Algorithm

Based on the biological RNA operations, DNA sequence selection, and mutation model, a RNA genetic algorithm (RNA-GA) algorithm is described in detail in this chapter. RNA molecules A, T, U, and C are utilized to encode the chromosome, and RNA molecular operations and DNA mutation model are combined to improve the crossover and mutation operators of SGA. The convergence of RNA-GA is analyzed using the Markov chain model. Five benchmark functions are applied to demonstrate the application process of the RNA-GA algorithm, and compare with SGA to effectively show the results by alleviating the premature convergence and improving the exploitation capacity of SGA.

2.1 Introduction

Since a computationally hard issue of the directed Hamiltonian path problem was first solved by DNA computing [1], many research results on different NP hard problems with fewer variables have been carried out [2–7]. Generally, there are three major steps in Adleman-style DNA computing: (1) a data pool of DNA molecules that represent all possible solutions is generated to the studied problem, (2) a series of biology laboratory techniques are utilized to exclude the DNA strands that don't match the logic constraints of the problem, (3) the surviving DNA molecules are collected for the process of answer readout. Obviously, the size of the initial data pool in DNA computing will increase exponentially with the increasement of the variables to be solved, it is a brute-force method in nature. In fact, the difficulty of DNA computing is not the absence of correct strands, but vast contaminating DNA sequences. To overcome the shortcomings of the brute-force method and implement the DNA operations with an existing digital computer, several DNA computing methods [7–9] and electronic DNA computing algorithms have been presented [10]. Because laboratory implements of DNA computing are highly difficult, inefficient,

J. Tao et al., *DNA Computing Based Genetic Algorithm*,
https://doi.org/10.1007/978-981-15-5403-2_2

un-scalable, and expensive, most of the DNA computings are carried out theoretically. Moreover, electronic DNA (EDNA) computing simulating a virtual test tube by digital computer was also proposed [11–13].

Holland first presented GA in 1975 [14], it is a global optimization algorithm based on the principle of survival of the fittest, similar partly to DNA computing. It may be possible to break the barrier of DNA computing and make it practical as the problem size scales up by combining GA. However, three operations of standard GA (SGA) in each chromosome are time-consuming, and the fixed mutation probability neglects the differences among various genes. SGA with too large mutation probability becomes a random search algorithm, on the contrary, it is prone to fall into local optimum with too small mutation probability.

Since the double helix structure of DNA strand cannot be directly combined with the chromosome of SGA, other genetic material needs to be considered. Recently, RNA computing was developed based on DNA computing [15–18]. Using the complementary oligonucleotides of DNA molecules, RNA strands inherit the DNA genetic information. Moreover, RNA computing with unique single-chain structure and various RNA operations is much easier to be combined with SGA. Enlightened by the RNA computing rules, a digital RNA-GA is proposed in this chapter [19] to improve the performances of SGA. The proposed algorithm may be applied to the biological computing to break the barrier of brute-force DNA computing.

First, crossover operators based on RNA operations and the mutation operators based on DNA sequence model are introduced to the SGA.

Secondly, the convergence of RNA-GA is analyzed based on the Markov chain model.

Finally, the parameter setting is illustrated by simulation on the typical test functions, and the application process is illustrated and compared with SGA.

2.2 RNA-GA Based on DNA Computing

2.2.1 Digital Encoding of RNA Sequence

There are four nucleotide bases of a RNA sequence, i.e., Adenine (A), Uracil (U), Guanine (G), and Cytosine (C), which are utilized to encode the solution of the given problem in RNA calculations. The type space for a RNA sequence is $E = \{A, U, G, C\}^l$, and l is the length of sequence. However, such RNA sequence cannot be processed by a digital computer. Because the 2-bit binary digital encoding (00, 01, 10, 11) can represent the structure, functional group, complementary relationship, and the number of hydrogen bonding of RNA nucleotide bases, it is used to encode four RNA nucleotide bases. There are totally $P_4^4 = 24$ possible encoding formats, among them, 0123/CUAG obtained in terms of the molecular weight of the nucleotide bases is selected as the best coding mode [17]. To be convenient of the mathematical and logical operations, the first bit is defined as a structure bit, and the second bit as the function bit, e.g., 1 × delegates the purine base, 0 × delegates the pyridine base, ×0 represents the keto group, and ×1 represents the amido group, where × is 0 or 1 [18]. Therefore, Cytosine (C) corresponds to 00, Uracil (U) corresponds to 01, Adenine (A) corresponds to 10, and Guanine (G) corresponds to 11.

A RNA sequence with a length of $L = 6$ used as the SGA chromosome can be expressed as $R_1 = CAUCGA$, whose quaternary number is coded as $R_1 = 021032$. All of the following discussions are based on this digital encoding.

2.2.2 *Operations of RNA Sequence*

There are many operations on RNA sequences. However, the genetic operators changing the length of RNA sequence are not introduced considering that the length of a single chromosome remains unchanged in SGA. We choose three main operations that can be applied to a single RNA sequence, i.e., namely translocation, transformation, and permutation.

(1) Translocation operator
The subsequences of RNA sequence are transferred to the new locations. For example, the original RNA sequence is $R = R_5R_4R_3R_2R_1$, where R_i ($i = 1, 2, \ldots, 5$) is the subsequence of RNA sequence (R) composed of four nucleotide bases (0123/CUAG), after translocation operation, the new RNA sequence becomes $R' = R_5R_2R_4R_3R_1$.
(2) Transformation operator
Two segments of RNA sequence exchange their locations mutually. For example, after transformation operator, the sequence $R = R_5R_4R_3R_2R_1$ becomes $R' = R_5R_2R_3R_4R_1$ by exchanging R_4 with R_2.
(3) Permutation operator
One subsequence of an RNA sequence is replaced by another one. For example, R_2' subsequence obtained from other RNA sequence or the same one is used to replace R_2, the new sequence is $R' = R_5R_4R_3R_2'R_1$.

Because of four types of elements in RNA sequence, three mutation operators of nucleotide bases are designed as follows:

(1) Reversal operator
The function bit of RNA nucleotide base is reversed while keeping its structure bit invariable, i.e., the premier digit of RNA encoding is reversed. There are totally four cases: $C \leftrightarrow A$, $U \leftrightarrow G$, i.e., $0 \leftrightarrow 2$, $1 \leftrightarrow 3$.
(2) Transition operator
Contrary to the reversal operator, the structure bit of RNA nucleotide base is inverted while keeping its function bit invariable. There also exist four cases: $C \leftrightarrow U$, $A \leftrightarrow G$, i.e., $0 \leftrightarrow 1$, $2 \leftrightarrow 3$.

(3) Exchange Operator
Both structure bit and function bit of RNA nucleotide base are transformed, i.e., both the first bit and the second bit are reversed, and then a complementary sequence of the given RNA sequence is then yielded. The four cases are as follows: $A \leftrightarrow U$, $C \leftrightarrow G$, i.e., $2 \leftrightarrow 1$, $0 \leftrightarrow 3$.

2.2.3 *Encoding and Operators in RNA-GA*

According to DNA sequence model under selection and mutation operations [20], some sequences are classified as harmful ones and the rest are classified as neutral ones. Analogously, RNA sequences are divided into the above two classes in terms of the value of fitness function of the individual, e.g., there are N individuals in the population at the current generation, the former <N/2> individuals are neutral and the left are deleterious, where $< \cdot >$ denotes the reduction (an exact figure) to the nearest integer. The crossover operator is executed among the neutral sequences to keep the excellent parents and produce better offspring, while the mutation operator is implemented in all sequences to keep the diversity of individuals. Thus, the encoding/decoding and RNA operators with SGA are given as follows:

(1) The encoding and decoding

In terms of the digital encoding method described in Sect. 2.2.1, the encoding space for a RNA sequence is $E = \{0, 1, 2, 3\}^l$ with sequences of length l. The chromosomes with population size N are shown in Fig. 2.1, which is an $nl \times N$ matrix, that means Nchromosomes in the population for the optimization problem with n variables.

MATLAB code for generating the chromosome in a population is shown as follows:

```
function [chromosome] = createchromo(Size,CodeL)
  % Size is the population size
  % CodeL is the length of Bases
  for i = 1:Size
    for j = 1:CodeL
      chromosome(i,j) = round(rand*3);
    end
  end
```

Fig. 2.1 RNA encoding of one population

$$\begin{bmatrix} \underbrace{c_{11}}_{l} \cdots \underbrace{c_{1n}}_{l} \\ \vdots \quad\quad \vdots \\ \underbrace{c_{N1}}_{l} \cdots \underbrace{c_{N_n}}_{l} \end{bmatrix} = \begin{bmatrix} q_{1l_1} \cdots q_{11_1}, \cdots, q_{1l_n} \cdots q_{11_n} \\ \vdots \quad\quad\quad\quad \vdots \\ q_{Nl_1} \cdots q_{N1_1}, \cdots, q_{Nl_n} \cdots q_{N1_n} \end{bmatrix}_{nl \times N}$$

The decoding process of the ith chromosome is to take a decimal number and map it into the given range.

$$
\begin{aligned}
c_{ij} &= q_{il-1}4^{l-1} + q_{il-2}4^{l-2} + \cdots q_{i0}4^{0} \\
x_{ij} &= x_{i\min} + \frac{c_{ij}}{4^{l}-1}(x_{i\max} - x_{i\min})
\end{aligned}
\tag{2.1}
$$

There are three MATLAB subfunctions to finish the decoding and mapping functions of the chromosome, which are listed as follows:

```
function [genepara] = uncode(chromosome, howlong)
% howlong is the number of the Bases for 1 variable
m = size(chromosome,1);% m is the population size
n = size(chromosome,2);% n is the whole length of chromosome
for i = 1:m
  for j = 1:n/( howlong)
partofchromosome = chromosome(i,((j-1)* howlong + 1):j * howlong);
genepara(i,j) = onegenepart (partofchromosome, howlong);
  end
end
function [genepara] = onegenepart(partofchromosome, howlong)
    % Calculate 1 variable of the optimal problem
      ii = howlong;
        sum = 0;
        jj = 0;
        while ii >=1
              sum = sum + partofchromosome (ii)*4^jj;
          ii = ii-1;
          jj = jj + 1;
        end
      genepara = sum;

function [unipara] = uniformity(genepara,umax,umin, paragennum)
% Map the value of chromosome to the given range [umin, umax];
m = size(genepara,1);
n = size(genepara,2);
for i = 1:m
  for j = 1:n
    unipara(i,j) = (umax-umin)/(4^ paragennum -1)*genepara(i,j) + umin;
  end
end
```

(2) The crossover operator based on RNA operations

The <N/2> sequences with the former <N/2> smaller fitness value (f) are defined as the neural ones and the left are the deleterious ones. The crossover operators are then performed in the neural ones, including translocation operator, transformation operator, and permutation operator. The permutation probability is set as 1, subsequence R_2 is produced randomly among the range of [1, l] in the current sequence, and subsequence R_2' possessing the same length of R_2 is generated from the other sequences. The translocation probability is chosen as 0.5, when implemented, we randomly obtain the subsequence R_2 in the range of $[1, l/2]$, and randomly find a new position in the range of $[R_{2h} + l/2, l]$, where R_{2h} is the higher location of R_2 in the crossover sequence. If the translocation operator is not carried out, the transformation operator is performed. The subsequence R_2 is selected in the first half part of the crossover sequence, while R_4 with the same length as R_2 is located in the second half part. After implementing the crossover operator in the neural sequences, N offspring are produced in terms of $N/2$ parents. MATLAB code of crossover operation is given as follows:

```
% ***** Crossover Operation ************
for i = 1:1:Size/2
        temp = rand;
        %permutation operator is implemented with probability 1
        if temp <=1
            n11(1) = ceil(CodeL*rand);
            n11(2) = ceil(CodeL*rand);
            while n11(2) ==n11(1)
              n11(2) = ceil(CodeL/2*rand);
            end
            location = ceil(rand*Size/2);
            while location ==i
              location = ceil(rand*Size/2);
            end
            TempE1(i,n11(1):n11(2)) = TempE(location,n11(1):n11(2));
        end
        % translocation operator
        if temp <=0.5
            n11(1) = ceil(CodeL/2*rand);  % define the R in the left part
            n11(2) = ceil(CodeL/2*rand);
            while n11(2) ==n11(1)
              n11(2) = ceil(CodeL/2*rand);
            end
            [n11,n] = sort(n11);
            n21(2) = CodeL/2 + ceil((CodeL/2-n11(2) + n11(1)-1)*rand) + n11(2)-n11(1);
            n21(1) = n21(2)-n11(2) + n11(1);
            TempE2(i,n21(1):n21(2)) = TempE(i,n11(1):n11(2));
            TempE2(i,n11(1):n11(2)) = TempE(i,n21(1):n21(2));
        end
        %transformation operator
        if temp > 0.5
            % define the R in the left part
            n11(1) = ceil(CodeL/2*rand);
            n11(2) = ceil(CodeL/2*rand);
            while n11(2) ==n11(1)
              n11(2) = ceil(CodeL/2*rand);
            end
            [n11,n] = sort(n11);
            nmove = CodeL/2 + n11(2)-n11(1) + 1+round((CodeL/2-n11(2) + n11(1)-1)*rand);
            % define the sequence moved
            tempa1(1,:) = TempE3(i,:);
            tempa2 = [];
```

```
            tempa2 = [tempa2,TempE3(i,1:n11(1)-1),TempE3(i,n11(2) + 1:CodeL)];
            [tempm,tempn] = size(tempa2);
            tempa1(1,nmove-n11(2) + n11(1):nmove) = TempE3(i,n11(1):n11(2));
            tempa1(1,1:nmove-n11(2) + n11(1)-1) = tempa2(1,1:nmove-n11(2) + n11(1)-1);
            if nmove ~=CodeL
              tempa1(1,nmove + 1:CodeL) = tempa2(1,nmove-n11(2) + n11(1):tempn);
            end
          end
end
```

(3) The RNA mutation operators

The RNA mutation is to keep the diversity of the population and generate new genetic material of RNA sequence, it is performed among the offspring produced by crossover operator among the neural ones, i.e., *N* offspring, and the <*N*/2> deleterious ones. Hence, there are totally 3*N*/2 sequences for mutation operators. There exist "hot spots" and "cold spots" in the DNA sequence model [20]. The mutations of nucleotide bases in the "cold spots" are slower than those in the "hot spots", that is consistent with the fact that the spots in different bit positions have different effects on the solutions of the problem. Therefore, at the beginning stage of the evolution, higher mutation probability is assigned to the RNA nucleotide bases in the higher bit positions ("the hot spots") to explore the larger feasible region. When a globally optimal region is found, the probabilities of mutation operators of RNA nucleotide bases in the higher bit positions will be reduced to prevent better solutions from disruption. The hot spots are converted into the cold spots accordingly. Set the nucleotide bases of RNA sequence between 1 and <*L*/2> to the low bit position, and the left as the high bit position, two kinds of mutation probability p_{mh} and p_{ml} are described as follows:

$$p_{\mathrm{mh}} = a_1 + \frac{b_1}{1 + \exp[aa(g - g_0)]} \tag{2.2}$$

$$p_{\mathrm{ml}} = a_1 + \frac{b_1}{1 + \exp[-aa(g - g_0)]} \tag{2.3}$$

where a_1 denotes the initial mutation probability of p_{ml}, b_1 is the range of trans-mutability, g is the evolution generation, g_0 delegates the generation where great change of mutation probability occurs, and aa denotes the speed of change. The p_{mh} and p_{ml} curves changing with evolution generation are shown in Fig. 2.2. The coefficients of Eqs. 2.2 and 2.3 are selected as follows: $a_1 = 0.02$, $b_1 = 0.2$, $g_0 = G/2$, and $aa = 20/G$.

After calculating the mutation probability, *nl* decimal numbers between 0 and 1 are produced and compared with the above changing probability. If the mutation probability is greater than the corresponding decimal number, the nucleotide base is replaced by one of another three integers, i.e., a random integer between 0 and 3 besides the nucleotide base itself. Therefore, three mutations in the nucleotide base of Sect. 2.2.2 are achieved. MATLAB code for mutation operator is shown as follows:

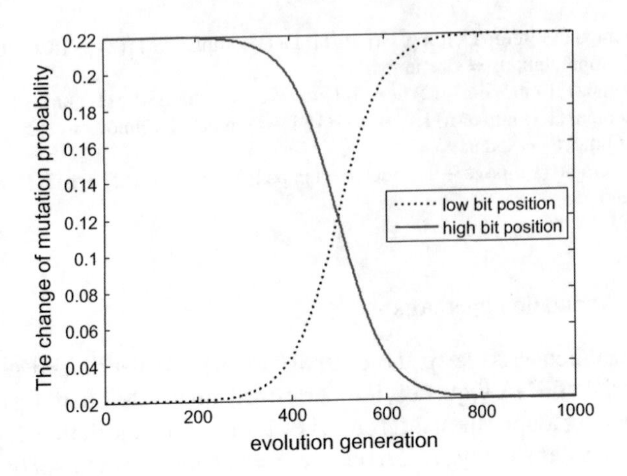

Fig. 2.2 Curves of p_{mh} and p_{ml}

```
%***** Step 4: Mutation Operation **************
  pmh = a1 + b1/(1 + exp(-aa*(kg-g0)));
  pml = a1 + b1/(1 + exp(aa*(kg-g0)));
  alphatable = [0,1,2,3];
  for i = 1:1:1.5*Size
    for ij = 1:num
      for j = (ij-1)*howlong + 1:(ij-1)*howlong + howlong/2 % the cold part
          temp = rand;
          if pml > temp          %Mutation Condition
            tmpa = TempE5(i,j);
            randnum = round(rand*3);          %mutation
            while(randnum ==tmpa)
              randnum = round(rand*3);
            end
             br = randnum + 1;
             TempE5(i,j) = alphatable(br);
          end
      end
      for j = (ij-1)*howlong + howlong/2 + 1:ij*howlong% the hot spots
          temp = rand;
          if pmh > temp          %Mutation Condition
              tmpa = TempE5(i,j);
              randnum = round(rand*3);          %mutation
              while(randnum ==tmpa)
                 randnum = round(rand*3);
              end
              br = randnum + 1;
              TempE5(i,j) = alphatable(br);
          end
      end
    end
  end
```

(4) The selection operator

There are now totally $3N/2$ sequences, the new population is constructed by selecting the best $N/2$ sequences and the worst $N/2$ sequences. According to DNA computing, DNA molecules are usually amplified by polymerase chain reaction (PCR) technology, where DNA sequences with higher concentration or melting temperature (Tm) can be copied with more chances. It can be seen that the PCR technology in DNA computing is analogous to the proportionate selection of SGA in some degree; hence, the same selection operator is adopted by RNA-GA. The proportional selection operator is then performed in the population with N sequences, which is made up of the best $N/2$ sequences and the worst $N/2$ sequences. The number of sequences is reproduced as follows:

$$n_r = \left\langle \left(\frac{J_i}{\sum_{i=1}^{N} J_i} \right) \right\rangle * N \tag{2.4}$$

where $J_i = F_{\max} - f_i$, and $F_{\max}$ is a constant chosen to guarantee $J_i > 0$. Thus, N sequences can be reproduced, and $N/2$ neural sequences as the parents of the crossover operator. Note that if n_r is equal to zero, the sequence will still be reproduced because $< \cdot >$ is used and the decimal number less than 1 is set to one. If $n_r > 1$, n_r sequences will be generated. The corresponding MATLAB code is given as follows:

```
% calculate the fitness function
Ji = Functionmax-fi;    %turn the min to max problem
[Oderfi,Indexfi] = sort(Ji);    %Arranging Ji from small to big

  % Select and Reproduct Operation
  fi_sum = sum(Ji);
  fi_Size = (Oderfi/fi_sum)*Size;
  fi_S = round(fi_Size);     %Selecting Bigger Ji value

  kk = 1;
  for i = Size:-1:1
    if fi_S(i) ==0
      TempE(kk,:) = E1(Indexfi(i),:);
      kk = kk + 1;
    else
      for j = 1:1:fi_S(i)     %Select and Reproduce
        TempE(kk,:) = E1(Indexfi(i),:);  % copy the chromosome
        kk = kk + 1;              %kk is used to reproduce
      end
    end
  end
```

2.2.4 *The Procedure of RNA-GA*

The processes of function evolution, selection, crossover, and mutation operators are described for RNA-GA as follows:.

Step 1: Initialize the maximum evolutionary generation G, chromosome encoding length $L = nl$, and population size N. Generate N RNA chromosomes randomly in the search space.
Step 2: Decode and calculate the fitness function value f for each individual in the population.
Step 3: Select the chromosomes to generate the parents of the next generation according to selection operator according to Eq. 2.4.
Step 4: In the former $N/2$ neural individual, three crossover operations are performed as follows: the permutation operator is executed with probability 1, and the transposition operator is executed with probability 0.5, while the transposition operator is executed if the transposition operator does not execute. After the crossover operation, N individuals can be generated.
Step 5: N individuals generated in step 4 and the remaining $N/2$ individuals in the crossover operator are used as the parents of the mutation operation, and the adaptive dynamic mutation probability is performed according to Eqs. 2.2 and 2.3.
Step 6: For the individuals generated in step 5, the best $N/2$ sequences and the worst $N/2$ sequences are selected as the parents of the selection operation.
Step 7: Repeat steps 2–6 until a termination criterion is satisfied. This can be the set maximum number of evolutions, the minimum improvement of the best performance of successive generations, or a known global optimum. In addition, Elitism is used throughout the process to include the best current individuals into the offspring.

2.3 Global Convergence Analysis of RNA-GA

Li et al. made a summary of conditions guaranteeing the convergence of GA with mutation operator for the global optimization problem, which is listed as follows [21]:

Assumption 2.1 At each generation t, if every individual (x) in the population and a random individual (y) satisfy $x \neq y$, then there exists $p(t) > 0$, where $p(t)$ is the probability changing x into y by one mutation operator.

Theorem 2.1 *If GA with elitist keeping strategy satisfies Assumption* 2.1, *it will converge in probability to the optimal solution of the problem. Moreover, its convergence is irrelevant to the distribution of the initial population. If the optimal chromosome of the population is monotonic in time and any individual in X can be reached* *by mutation and reorganization in a limited number of steps, GA will converge to the optimal solution with probability 1.*

Based on the above assumption and the theorem, the convergence of the proposed RNA-GA is analyzed as follows:

As defined in RNA-GA, RNA strands are essentially the quadruple sequences with length of L. Its encoding space is defined as $S = \{0, 1, 2, 3\}^L$, i.e., $|S| = 4^L$. Set N as the population size of individuals, Ω as the sequence set of RNA, X as the population composed of the elements in Ω, and Ω^N as all possible populations in Ω, $\Omega^N = \{X_1, X_2, \ldots X_N\}$.

The transition probability matrix (P) can be decomposed into the product of three probability matrices [21]: crossover (C), mutation (M), and selection (S), i.e., $P = CMS$. Matrices C, M, *and* S have the following properties:

$$\sum_{j=1}^{n} c_{ij} = 1 \tag{2.5}$$

where c_{ij} is the probability of changing sequence i into sequence j by the crossover operator.

$$m_{jk} = \prod\nolimits_{i=1}^{N} \left(\frac{p_m}{C-1}\right)^{H_i} (1 - p_m)^{L-H_i} \quad k \in \{1, 2 \ldots n\} \tag{2.6}$$

where m_{jk} denotes the probability of turning sequence j into sequence k by mutation, $C = 4$, and H is the individual Hamming distance between sequence j and sequence k.

Because at least one individual is selected, the following inequality holds:

$$\sum_{k=1}^{n} s_{kq} > 0 \quad q \in \{1, 2 \ldots n\} \tag{2.7}$$

where s_{kq} denotes the probability of changing sequence k into sequence q by the selection operator. Let

$$\rho = \left[\min\left(\frac{p_m}{3}, 1 - p_m\right)\right]^{NL} \tag{2.8}$$

Then, the following inequality can be obtained based on Eqs. 2.6 and 2.8:

$$m_{jk} > \rho \tag{2.9}$$

The equation is then derived in term of the running processes of RNA-GA

$$P = CMS = \sum_{k=1}^{n} \left(\sum_{j=1}^{n} c_{ij} m_{jk} \right) s_{kq} \tag{2.10}$$

Substituting Eqs. 2.9 into 2.10, the following inequality is obtained:

$$P \geq \sum_{k=1}^{n} \left(\sum_{j=1}^{n} c_{ij} \rho \right) s_{kq} \tag{2.11}$$

Equation 2.11 can be rewritten by substituting Eq. 2.5 into it:

$$P \geq \rho \sum_{k=1}^{n} s_{kq} \tag{2.12}$$

From Eqs. 2.7–2.12, the ultimate answer is

$$P > 0 \tag{2.13}$$

Since Eq. 2.13 is satisfied, the finite states homogeneous Markov chain constructed at each generation is ergodic, which is consistent with the fact that the initial population of RNA-GA is finite. Thus, any point in X can be reached from limited steps by mutation and reorganization operations.

Denote f^* as the global optimal solution for the given optimization problems, and define $Z_t = \max\{f(X_k^{(t)})|k = 1, 2, \ldots n\}$ as the state of optimal fitness value attainable among t generations, and p_i^t as the probability when the individual reaches Z_t. It is obvious that

$$P\{Z_t \neq f^*\} \geq p_i^t \tag{2.14}$$

Thus, Eq. 2.15 can be obtained as follows:

$$P\{Z_t = f^*\} \leq 1 - p_i^t \tag{2.15}$$

with time t approaches to infinity:

$$p_i^{\infty} > 0 \tag{2.16}$$

Hence,

$$\lim_{t \to \infty} P\{Z_t = f^*\} \leq 1 - p_i^{\infty} < 1 \tag{2.17}$$

It should be noted that the elite maintaining strategy is not used in the derivation of Eq. 2.17. In fact, once a better individual is generated in X, the elite maintaining strategy can retain it in the offspring. After the finite-time state transition, the best solution can always be obtained, which makes the optimal fitness sequence monotonous to time and satisfy the condition of Theorem 2.1. Therefore, when elite maintaining strategy is adopted, RNA-GA can converge to the global optimal solution with probability 1.

2.4 Performance of the RNA-GA

2.4.1 Test Functions

In order to given the application procedure of RNA-GA algorithm, a test environment is provided in the form of several typical optimization functions. Choosing a set of representative functions is not an easy task, since any particular combination of properties represented by a test function doesn't allow for the generalized performance statements. Five commonly used test functions are shown in the Appendix in the chapter, which represent a group of landscape classes with various characteristics, i.e., large search space, numerous local minima, and fraudulence. All the functions are two-dimensional, which makes it easy to visualize them and to see the action of algorithm.

The global optimum of the Rosenbrock function f_1 is located in a very narrow valley with a flat bottom. Moreover, the non-separable characteristic of this quadratic function with different eigenvalues further increases the difficulty of solving the problem. Needle-in-haystack (NiH) function f_2 has four local optima with the value of 2748.78, and they are close to the global optimum (3600). In addition, the degree of fraudulence will change when selecting different coefficients of NiH function. Schwefel's function f_3 is symmetric, separable, and multimodal, and its global minimum is close to the domain boundary and geometrically distant from the second minimum points. Therefore, the search algorithms are prone to be cheated and converge to the wrong direction. Rana function f_4 is a non-separable and highly multimodal function. Its best solutions are located at the corner of the domain. The Griewank function f_5 is highly multimodal with thousands of widespread local minima, which are distributed regularly. To get a topological impression, the landscapes of the two-dimensional functions from f_1 to f_5 are plotted in Figs. 2.3, 2.4, 2.5, 2.6, and 2.7, respectively. These functions are hard to be optimized using classical methods as well as most evolutionary algorithms. The successful search of the optimal solutions can only be derived by methods with effective anti-deceptive search capability.

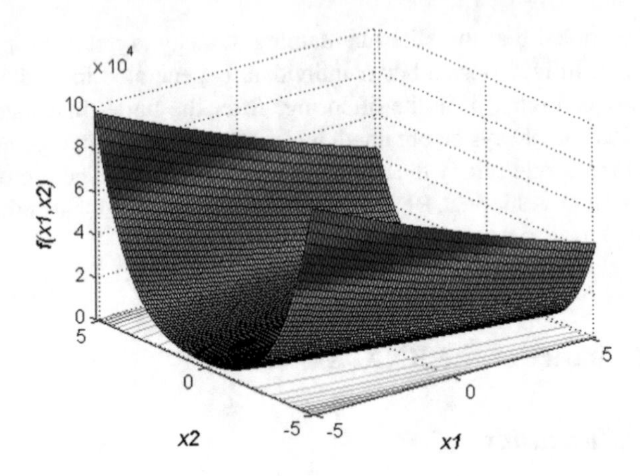

Fig. 2.3 The two-dimensional Rosenbrock function f_1

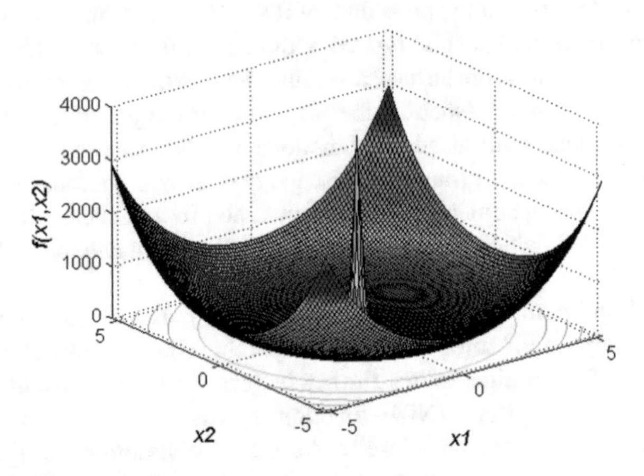

Fig. 2.4 The two-dimensional NiH function f_2

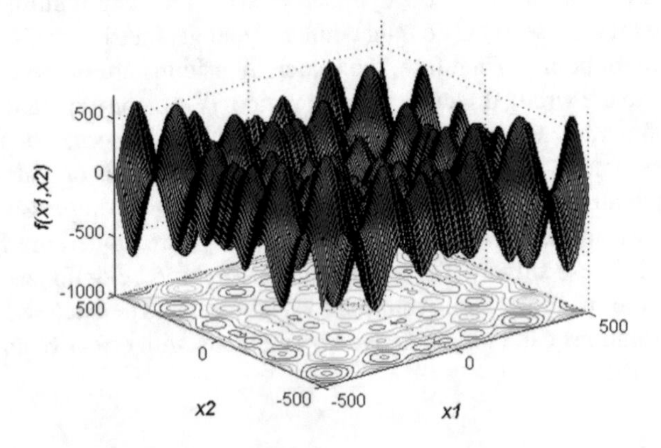

Fig. 2.5 The two-dimensional Schwefel function f_3

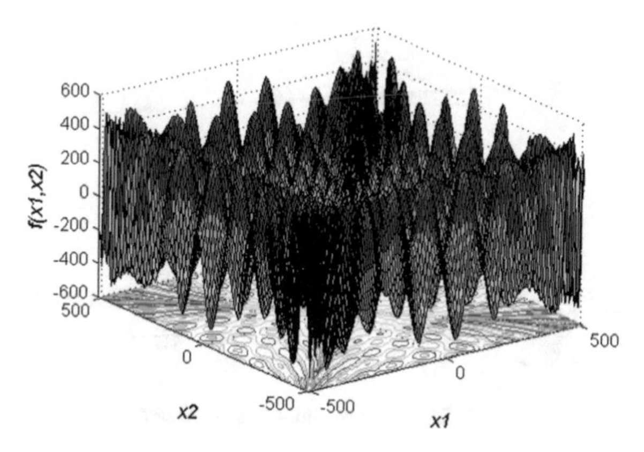

Fig. 2.6 The two-dimensional Rana function f_4

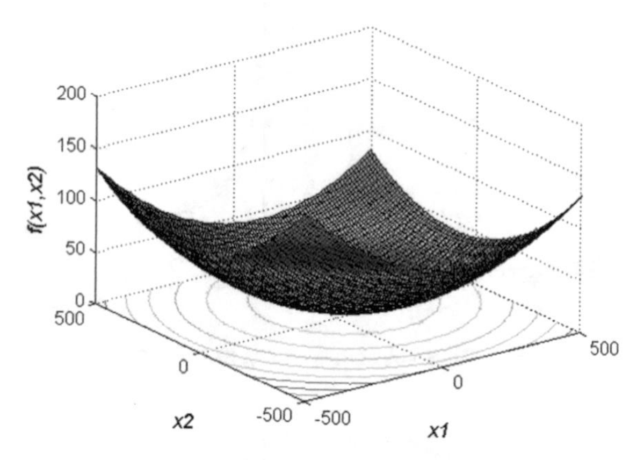

Fig. 2.7 The two-dimensional Griewank function f_5

2.4.2 *Adaptability of the Parameters*

How to set the parameters is critical to the application of RNA-GA, and the adaptation ability of the main parameters in RNA-GA is then illustrated by applying to optimize the Rana model f_4 as well as the Griewank function f_5. In the first group of experiments, the coefficients $aa = 20/G$ and $a_1 = 0.02$ are fixed, but the value of b_1 is varying. The second uses fixed $b_1 = 0.2$ and $a_1 = 0.02$, but various aa; Similarly, the third uses fixed coefficients $aa = 20/G$ and $b_1 = 0.2$, but various a_1. In Figs. 2.8, 2.9, and 2.10, the experiment results are given in a trend curves of the average best-so-far objective function values over 50 independent runs.

The linear convergence can be clearly observed in these figures, and the adaptation mechanism of RNA-GA can operate effectively. Moreover, it can be seen that

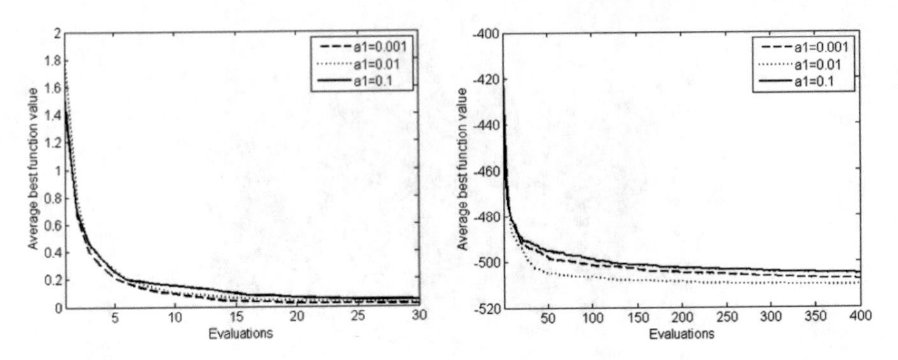

Fig. 2.8 Convergence curves of RNA-GA corresponding to different settings of b_1 for f_4 and f_5

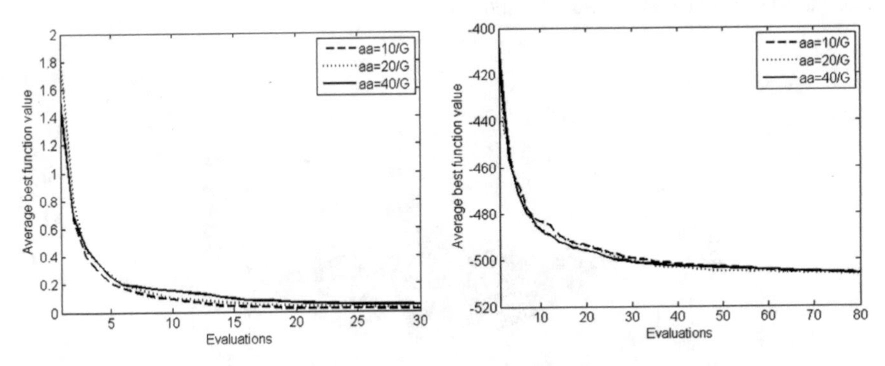

Fig. 2.9 Convergence curves of RNA-GA corresponding to different settings of aa for f_4 and f_5

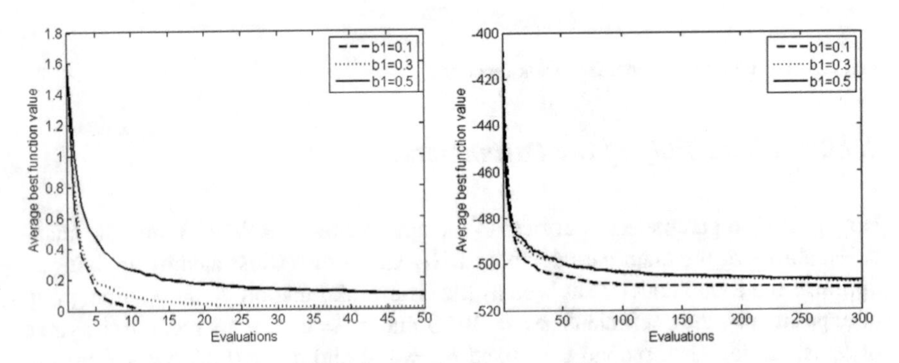

Fig. 2.10 Convergence curves of RNA-GA corresponding to different settings of a_l for f_4 and f_5

the convergence rate is more sensitive to b_1 and a_1 than *aa*, and larger b_1 and a_1 lead to faster convergence of RNA-GA. However, for the sake of robustness, the more moderated settings of $a_1 = 0.001 - 0.05, b_1 = 0.1 - 0.3$, and $aa = 20/G$ are recommended, since too large b_1 and a_1 tend to become random search, while too small

b_1 and a_1 are prone to converge prematurely. The feasibility of the recommended parameters setting has also been empirically verified by the successful optimizations of the functions f_1 to f_3.

2.4.3 *Comparisons Between RNA-GA and SGA*

To understand how RNA-GA converge to the global optimal value, the distribution of the individuals in the RNA-GA population on contour plots of the functions to be optimized are shown in this section. When the RNA-GA algorithm is implemented, the maximum evolution generation is limited to 1000, the population size N is set to 60, and the individual length is set to 40. Other parameters are kept unchanged as shown in Sect. 2.4.2. To be convenient for comparison, SGA in the GA toolbox of MATLAB 7.1 is adopted to optimize these functions. GA uses proportional selection, adaptation mutation, and two-point crossover with the crossover probability 0.8. Elitist reservation mechanism is utilized to ensure the monotony of the best-so-far individual in population. The runs of the algorithms terminate when the predetermined maximal evolution generation is performed or the inequality $|F_b - F^*| < \Delta$ is satisfied, where F_b denotes the objective function value of the best-so-far individual, and F^* the global optimum, Δ is a precision requirement of the optimal solution, as set to 0.0001. The behavior of RNA-GA and SGA are shown in Figs. 2.11, 2.12, 2.13, 2.14, 2.15, 2.16, 2.17, 2.18, 2.19, and 2.20. It is obvious that RNA-GA possesses an improved population diversity. With a similar initial population, RNA-GA is capable to explore more search space than SGAs during the in-process population. Even at the end of the evolution, the population of RNA-GA still remains in diversity, while the population of SGA is prone to converge to one point, which makes SGA easy to trap into local minima.

In order to obtain statistically significant data, a sufficiently large number (R) of independent runs must be implemented. The performance of the convergence speed is measured by the average evaluation number $\bar{E}$, the minimum and maximum evaluation number $E_{\min}$, and $E_{\max}$ over R runs, where $\bar{E} = \frac{1}{R}\sum_{i=1}^{R} E_i$, and $E_i(i \in 1, \ldots, R)$ are the actual evaluation generations satisfying terminate conditions. The corresponding data obtained by $R = 50$ are given in Table 2.1.

The global search capability is measured by $F_{\min}$, $F_{\max}$, and $\bar{F}$, denoted as the minimum, maximum, and average optimal value of the test functions over R runs, respectively. The rate of the runs accurately reaching the global optimum is also demonstrated by Suc.rate. The corresponding statistical results are listed in Table 2.2.

The statistic results in Table 2.2 show that RNA-GA can overcome the fraud of test functions with fewer local optima values, such as f_2 and f_3, and has been succeeded in finding the solutions in all tests. As for f_5, because there exist thousands of local optima with regularity, RNA-GA also traps into local optima. However, most numerical experiments can find the global optimum. As for the Rana function f_4, RNA-GA has found the global optimum located at (−488.63, 512) with a value of −511.7329, which is different from the optimum in the literature, that is (512, 512)

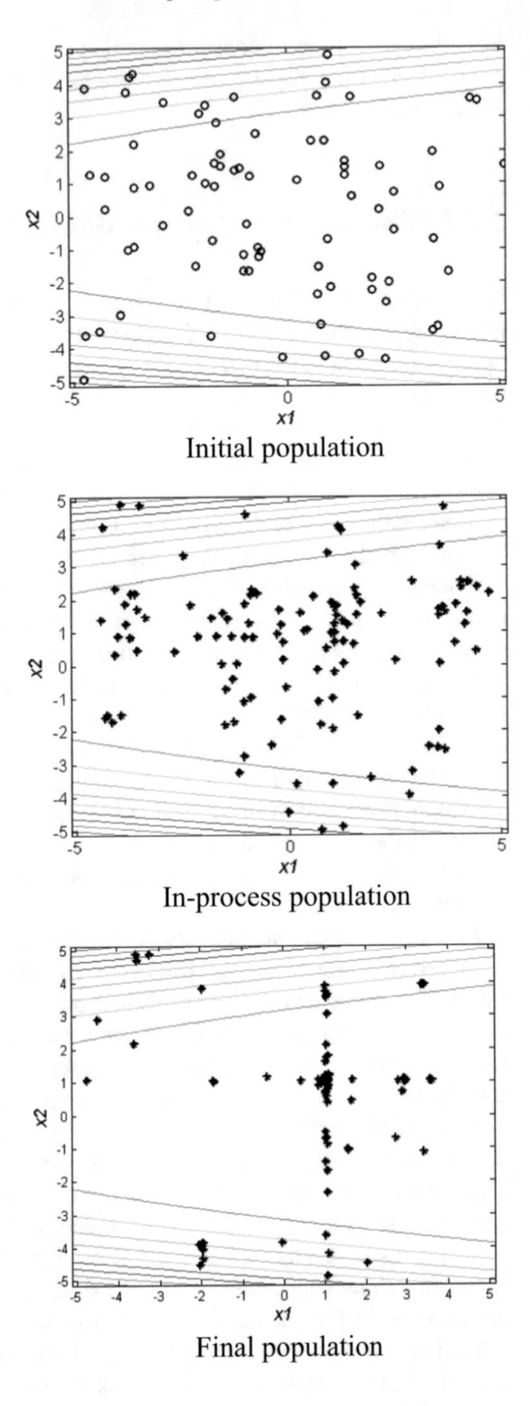

Fig. 2.11 Behavior of RNA-GA in the Rosenbrock problem

Fig. 2.12 Behavior of SGA in the Rosenbrock problem

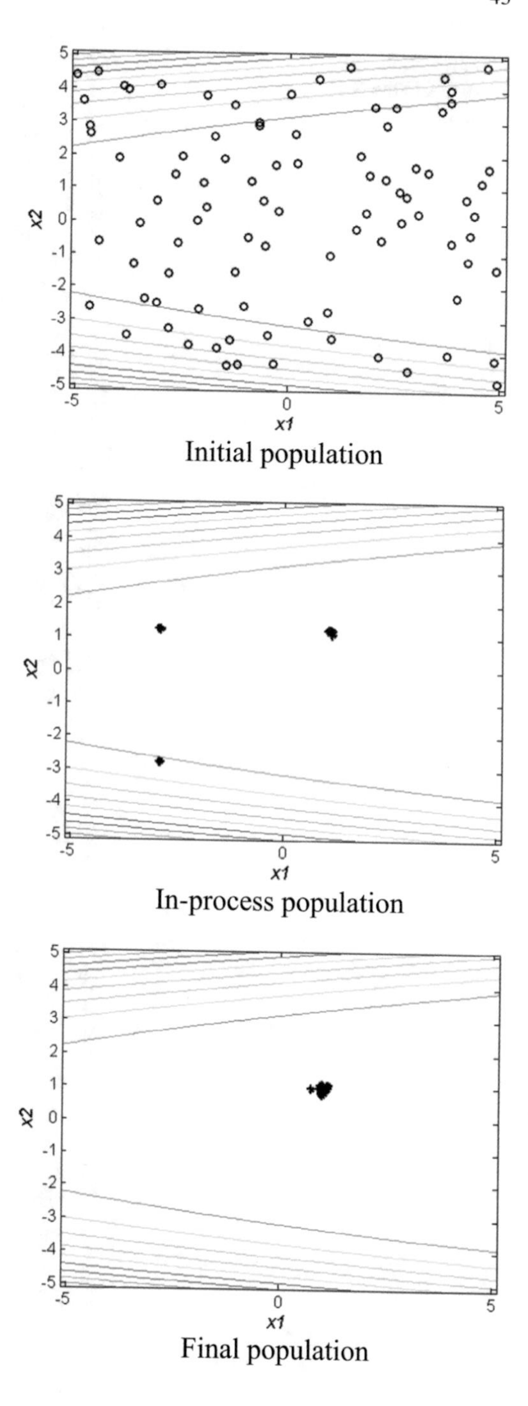

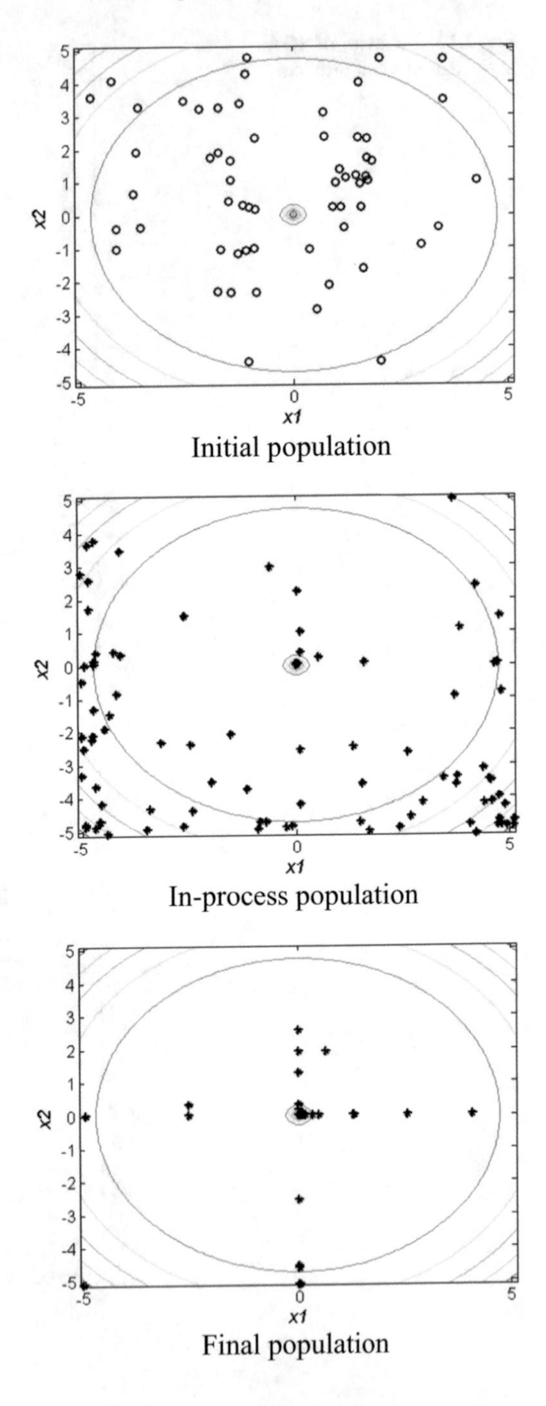

Fig. 2.13 Behavior of RNA-GA in the NiH problem

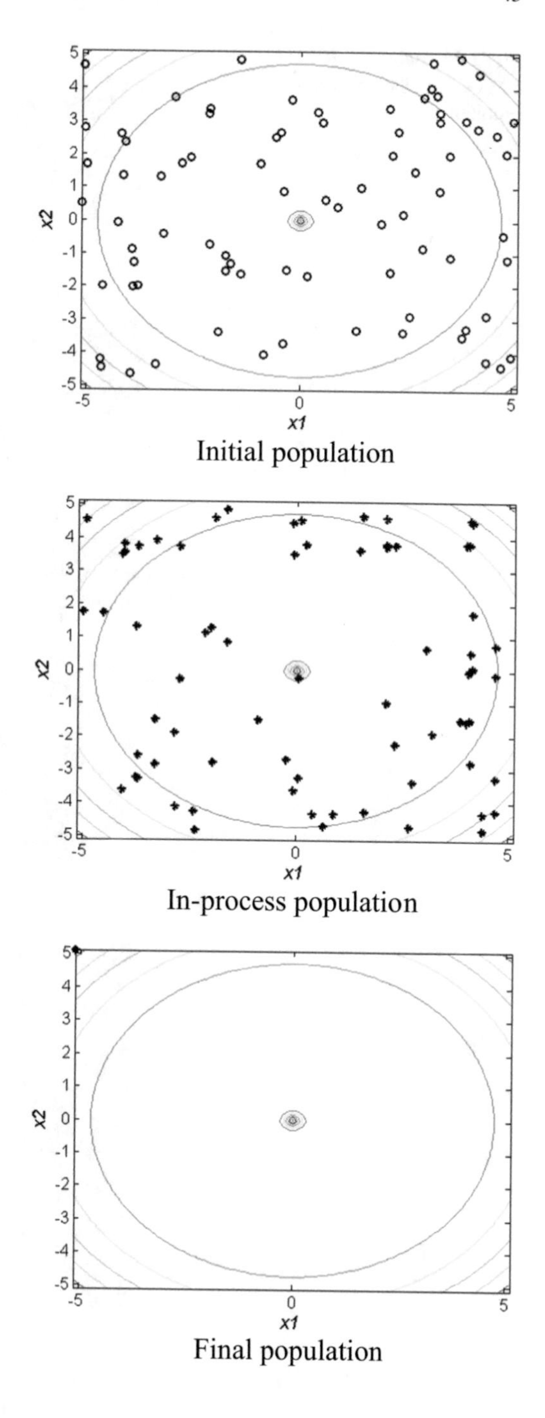

Fig. 2.14 Behavior of SGA in the NiH problem

Fig. 2.15 Behavior of RNA-GA in the Schwefel problem

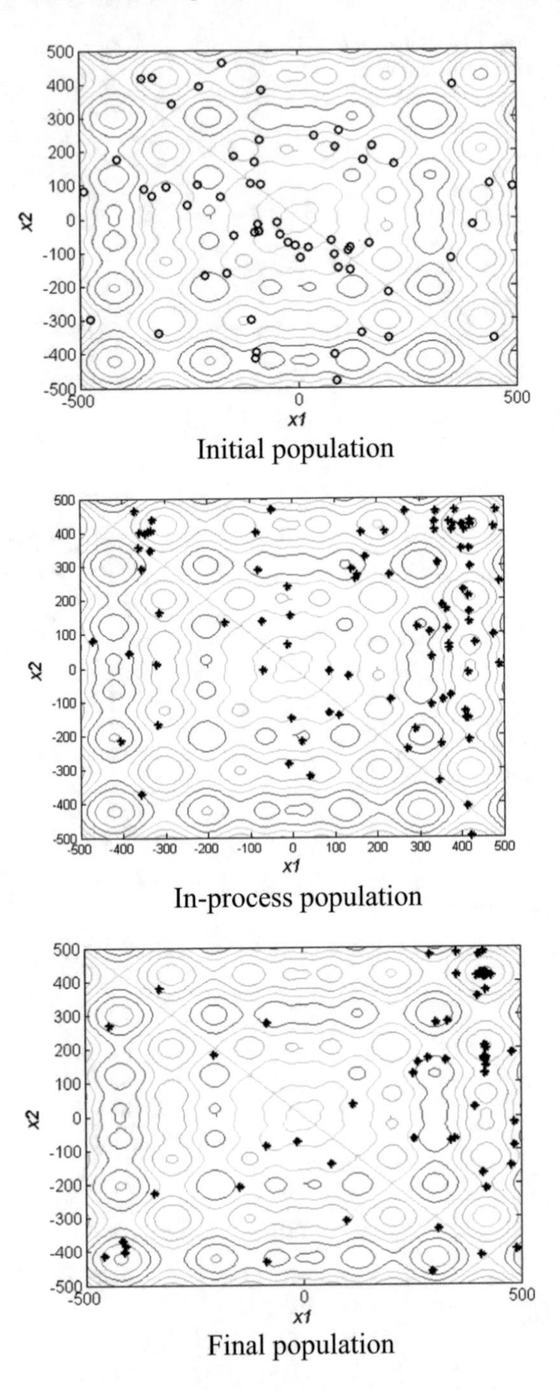

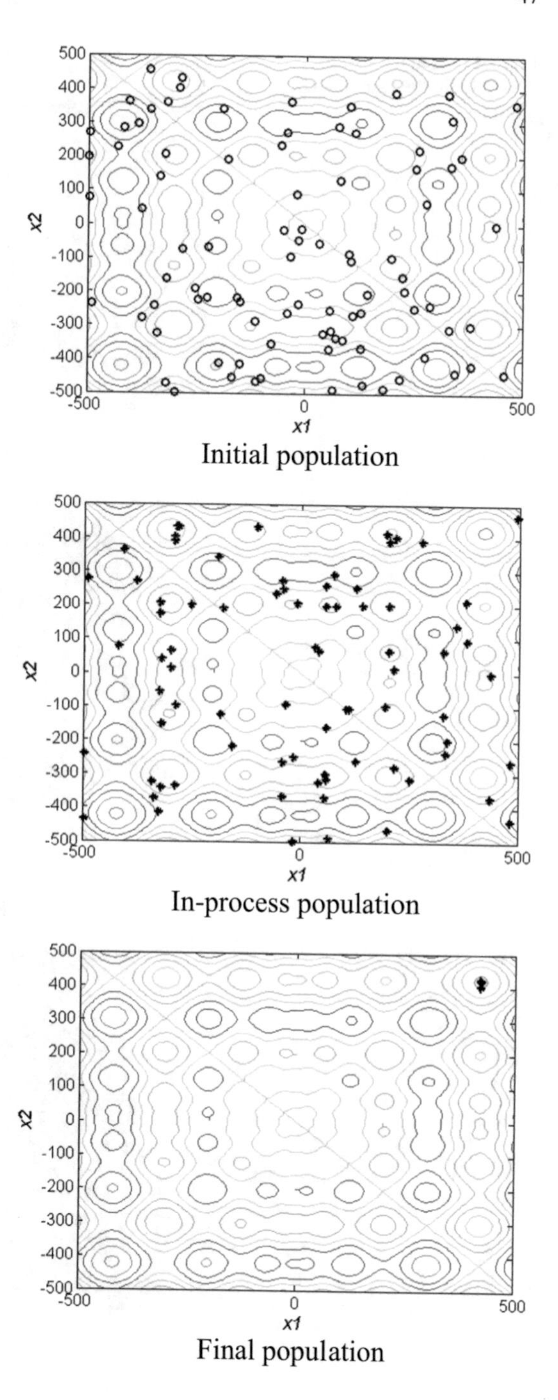

Fig. 2.16 Behavior of SGA in the Schwefel problem

Fig. 2.17 Behavior of RNA-GA in the Rana problem

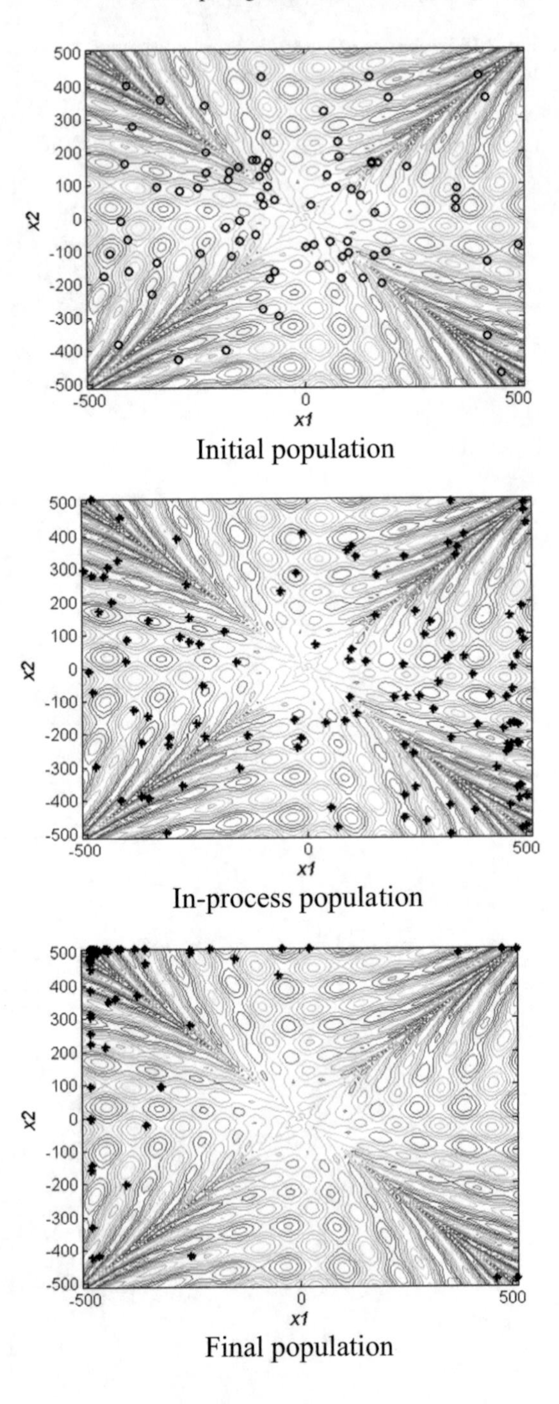

Fig. 2.18 Behavior of SGA in the Rana problem

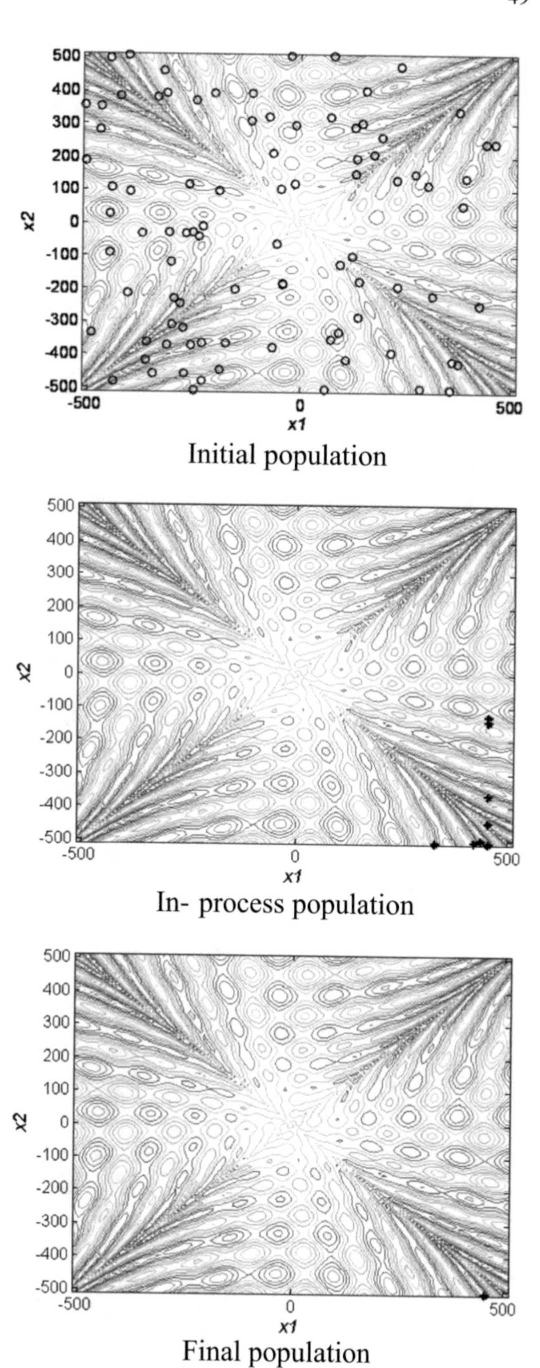

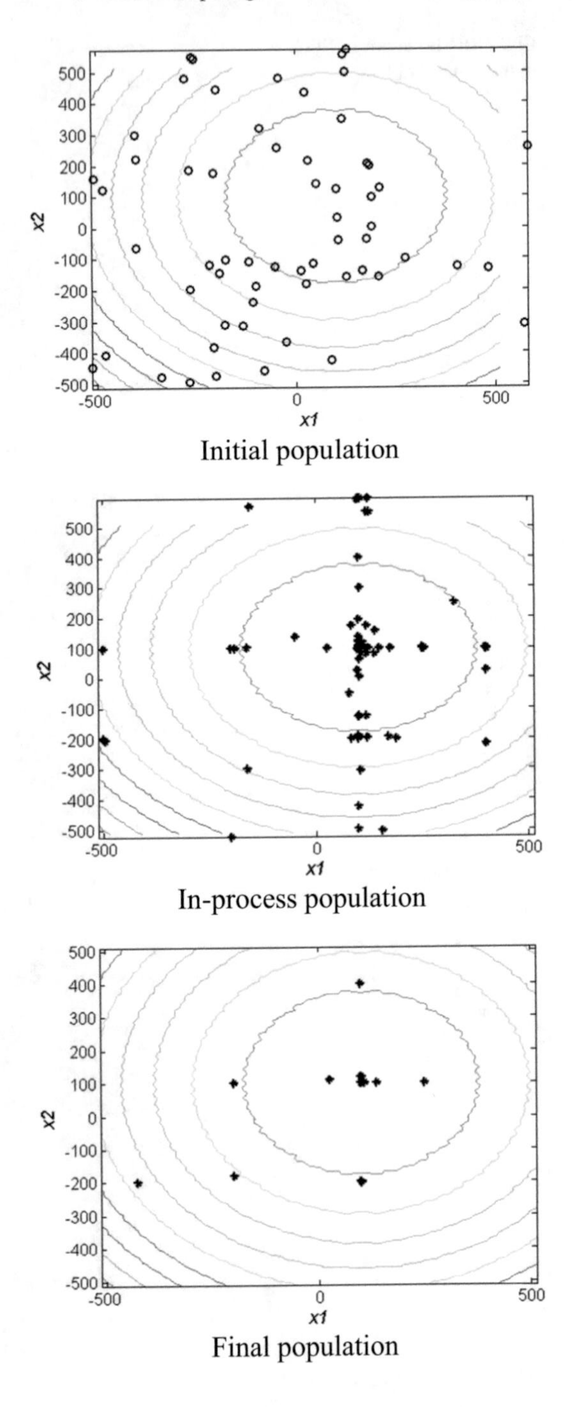

Fig. 2.19 Behavior of RNA-GA in the Griewank problem

Fig. 2.20 Behavior of SGA in the Griewank problem

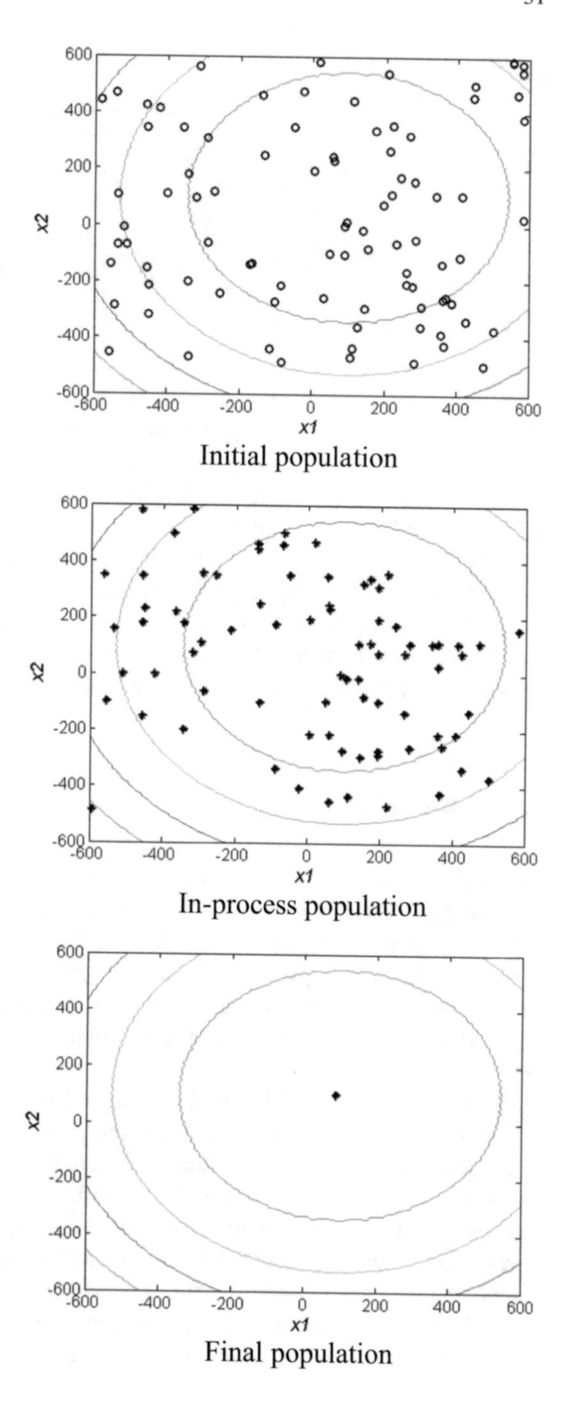

Table 2.1 Comparison of the convergence speed by RNA-GA and SGA over 50 runs

Test functions	RNA-GA			SGA		
	$\bar{E}$	$E_{\min}$	$E_{\max}$	$\bar{E}$	$E_{\min}$	$E_{\max}$
f_2	614.45	109	1000	311.3	77	1000
f_2	489.5	327	577	941.18	724	1000
f_3	323.5	117	603	240.95	201	1000
f_4	760.1	493	1000	961.5	211	1000
f_5	497.0667	5	1000	859.25	89	1000

with a value of -511.7011. However, the success rate is still quite small because of thousands of local optima, which is only 32%. Compared with the statistic results of SGA, such as the success rate of f_4 is as low as 5%, the global search capability of RNA-GA is greatly improved. As for f_1 and f_3, SGA has similar success rate as RNA-GAs in Table 2.2, but the convergence rate of SGA is faster than RNA-GA in Table 2.1, because there is a turning point of mutation probability at evolution generation g_0, and the rapidity of convergence is sacrificed to keep the diversity of population. As for the single model function f_1, though both SGA and RNA-GA can find the best results with 100% success rate, the results of RNA-GA are better than SGAs. Therefore, RNA-GA has obtained better search performance than SGA.

2.5 Summary

In this chapter, a RNA-GA framework for complex function optimization is described by combining RNA operations and DNA sequence models with genetic algorithms. Numerical simulation results demonstrate the effectiveness of the hybridization, especially the advantages of RNA-GA in optimizing quality, efficiency, and initial conditions. The superiority of the RNA-GA lies in the combination of a DNA sequence model with variable mutation probability and the combination of multiple RNA operators.

When the biological RNA molecule calculation is used instead of the electronic RNA-GA calculation process, even if the good resolution in the RNA molecule calculation is accidentally eliminated, a better solution can still be obtained after the next recombination of left solutions. After a limited number of repetitions, the optimal solutions or near-optimal solutions of RNA sequences will increase greatly, and the correct RNA sequence can be obtained relatively simply. In theory, since all the operators in RNA-GA are obtained through the operation of RNA molecules, the algorithm can be simply converted and applied to actual biochemical reactions, and it can break the limitation of the brute-force method of DNA calculation.

Table 2.2 Comparison of the global research ability by RNA-GA and SGA over 50 runs

Test function	RNA-GA				SGA			
	$\bar{F}$	F_b	F_w	Suc.rate (%)	$\bar{F}$	F_b	F_w	Suc.rate (%)
f_2	8.570e−7	8.336e−9	3.4793e−7	100	4.598e−6	3.254e−9	4.065e−5	100
f_2	3.600e+3	3.600e+3	3.600e+3	100	3.316e+3	3.600e+3	2.7488e+3	66.67
f_3	−837.966	−837.966	−837.966	100	−833.228	−837.966	−719.527	96
f_4	−511.613	−511.733	−510.585	32	−500.406	−511.733	−463.419	5
f_5	0.0020	4.156e−8	0.0074	76	0.0102	5.849e−10	0.0271	21

Appendix

Five test functions:

Test functions	Optimal solution	Optimal value
$f_1(\mathbf{x}) = 100(x_2 - x_1^2)^2 + (1 - x_1)^2$ $x_1, x_2 \in [-5.12, 5.12]$	(1,1)	0
$\max f_2(\mathbf{x}) = \left(\frac{a}{b+(x_1^2+x_2^2)}\right)^2 + (x_1^2 + x_2^2)^2$ $a = 3.0, b = 0.05, x_1, x_2 \in [-5.12, 5.12]$	(0,0)	3600
$f_3(\mathbf{x}) = \sum_{i=1}^{2} -x_i \sin(\sqrt{\lvert x_i \rvert})$ $x_1, x_2 \in [-500, 500]$	(420.9687, 420.9687)	−837.9658
$f_4(\mathbf{x}) = x_1 \times \sin(\sqrt{\lvert x_2 + 1 - x_1 \rvert}) \times \cos(\sqrt{\lvert x_1 + x_2 + 1})$ $+ (x_2 + 1) \cos(\sqrt{\lvert x_2 + 1 - x_1 \rvert}) \times \sin(\sqrt{\lvert x_1 + x_2 + 1 \rvert})$ $x_1, x_2 \in [-512, 512]$	(−488.63, 512)	−511.7329
$f_5(\mathbf{x}) = ((x_1 - 100)^2 + (x_2 - 100)^2)/4000$ $- \cos(x_1 - 100) \cos((x_1 - 100)/\sqrt{2} + 1$ $x_1, x_2 \in [-600, 600]$	(100,100)	0

References

1. Adleman, L.M. 1994. Molecular computation of solutions to combinatorial problems. *Science* 266 (1): 1021–1024.
2. Braich, R.S., et al. 2002. Solution of a 20-variable 3-SAT problem on a DNA computer. *Science* 296: 499–502.
3. Dan, B., C. Dunworth, and R.J. Lipton.1995. Breaking DES using a molecular computer. In *Proceedings of a DIMACS Workshop, Series in Discrite Mathematics Theoretical Computer Science*. Princeton University.
4. Ji, Y.L., et al. 2004. Solving traveling salesman problems with DNA molecules encoding numerical values. *Biosystems* 78 (1): 39–47.
5. Ouyang, Q., et al. 1997. DNA solution of the maximal clique problem. *Science* 278: 446–449.
6. El-Seoud, S.A., R. Mohamed, and S. Ghoneimy. 2017. DNA computing: challenges and application. *International Journal of Interactive Mobile Technologies* 11 (2): 74–87.
7. Ren, J., and Y. Yao. 2018. DNA computing sequence design based on bacterial foraging algorithm. In *Proceedings of the 2018 5th International Conference on Bioinformatics Research and Applications*. ACM.
8. Yamamoto, M., et al. 2001. Solutions of shortest path problems by concentration control. *Lecture Notes in Computer Science* 2340 (40): 231–240.
9. Yang, C.N., and C.B. Yang. 2005. A DNA solution of SAT problem by a modified sticker model. *Biosystems* 81 (1): 1–9.

10. Boruah, K., and J.C. Dutta. 2016. Development of a DNA computing model for Boolean circuit. In *2016 2nd International Conference on Advances in Electrical, Electronics, Information, Communication and Bio-Informatics (AEEICB)*. IEEE.
11. Garzon, M., R.J. Deaton, and J.A. Rose. 1999. Soft molecular computing. DNA Based Computers 91–100.
12. Hartemink, A.J., T.S. Mikkelsen, and D.K. Gifford. 1999. Simulating biological reactions: A modular approach. In *International Meeting on Preliminary*.
13. Li, Y., C. Fang, and Q. Ouyang. 2004. Genetic algorithm in DNA computing: A solution to the maximal clique problem. *Chinese Science Bulletin* 49 (9): 967–971.
14. Holland, J.H. 1992. *Adaptation in Natural and Artificial Systems*, vol. 6, issue 2, 126–137.
15. Cukras, A.R., et al. 1999. Chess games: A model for RNA based computation. *Biosystems* 52 (1–3): 35–45.
16. Faulhammer, D., et al. 2000. Molecular computation: RNA solutions to chess problems. In *Proceedings of the National Academy of Sciences of the United States of America*.
17. Li, S., and J. Xu. 2003. Digital coding for RNA based on DNA computing. *Computer Engineering Applications* 39 (5): 46–47.
18. Shu chao, L.I. 2003. Operational rules for digital coding of RNA sequences based on DNA computing in high dimensional space. *Bulletin of Science Technology*.
19. Tao, J., and N. Wang. 2007. DNA computing based RNA genetic algorithm with applications in parameter estimation of chemical engineering processes. *Computers Chemical Engineering* 31 (12): 1602–1618.
20. Neuhauser, C., and S.M. Krone. 1997. The genealogy of samples in models with selection. *Genetics* 145 (2): 519.
21. Hu, J., Y. Sun, and Q. Xu. 2010. The theory and application of genetic algorithm. In *International Conference on Computer & Communication Technologies in Agriculture Engineering*.

Chapter 3
DNA Double-Helix and SQP Hybrid Genetic Algorithm

By utilizing the global exploration of GA and local exploitation characteristics of sequential quadratic programming (SQP), a hybrid genetic algorithm (HGA) is proposed in this chapter for the highly nonlinear constrained functions. Thereafter, the theoretical analysis for the convergence of the HGA is then made. In the global exploration phase, the Hamming cliff problem is solved by DNA double-helix structure, and DNA computing inspired operators are introduced to improve the global searching capability of GA. When the feasible domains are located, the SQP method is executed to find the optimum quickly, in the meantime, it can achieve better solution accuracy. Six benchmark functions are applied to demonstrate the application process of the hybrid algorithm and compare with GA to effectively show the results by alleviating the premature convergence and improving the exploitation capacity of the constrained optimization algorithm.

3.1 Introduction

Most of the problems in theoretical research and engineering practice are the nonlinear constrained optimization problems. Although GAs perform well for unconstrained or simple constrained optimization problems, such as box constraint and spherical constraint, they may have trouble in solving highly nonlinear constrained optimization problems [1]. Standard GAs have several shortcomings: the Hamming cliffs in binary encoding format, the premature convergence, the weak local search capability, and the blindness to the constraints [2]. Enlightened from the DNA double-helix structure, which is complementary and symmetry, it is utilized to overcome the Hamming cliff problem in SGA. In addition, the new operators in DNA computing are introduced to improve its global search capability of SGA. Since biological DNA computing was first introduced to solve a directed Hamilton path problem in 1975, it has been used to solve various different NP hard problems with fewer variables [3–5].

J. Tao et al., *DNA Computing Based Genetic Algorithm*,
https://doi.org/10.1007/978-981-15-5403-2_3

Moreover, research results on DNA computing have been extended to other artificial intelligent fields, such as evolutionary computation, neural networks, and fuzzy control. Ren et al. presented a DNA-GA to find the rule sets of fuzzy controller in a Mamdani fuzzy system [6]. Yoshikawa et al. combined the DNA encoding method with the pseudo-bacterial genetic algorithm [7]. Our group presented an RNA-GA by utilizing RNA computing based operators and the DNA sequence model [8].

The Newton method, interior point method, SQP method, and other traditional calculus-based optimization techniques [9–11] can quickly derive the optimal solutions by several iterations from a starting location; however, all of them are local search algorithms and may not obtain the solutions for complex optimization problems. GA is a global random searching algorithm, which has been applied as a practical optimization tool in many disciplines. Therefore, combining the traditional calculus-based optimization algorithm with GA to construct a hybrid GA is regarded as efficient for solving the highly nonlinear constrained optimization problems [12].

In this chapter, DNA complementary double-helix structure is utilized to overcome the Hamming cliffs in binary encoding format, and DNA single-strand based genetic operations are adopted to avoid the premature convergence and improve the global exploration capability of SGA; thus, it can quickly find the feasible regions of the given problem [13]. Moreover, a self-adaptive constraint handling approach is designed to avoid selecting the penalty parameters that will directly affect the solution of the given problem. When the starting points located in the feasible domain are found, SQP method is used to accelerate the convergence speed and improve the solution accuracy. The algorithm is carried out to solve six benchmark functions and compared with the optimal or the best-known solutions. Alongside solving the numerical function problems, the convergence rate is analyzed theoretically.

3.2 Problem Description and Constraint Handling

Without loss of generality, the nonlinear constrained optimization problem can be expressed as follows:

$$\begin{aligned} & \min \quad f(\mathbf{x}) \\ s.t. \quad & l_m \le g_m(\mathbf{x}) \le u_m, \quad m = 1, 2, \ldots, M \\ & x_i^l \le x_i \le x_i^u, \quad i = 1, 2, \ldots, z \end{aligned} \tag{3.1}$$

where $\mathbf{x} = [x_1, x_2, \ldots x_z]$ represents the solution vector, the ith variable x_i is in the range $[x_i^l,\ x_i^u]$, and $g_m(\mathbf{x})$ $m = 1, 2, \ldots, M$ are the inequality constraints that define the feasible regions, $l_m, u_m \in R$.

In order to solve the above nonlinear constrained optimization problem, the constraints should be handled efficiently. There exist many constraint handling approaches [14], of which the penalty function method was the most popular one. However, penalty parameters are required to be set for each constraint, which are difficult to be obtained. Moreover, different penalty parameters will derive different optimal solutions for the same problem. To avoid setting the penalty parameters, the following criteria were proposed [15]: (1) Any feasible solution was preferred to any infeasible solution. (2) Among two feasible solutions, the one having a better objective function value was preferred. (3) Among two infeasible solutions, the one having a smaller constraint violation was preferred. Following the above criteria, a constraint handling approach without penalty parameter is derived, and the new fitness function is given as follows:

$$F(\mathbf{x}) = \begin{cases} \text{violation} = \sum_{j=1}^{M} \left(\left| \min(0, g_j(\mathbf{x}) - l_m) \right| + \left| \max(0, g_j(\mathbf{x}) - u_m) \right| \right) & \text{all solutions are infeasible} \\ \left. \begin{array}{ll} f(\mathbf{x}) & \text{if feasible} \\ f\max(k)\,(1 + sign\,(f\max(k) \cdot violation)) & \text{if infeasible} \end{array} \right\} & \text{otherwise} \end{cases} \tag{3.2}$$

As a minimization problem in Eq. 3.1, $f\max(k)$ records the worst value of the objective function in the feasible solutions for k evolution generations, $f_{\max}$ (k) $\neq 0$, it can update with the evolution processes and act as a penalty parameter needless to be selected in advance. *sign* is a sign function utilized to guarantee that the constraint violation is added to the fitness function as a penalty term without considering the symbol of $f\max(k)$. In Eq. 3.2, if all solutions in a population are infeasible, only the offset between the upper and lower bound is calculated because the infeasible solutions cannot be implemented in practice. If some solutions are feasible and some are infeasible, the infeasible ones are punished by adding the violation to the current maximal value of the fitness function.

3.3 DNA Double-Helix Hybrid Genetic Algorithm (DNA-DHGA)

3.3.1 DNA Double-Helix Encoding

Four nucleotide bases: Adenine (A), Thymine (T), Guanine (G), and Cytosine (C) are existed in a DNA double-helix chromosome, and ranked according to Watson–Crick complementary criterion, which are shown in the upper part of Fig. 3.1.

We want to utilize DNA chromosome to encode the solution vector in GA. However, it is obvious that such A, T, G, and C DNA double-helix chromosome cannot be directly processed in the digital computer. 0(00), 1(01), 2(10),

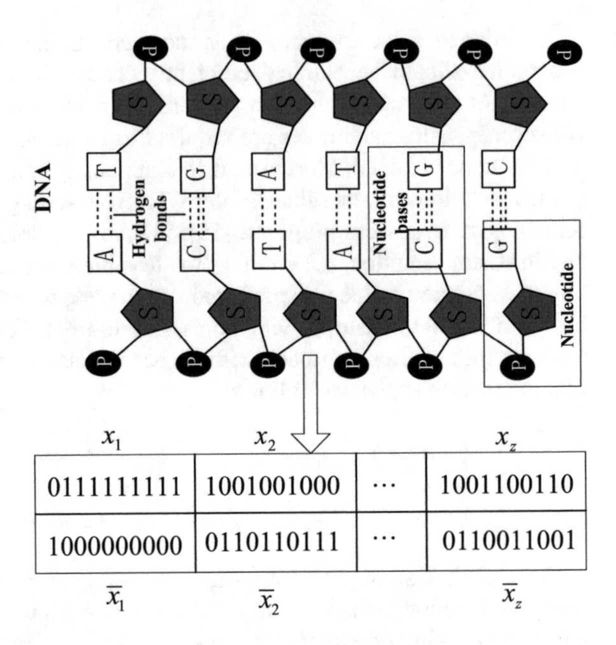

Fig. 3.1 DNA double-helix based binary encoding

and 3(11) are then substituted for A, T, G, and C, which are similar to those in Chap. 1. Till now, binary encoding is still extensively applied in evolutionary algorithm due to its simplicity and easy operation but there exists the Hamming cliff in the traditional binary encoding format. When the DNA double-helix chromosome is combined, the DNA double-helix encoding structure actually contains the complementary binary chromosomes as shown in the lower part of Fig. 3.1., e.g., x_1 and its complement representation $\bar{x}_1$ coexist in one DNA chromosome. Once a DNA double-helix chromosome is produced, there are actually two solutions: $\mathbf{x} = [x_1, x_2, \ldots, x_z]$, $\bar{\mathbf{x}} = [\bar{x}_1, \bar{x}_2, \ldots, \bar{x}_z]$. The Hamming cliff problem is then solved naturally since the complementary DNA single strands coexist. For example, there are Hamming cliff problem in 10000 and 01111, they are coexisted in the double-helix chromosome; thus, the Hamming cliff problem is solved naturally. To be compatible with A T G C, the quanternary encoding (0 1 2 3) is adopted, and MATLAB code for generating DNA double-helix chromosome is shown as follows:

```
function [chromosome]=createchromosome(Size,CodeL)
   % Size is the number of chromosome in the population
   % CodeL is the length of DNA chromosome
   for i = 1:2:2*Size    % double-helix
    for j = 1:1:CodeL
         k=round(rand*3); % generate 0,1,2,3 randomly
         chromosome(i,j) = k;
         switch k                    %create the double helix structure
              case 0
              chromosome(i+1,j) =3;
              case 1
              chromosome(i+1,j) =2;
              case 2
              chromosome(i+1,j) =1;
              case 3
              chromosome(i+1,j) =0;
         end
    end
   end
```

The decoding process is the same as Sect. 2.2.3, readers can refer to the corresponding code in Chap. 2.

3.3.2 DNA Computing Based Operators

Genetic operators are widely used for the application of GAs [16], and they are critically verified as a successful way. In this chapter, except for DNA double-helix chromosome, more complicated gene-level operators enlightened by DNA computing and RNA computing are introduced to enhance the searching capability of SGA.

3.3.2.1 Selection Operator

The value of $F(\mathbf{x})$ is required to calculate twice for one DNA chromosome because of the double-helix structure. Two complementary solution vectors $\mathbf{x},\bar{\mathbf{x}}$ will lead to a double increasement of the computational complexity when the new encoding format is directly applied to the crossover and mutation operators. To reduce the calculation complexity, the strand with a better value of fitness function denoted as $\mathbf{x}'$ is selected as the representation of DNA double-helix chromosome before implementing the genetic operators, which can be obtained according to Eq. 3.3:

$$\mathbf{x}' = \begin{cases} \mathbf{x} & F(\mathbf{x}) \le F(\overline{\mathbf{x}}) \\ \overline{\mathbf{x}} & F(\mathbf{x}) > F(\overline{\mathbf{x}}) \end{cases} \tag{3.3}$$

where the value of fitness function F can be calculated in terms of Eq. 3.2.

MATLAB code for the constraint handling in Sect. 3.2 is shown as follows, here, f_4 in the appendix is taken as an example.

```
%calculate the constraint and function value, x1 is the decoded variables
    kk=1;tempF=[];violation=[];
     for i=1:size     % size is the population size
          [funci(i),violation(i,:)]=my_funcCon5(x1(i,:),5);
%judge whether all of the constraints are satisfied
          tempflag(i)=violation(i,1)==0 && violation(i,2)==0 &&
                    violation(i,3)==0 && violation(i,4)==0 && violation(i,5)==0;
          if tempflag(i)
               tempF(kk)=funci(i);
               kk=kk+1;
          end
     end
     flagfail=0;
    if size(tempF,1)==0
         fmax=0;
         flagfail=1;
    else
          tempfmax=max(funci);
         if tempfmax>fmax
              fmax=tempfmax;
         end
    end
% constraint handling in terms of Eq.3.2 %
    for i=1:size
         if flagfail==1
              F(i)=sum(violation(i,:));
         else
              if tempflag(i)
                  F(i)=funci(i);
              else
                   F(i)=fmax*(1+sign(fmax)*sum(violation(i,:)));
              end
         end
    end
```

Once the constraint handling and fitness function calculating are finished, $\mathbf{x}'$ is easy to be derived according to Eq. 3.3.

To produce the parents of crossover and mutation operators, the proportional selection operator is implemented based on $\mathbf{x}'$, and the number of selected individuals is calculated as follows:

$$S_N = < J(\mathbf{x}')N \Big/ \sum_{i=1}^{N} J(x_i') > \tag{3.4}$$

where $J(\mathbf{x}') = F_{\max} - F(\mathbf{x}')$, $F_{\max}$ is a positive number which is set to guarantee $J(\mathbf{x}') > 0$, $< \cdot >$ denotes the round-off operation symbol, and N is the population size. For MATLAB code of selection operator, refer to Sect. 2.2.3.

After calculating Eq. 3.4 for each $\mathbf{x}'$, there will produce at most $N + 1$ strands due to the round-off operation, and the former N strands will be kept as the parents of the crossover and mutation operations.

3.3.2.2 Crossover Operator

After the selection operation, the crossover operator based on DNA computing for single strand can then be implemented [17]. By application of various enzymes, there are various operations for DNA sequences. However, the genetic operations that change the length of the DNA sequence are not appropriate for crossover operator since the length of the chromosome keeps invariable in GA. On a single strand of DNA sequence, there are three main operations: translocation, transformation, and permutation, which have been described in Sect. 2.2.2, and obtained better performance than SGA. MATLAB code for the crossover operation can be derived in Chap. 2.

3.3.2.3 Mutation Operator

Mutation operator is necessary for GA to explore better searching domain. However, the mutation probability of SGA is fixed during the whole evolution process. Since GA with too large mutation probability will become a random searching algorithm. And it is prone to trap into local optimum with too small mutation probability, a dynamical mutation probability in Chap. 2 is then adopted.

Actually, various bit positions in a chromosome have different effects on the solution at different evolution stages, e.g., the most significant bit and the least significant bit. Hence, at the beginning of evolution, the larger mutation probability is assigned to the higher bit positions to explore the larger feasible region. Then, if the global optimum located region is found, the mutation probability in the higher bit positions is decreased to prevent disrupting the better solution, while the mutation probability in the lower bit positions is increased to improve the solution accuracy. The quaternary strand between 1 and $L/2$ is defined as the low bit position, and the left $[L/2+1, L]$ as the high bit position. Accordingly, there are two kinds of mutation probabilities: p_{mh} and p_{ml}, which will change dynamically with the evolution of generations. The dynamic probabilities are given in Eqs. 2.2and 2.3. If p_{mh} or p_{ml} is satisfied, then the quaternary base in the relevant position is reversed; thus, the mutation operation is finished. For MATLAB code for the dynamic mutation operation, refer to Sect. 2.2.3.

3.3.3 Hybrid Genetic Algorithm with SQP

SQP is an iterative optimization method, which is starting from an initial point for the constrained nonlinear optimization problem where the constraints and the objective function would be required twice continuously differentiable. It actually solves a sequence of optimization sub-problems. Each optimization sub-problem optimizes a quadratic model of the objective subjecting to the linearization of the constraints. If the optimization problem is unconstrained, SQP is then reduced to Newton's method to find a point where the gradient of the objective function vanishes. If the optimization problem has only equality constraints, SQP is then equivalent to apply Newton's method to the first-order optimality conditions, or Karush–Kuhn–Tucker conditions of the problem. Since SQP is a local search method, its starting point will affect the optimal solutions greatly. Without an appropriate starting point, SQP algorithm may trap into the local optimum and cannot obtain the optimal solution. To obtain a good starting point, GA is applied to optimize the starting point of SQP. When the current optimum of starting point keeps invariable for 100 successive generations in GA or all constraints are met, the point is regarded as a good one, and SQP is implemented. The new solution solved by SQP algorithm is changed as DNA quaternary single-strand chromosome to perform the above genetic operations.

Based on the DNA double-helix chromosome and DNA computing single-strand operators, the procedure of the SQP hybrid GA is summarized as follows:

Step 1: Initialize the maximum generation G, the length of chromosome L, and the population size N. Generate the complementary quaternary double-helix chromosomes for the population initialization.

Step 2: Decode the DNA double-helix chromosome and calculate the value of fitness function in terms of Eq. 3.2.

Step 3: Select the representation of the DNA double-helix chromosome by using Eq. 3.3. Sort the representations and choose the best $3N/4$ and the worst $N/4$ to form N individuals.

Step 4: Judge whether the starting point is appropriate to implement SQP algorithm. If no, go to step 5. If yes, obtain the solution derived by SQP algorithm, then change it into quaternary encoding format to perform the genetic operators.

Step 5: Execute the selection operator to obtain N individuals according to Eq. 3.4 as the parents of crossover and mutation operators.

Step 6: Implement the crossover operator in the best $N/2$ individuals, where the permutation operator is implemented with probability 1 and the transformation operator is implemented with probability 0.5. If the transformation operator is not performed, the translocation operator is then implemented. N offspring are then generated by the crossover operator.

Step 7: Carry out the mutation operator in the left $N/2$ individuals and the new N offspring; thus, there are totally 3 $N/2$ individuals generated.

Step 8: Generate the complementary quaternary strand to construct the DNA double-helix chromosome.

Step 9: Repeat steps 2–8 until a termination criterion is met. We set 3 termination conditions, i.e., the set maximal number of evolutions, the set minimum improvement of the best performance in successive generations, and the set minimum error between the obtained solution and the global optimum.

The program framework is shown in Fig. 3.2.

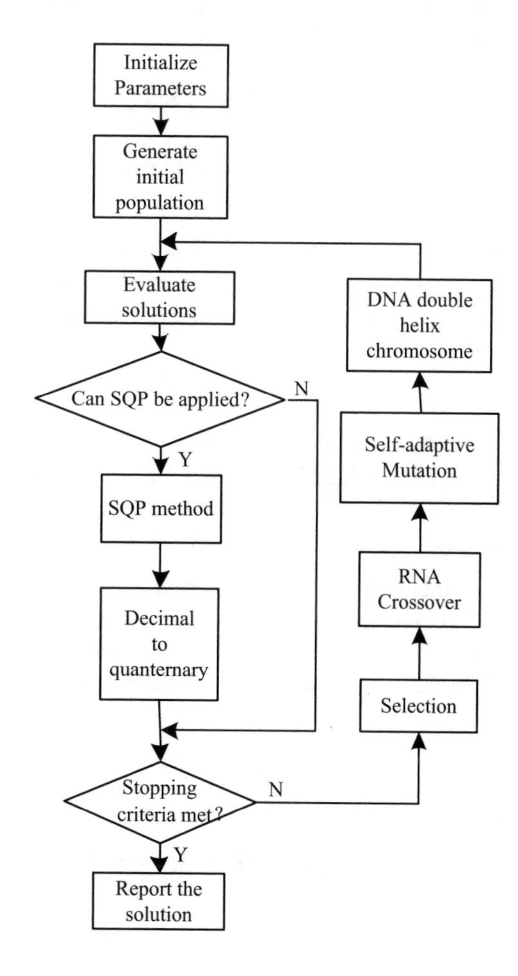

Fig. 3.2 SQP hybrid GA based on DNA double-helix structure

3.3.4 Convergence Rate Analysis of DNA-DHGA

Essentially, DNA double-helix chromosomes are binary strings with length of L. Its encoding space can be defined as $S = \{0, 1\}^L$, i.e., $|S| = 2^L$. Set Ω as the single-strand set of DNA chromosome, X as the population composed of the elements in Ω, and Ω^N as all possible population in Ω, $\Omega^N = \{X_1, X_2 \ldots X_n\}$.

For the purpose of analyzing the convergence rate of DNA-DHGA, lemma 3.1 and proposition 3.1 are given firstly as follows:

Lemma 3.1 Suppose Markov chains meet the following inequality:

$$P^{k_0}(\mathbf{x}, A) \geq \beta\xi(A), \quad \mathbf{x} \in R, \ A \subseteq \Omega^N \tag{3.6}$$

where k_0 is a positive integer, R is the subset of Ω^N, $\beta > 0$, and $\xi(\cdot)$ is a probability distribution on Ω^N. Equation 3.6 is a minorization condition of the Markov chain [18]. Suppose $k_0 = 1$, the following inequality holds:

$$P(\mathbf{x}, A) \geq \beta\xi(A), \quad \mathbf{x} \in R, \ A \subseteq \Omega^N \tag{3.7}$$

Given the above minorization condition, Rosenthal gave an upper bound of the Markov chain, as shown in the following proposition [19]:

Proposition 3.1 Suppose Markov chain meets the minorization condition and converges to the distribution π, then, there should be $\beta \in [0, 1]$, and the following inequality holds for any random initial distribution π_k:

$$\|\pi_k - \pi\| \leq (1 - \beta)^k \tag{3.8}$$

Set the probability transition matrix $P(\cdot, \cdot)$ as the Markov chain of the population $\{X(k), k \geq 0\}$, and define the probability transition matrix from the population at k generation to the population at $k + 1$ generation as $P(X, Y) = P\{X(k + 1) = Y|X(k) = X\}$, $X, Y \in \Omega^N$. In the discrete space, let $\beta = \sum\limits_{Y \in R} \min\limits_{X \in R} P(X, Y)$, and the maximum of Eq. 3.8 can be achieved for $P(X, \cdot)$.

The mentioned transition probability matrix (P) can also be decomposed into three probability matrices: crossover (C), mutation (M), and selection (S), i.e., $P = CMS$. Moreover, Matrix C possesses the following property:

$$\sum_{j=1}^{n} c_{ij} = 1 \tag{3.9}$$

where c_{ij} is the probability of changing sequence i into sequence j through the crossover operator. And Matrix M possesses the following property:

$$m_{jz} = \prod_{i=1}^{N} \left(\frac{p_m}{C-1}\right)^H (1-p_m)^{L-H} \quad z \in \{1, 2, \cdots, n\} \tag{3.10}$$

where m_{jz} denotes the probability of transforming sequence j into sequence z through the mutation operator, and C is 2 and H is the individual Hamming distance between sequence j and sequence z.

Suppose $0 < p_m \leq \frac{1}{L+1}$, i.e., $Lp_m \leq 1 - p_m$, then

$$\begin{aligned} &\prod_{i=1}^{N} \left(\frac{p_m}{C-1}\right)^H (1-p_m)^{L-H} \geq \prod_{i=1}^{N} \left(\frac{p_m}{C-1}\right)^H (Lp_m)^{L-H} \\ &= \prod_{i=1}^{N} \left(\frac{1}{C-1}\right)^H (L)^{L-H} p_m^L \geq \prod_{i=1}^{N} \left(\frac{1}{C-1}\right)^L p_m^L = \left(\frac{p_m}{C-1}\right)^{NL} \end{aligned} \tag{3.11}$$

Likewise, suppose $\frac{1}{L+1} \leq p_m < 1$, i.e., $p_m \geq \frac{1-p_m}{L}$, then

$$\begin{aligned} &\prod_{i=1}^{N} \left(\frac{p_m}{C-1}\right)^H (1-p_m)^{L-H} \geq \prod_{i=1}^{N} \left(\frac{1}{C-1}\right)^H \left(\frac{1-p_m}{L}\right) (1-p_m)^{L-H} \\ &= \prod_{i=1}^{N} \left(\frac{1}{C-1}\right)^H \left(\frac{1}{L}\right)^{L-H} (1-p_m)^L \geq \prod_{i=1}^{N} \left(\frac{1}{LC-1}\right)^L (1-p_m)^L = \left(\frac{1-p_m}{LC-1}\right)^{NL} \end{aligned} \tag{3.12}$$

In terms of Eq. 3.11 and Eqs. 3.12, 3.10 can be rewritten as

$$m_{jk} \geq \begin{cases} \left(\frac{p_m}{C-1}\right)^{NL} & 0 < p_m \leq \frac{1}{L+1} \\ \left(\frac{1-p_m}{LC-1}\right)^{NL} & \frac{1}{L+1} < p_m \leq 1 \end{cases} \tag{3.13}$$

Since at least one individual is selected, Matrix S holds the following inequality:

$$\sum_{k=1}^{n} s_{kq} > 0 \tag{3.14}$$

where s_{kq} denotes the probability of changing sequence k into sequence q through the selection operator.

Set the parent population as $S^{(k)}$ and the in-process population as p^+. If $q \notin p^+$, q cannot become a new parent by selection operator, i.e., $s_{kq} = 0$. Hence, suppose $q \in p^+$, the following equation of s_{kq} can be obtained:

$$s_{kq} = \prod_{\mathbf{x}_k \in S^{(q)}} \left[\frac{f(\mathbf{x}^*)}{\sum\limits_{\mathbf{x}_j \in p^+} f(\mathbf{x}_j)}\right] \tag{3.15}$$

There always exists $\mathbf{x}_k$ satisfying $f(\mathbf{x}_k) \geq f(\mathbf{x}^*)$ for the minimization problem, where $f(\mathbf{x}^*)$ is the minimum statistic of the given problem. In terms of Eq. 3.15, the following inequality can be derived:

$$s_{kq} \geq \left[\frac{f(\mathbf{x}^*)}{\sum\limits_{\mathbf{x}_j \in p^+} f(\mathbf{x}_j)} \right]^N \tag{3.16}$$

According to the running process of DNA-DHGA and Chapman–Kolmogorov equation, the following equation can be obtained:

$$P = CMS = \sum_{k=1}^{n} \left(\sum_{j=1}^{n} c_{ij} m_{jk}\right) s_{kq} \tag{3.17}$$

Let $\rho = \begin{cases} \left(\dfrac{p_m}{C-1}\right)^{NL} & 0 < p_m \leq \dfrac{1}{L+1} \\ \left(\dfrac{1-p_m}{LC-1}\right)^{NL} & \dfrac{1}{L+1} < p_m \leq 1 \end{cases}$, and substitute Eq. 3.13 into Eq. 3.17, the following inequality is obtained:

$$P \geq \sum_{k=1}^{n} \left(\sum_{j=1}^{n} c_{ij} \rho\right) s_{kq} \tag{3.18}$$

Substituting Eq. 3.9 into Eq. 3.18, the following inequality can be achieved:

$$P \geq \rho \sum_{k=1}^{n} s_{kq} \tag{3.19}$$

According to Eqs. 3.16 and 3.19, the probability transition matrix P satisfies the following equation:

$$\min_{X, Y \in \Omega^N} P(X, Y) = \rho \left[\frac{f(\mathbf{x}^*)}{\sum\limits_{\mathbf{x}_j \in P^+} f(\mathbf{x}_j)} \right]^{nN} \tag{3.20}$$

Let $\beta = C^{NL} \cdot \rho \left[\frac{f(\mathbf{x}^*)}{\sum\limits_{\mathbf{x}_j \in p^+} f(\mathbf{x}_j)} \right]^{nN}$, $\xi(A) = C^{-NL}|A|$, $A \subseteq \Omega^N$, where $|A|$ represents the size of set A. For any X, A, the following inequality can be achieved:

$$P(X, A) = \sum_{Y \in A} P(X, Y) \geq \min_{X,Y \in S} P(X, Y)|A| = \beta\xi(A) \tag{3.21}$$

It can be seen from Eq. 3.21 that the above β and $\xi(\cdot)$ are specific minorization conditions for Markov chain. If there exists a steady-state distribution π to any initial distribution π_k according to proposition 3.1, Eq. 3.8 can then be established. Therefore, substituting β into Eq. 3.8, the following inequality can be derived:

$$\|\pi_k - \pi\| \leq \begin{cases} \left[1 - \left(\dfrac{p_m}{C-1}\right)^{NL} \left[\dfrac{f(\mathbf{x}^*)}{\sum\limits_{\mathbf{x}_j \in p^+} f(\mathbf{x}_j)}\right]^{nN}\right]^k & 0 < p_m \leq \dfrac{1}{L+1} \\ \left[1 - \left(\dfrac{p_m}{zLC-1}\right)^{NL} \left[\dfrac{f(\mathbf{x}^*)}{\sum\limits_{\mathbf{x}_j \in p^+} f(\mathbf{x}_j)}\right]^{nN}\right]^k & \dfrac{1}{L+1} \leq p_m < 1 \end{cases} \tag{3.22}$$

In terms of Eq. 3.22, it is obviously shown that the more the local minima found during the evolution process, the less the upper bound of $\|\pi_k - \pi\|$, also the faster the convergence of GA can be achieved. Hence, SQP method combined in GA will efficiently improve the local searching performance of GA and find more local minima; thus, the upper bound of $\|\pi_k - \pi\|$ decreases and the convergence rate becomes faster.

The bigger the population size N and the longer the length of the chromosome string L, the larger the upper bound of $\|\pi_k - \pi\|$ is obtained, then the slower the convergence speed of GA is achieved. Therefore, the length of the chromosome and the population size should be selected as small as possible to accelerate the convergence rate. However, the length of the chromosome should be determined according to the precision requirement of the given problem, and the population diversity should also be considered except for the convergence speed. Herein, the population with size $N = 40$ used in DNA-DHGA is similar to the population with size $N = 60$ in GA because of DNA double-helix structure and its special genetic operators. Therefore, the adoption of DNA gene-level operators can accelerate the convergence speed.

For $0 < p_m \leq \frac{1}{L+1}$, the bigger the mutation probability, the less the upper bound of $\|\pi_k - \pi\|$ becomes, and the faster the convergence speed of GA can be obtained; while for $\frac{1}{L+1} \leq p_m < 1$, the bigger mutation probability, the larger the upper bound of $\|\pi_k - \pi\|$ becomes, and the slower the convergence speed of GA can be gained. Hence, the mutation probability cannot be set too large.

Since p_{mh} decreases while p_{ml} increases gradually in the earlier evolution processes, the convergence speed of DNA-DHGA can be accelerated at this stage, and after that, the change of p_{mh} and p_{ml} may decelerate the convergence speed. Since the application of SQP method can greatly enhance the local search capability and speed up the GA convergence. Hence, the change of mutation probability at this stage has a small effect on the whole convergence speed.

In terms of the above analysis, it can be concluded that DNA-DHGA can accelerate the convergence speed and improve the solution accuracy theoretically.

3.4 Numeric Simulation

3.4.1 Test Functions

To investigate the efficiency of DNA-DHGA, the proposed GA is applied to six different nonlinear constrained optimization problems that have been studied in the literature [1, 20]; x^* and $f*$ delegate the optimal solution and the optimum values of the constraint problem. The details of six test functions for constraint optimization problems are given in the appendix.

3.4.2 Simulation Analysis

The parameters of the proposed algorithm are set as $L = 20$, $N = 40$, and $G = 1000$ by trial and error. A termination criterion is the set maximal 1000 generations or the set minimum error satisfying the precision requirement $\varepsilon \leq 1\%$, where $\varepsilon = |f - f^*|/f^*$, and f is the value of constrained optimization problem solved by the optimization algorithm. Since GA is a random searching algorithm, 50 runs are executed from randomly initialized populations in all problems. A comparison with the algorithms in [1, 20] is made and shown in Table 3.1, where Best denotes the best solution in 50 runs, similarly, Worst denotes the worst result obtained in 50 runs; avekg represents the average generations satisfying the termination criteria.

Table 3.1 Comparison of the solutions for test functions over 50 runs

f	Other methods*			DNA-DHGA		
	Best	Worst	G	Best	Worst	avekg
f_1	13.59085	117.02971	1000	13.59084	13.59084	80.0
f_2	680.6344	680.6508	5000	680.6300	680.6300	333.3
f_3	7060.221	10230.834	4000	7049.2840	7049.2840	531.5
f_4	-30665.537	-29846.654	5000	-30665.539	-30665.539	456.0
f_5	24.3725	25.0753	-	24.3062	24.3062	650.4
f_6	-0.8011	-0.745329	-	-1.1891	-1.1243	393.3

- represents no result in the subject

* represents that Deb's method is adopted for f_1–f_5 and Michalwicz's is utilized for f_6

In order to visualize easily, a two-dimensional constrained minimization problem f_1 is first selected. The unconstrained objective function of f_1 has a minimum solution located at (3, 2) with an optimal value $f^* = 0$. However, due to the presence of constraints in f_1, this solution is no longer feasible and the constrained optimum solution is $f^* = 13.59085$. The feasible region is a narrow crescent-shaped region located approximately 0.7% of the total search space, as shown in Fig. 3.3. The problem is relatively easy to solve because of low dimensions of the solution vector and the simplicity of the optimization problem. Hence, the average evolution generation is only 80 as listed in Table 3.1 and the evolution generations with the best solution over 50 runs are plotted in Fig. 3.4. It can be seen that the feasible region is gained at the initial population and the optimum is found soon with the process of population evolution. The best solution is located at(2.2468, 2.3819)with the function value 13.59084. After 50 runs, the distribution of solution vectors is given in

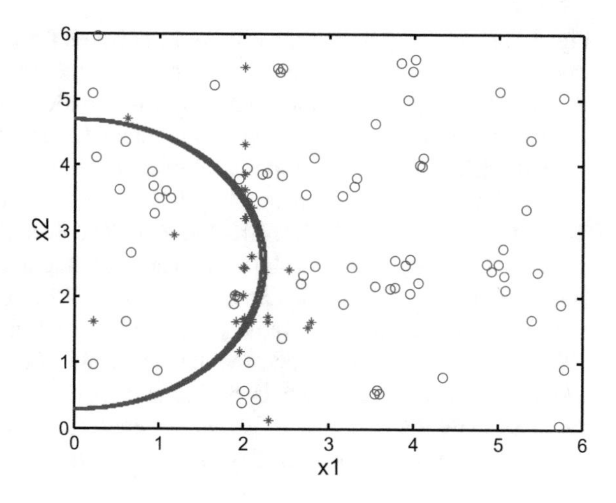

Fig. 3.3 Feasible search space and the initial population distribution (marked with open circles) and final population distribution (marked with *) on f_1

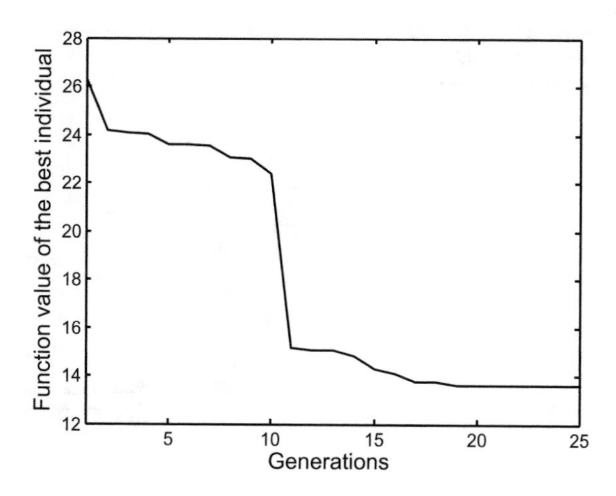

Fig. 3.4 The evolution process with the best solution over 50 runs on f_1

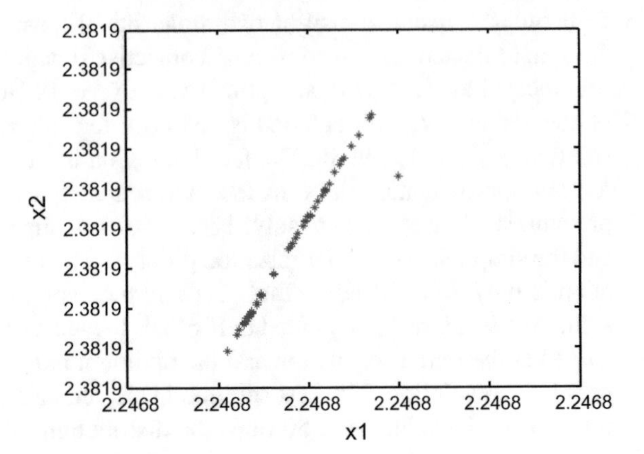

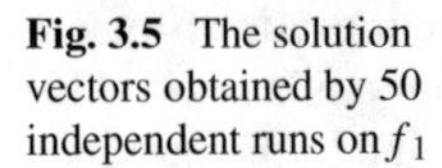
Fig. 3.5 The solution vectors obtained by 50 independent runs on f_1

Fig. 3.5, it can be seen that the difference between these solutions is so small that it can be ignored, which indicates the stability of the proposed algorithm. The results obtained by Deb's GA without niching and mutation operator are relatively worse in terms of the Best and Worst values in Table 3.1, and the comparison results also show the efficiency of the proposed algorithm.

The simulation results obtained from the best solution over 50 runs for f_2–f_6 are illustrated from Figs. 3.6, 3.7, 3.8, 3.9, 3.10. The feasible region of f_2 accounts for about 0.5% of the total search space. The best solution obtained by DNA-DHGA is located at $\mathbf{x} = (2.330498, 1.951372, -0.477540, 4.365726, -0.624487, 1.038137, 1.594228)$ with the function value $f(\mathbf{x}) = 680.630$. Deb has found the best solution 680.634, and Michalewicz reported the best result of 680.642 correspondingly. The evolution process in Fig. 3.6 is similar to that of Fig. 3.4. Figure 3.6 shows that the feasible region is gained at the first generation, and when the best solution in the feasible region is kept invariant for 100 continuous generations, SQP is implemented

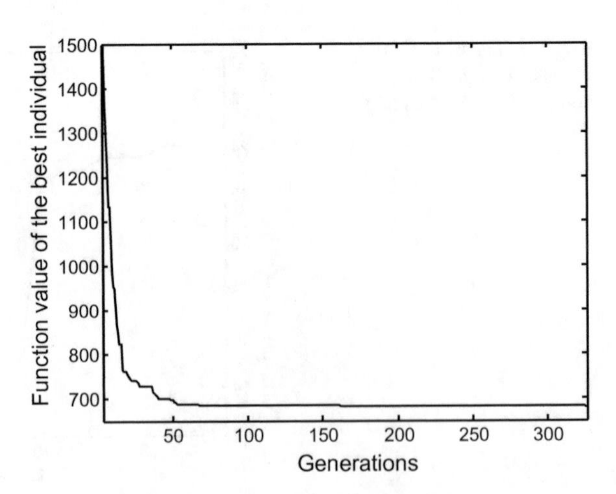

Fig. 3.6 The evolution process with the best solution over 50 runs on f_2

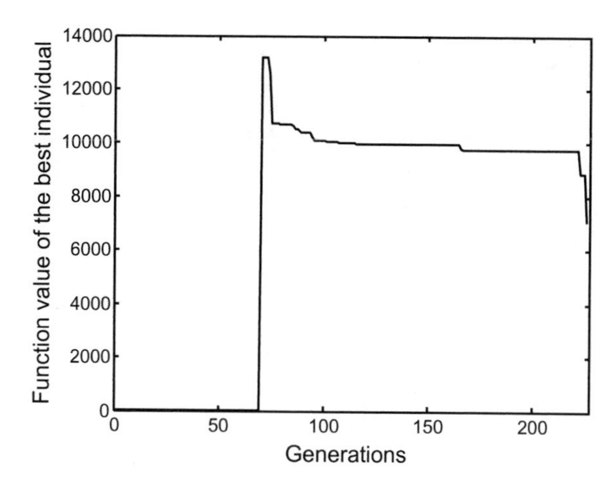

Fig. 3.7 The evolution process with the best solution over 50 runs on f_3

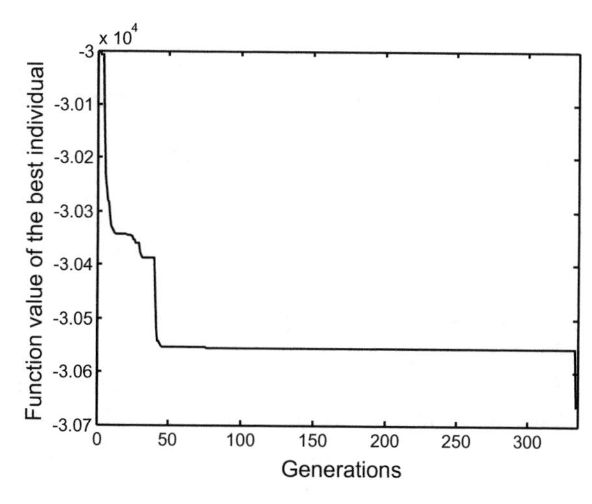

Fig. 3.8 The evolution process with the best solution over 50 runs on f_4

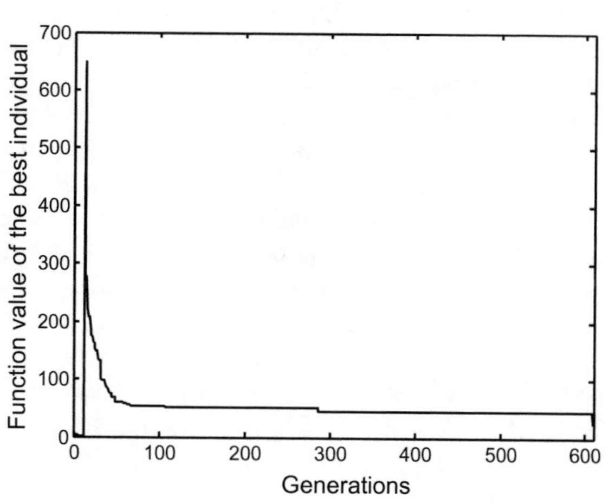

Fig. 3.9 The evolution process with the best solution over 50 runs on f_5

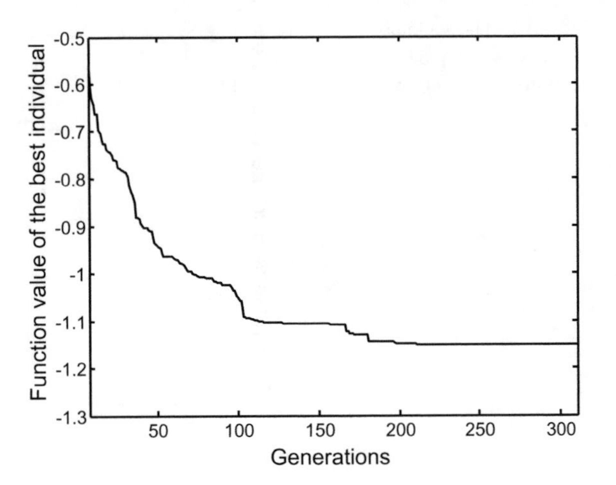

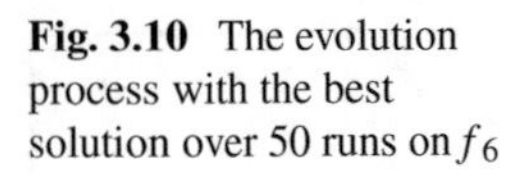
Fig. 3.10 The evolution process with the best solution over 50 runs on f_6

to get better solutions. The result derived by DNA-DHGA is obviously superior to Deb's and Michalewicz's because the Best solution of Deb's method is worse than the Worst solution of the proposed method.

For f_3, Michalewicz has experienced the difficulty of solving this problem. Deb obtained the best solution 7060.221 with niching and mutation with maximal generation 4000 and population size 80. The best solution solved by DNA-DHGA is located at $\mathbf{x} = (0.579.3067, 1359.9707, 5109.9706, 182.0177, 295.6012, 217.9823, 286.4165, 395.6012)$ with $f(\mathbf{x}) = 7049.2480$, which is superior to the current best solution $f^* = 7049.3309$. In Fig. 3.7, the feasible region cannot be found at the initial stage because of the difficulty of the problem, the population is then developed to satisfy all the constraints after 69 generations evolution. When the best solution located at the feasible domain cannot be improved for 100 continuous generations, SQP algorithm is executed and the problem is solved rapidly.

The function f_4 has 5 variables and 6 inequality constraints, and Deb solved this problem by GA employing a penalty function method without penalty coefficients, and obtained the best solution -30665.537. The best solution obtained by DNA-DHGA is located at $\mathbf{x} = (78, 33, 29.9953, 45, 36.7758)$ with $f(\mathbf{x}) = -30665.5387$. Though the Best value of the proposed method is a little inferior to Deb's, the stability of the solutions of the proposed method at different runs is superior to Deb's. The evolution process in Fig. 3.8 is quite similar to that of Fig. 3.6.

The function f_5 has 10 variables and 8 inequality constraints, and the optimum is still unknown, the best solution given in the literature is $f_5^* = 24.3062$ and regarded as the optimum of f_5. Eight inequality constraints obtained by Deb are 10^{-6} (0.0095, 0, 0.3333, 0.1006, 0, 0.4305, 0, 0), where the positive number denotes the degree of deviation from the constraint, and zero means that the constraint is satisfied. The best solution obtained by DNA-DHGA is $f(\mathbf{x}) = 24.3062$, located at $\mathbf{x} = (2.1720, 2.3637, 8.7739, 5.0960, 0.9907, 1.4306, 1.3216, 9.8287, 8.2801, 8.3759)$, where the deviation from the constraints are only $10^{-9}(0, 0, 0.0099, 0, 0, 0.1078, 0, 0)$. It is obvious that the degree of constraint deviation obtained by DNA-DHGA is less than Deb's. The evolution process in Fig. 3.9 is similar to that of Fig. 3.7.

The function f_6 is a heavily nonlinear and multimodal constrained optimization problem, and the global optimum is also unknown at present. Let n be 20, Michalwicz obtained the best solution -0.8036. By using DNA-DHGA, if the solution keeps invariant for 300 successive generations or the evolution generation reaches the set maximal value, the algorithm will be terminated and the solution is kept as the final result. The evolution process is illustrated in Fig. 3.10. It can be seen that for f_6 DNA-DHGA also trapped into the local optimum because of its multimodal characteristics. However, the solution in Table 3.1 is better than Michalwicz's, whose best solution is -0.8011. The best solution obtained by DNA-DHGA is located at $\mathbf{x}=$ (3.1520, 9.4137, 6.2666, 4.6805, 3.1172, 3.1085, 3.0997, 3.0910, 3.0821, 3.0733, 0.2080, 3.0554, 0.2003, 0.1965, 0.1929, 0.1894, 0.1859, 0.1826, 0.1793, 0.1762), with $f(\mathbf{x}) = -1.1891$.

In Table 3.1, the precision requirement with $\varepsilon \leq 1\%$ is met by DNA-DHGA for all test functions over 50 runs except for f_6, while the precision requirement is satisfied only for f_2 by other methods. Moreover, the Best and Worst solutions obtained by DNA-DHGA are obviously superior to those obtained by other methods. As to the comparison of convergence speed, since avekg cannot be obtained in the literature, G is substituted instead of avekg when using other methods. The number of avekg obtained by DNA-DHGA can be decreased greatly, and the result shows that its convergence has been speeded up greatly.

From Figs. 3.5, 3.6, 3.7, 3.8, 3.9, 3.10 and the comparison results in Table 3.1, it can be seen that the DNA-DHGA algorithm can be performed efficiently in solving the heavily constrained optimization problem, it has obtained the best-known solution and become more robustness according to the number of successful finding solution close to the best-known solution than previous methods.

3.5 Summary

DNA computing based genetic operators and DNA complementary double-helix binary encoding can efficiently improve the global exploration performance and help to determine the feasible regions of the nonlinear constrained problem, and the SQP method can accelerate the convergence speed and increase the solution precision. Therefore, the DNA-DHGA keeps a good balance between the local exploitation and the global exploration and guarantees an accurate solution with more rapid convergence speed. The convergence analysis based on the Markov model has shown the efficiency of DNA-DHGA theoretically. Moreover, six typical test functions of nonlinear constrained optimization problems are selected, and the optimal results show that DNA double-helix structure, gene-level operators, and SQP method play an important role in improving the efficiency, convergence speed, and solution accuracy of GA.

Appendix

Six heavily constrained test functions:

f_1:

$$\begin{aligned} \min \quad & f(\mathbf{x}) = (x_1^2 + x_2 - 11)^2 + (x_1 + x_2^2 - 7)^2 \\ s.t. \quad & g_1(\mathbf{x}) \equiv 4.84 - (x_1 - 0.05)^2 - (x_2 - 2.5)^2 \geq 0, \\ & g_2(\mathbf{x}) \equiv x_1^2 + (x_2 - 2.5)^2 - 4.84 \geq 0, \\ & 0 \leq x_1 \leq 6, \quad 0 \leq x_2 \leq 6. \end{aligned}$$

The optimal solution: $\mathbf{x}^* = (2.246826, 2.381865), f^* = 13.59085.$

f_2:

$$\begin{aligned} \min \quad & f(\mathbf{x}) = (x_1 - 10)^2 + 5(x_2 - 12)^2 + x_3^4 + 3(x_4 - 11)^2 + 10x_5^6 + 7x_6^2 + x_7^4 \\ & \quad -4x_6x_7 - 10x_6 - 8x_7 \\ s.t. \quad & g_1(\mathbf{x}) \equiv 127 - 2x_1^2 - 3x_2^4 - x_3 - 4x_4^2 - 5x_5 \geq 0, \\ & g_2(\mathbf{x}) \equiv 282 - 7x_1 - 3x_2 - 10x_3^2 - x_4 + x_5 \geq 0, \\ & g_3(\mathbf{x}) \equiv 196 - 23x_1 - x_2^2 - 6x_6^2 + 8x_7 \geq 0, \\ & g_4(\mathbf{x}) \equiv -4x_1^2 - x_2^2 + 3x_1x_2 - 2x_3^2 - 5x_6 + 11x_7 \geq 0, \\ & -10 \leq x_i \leq 10, \quad i = 1, \ldots 7. \end{aligned}$$

The optimal solution:

$$\begin{aligned} \mathbf{x}^* =& (2.330499, 1.951372, -0.4775414, 4.365726, -0.6244870, 1.038131, 1.594227) \\ f^* =& 680.6300573 \end{aligned}$$

f_3:

$$\begin{aligned} \min \quad & f(\mathbf{x}) = x_1 + x_2 + x_3 \\ s.t. \quad & g_1(\mathbf{x}) \equiv 1 - 0.0025(x_4 + x_6) \geq 0 \\ & g_2(\mathbf{x}) \equiv 1 - 0.0025(x_5 + x_7 - x_4) \geq 0 \\ & g_3(\mathbf{x}) \equiv 1 - 0.01(x_8 - x_5) \geq 0 \\ & g_4(\mathbf{x}) \equiv x_1x_6 - 833.33252x_4 - 100x_1 + 83333.333 \geq 0 \\ & g_5(\mathbf{x}) \equiv x_2x_7 - 1250x_5 - x_2x_4 + 1250x_4 \geq 0 \\ & g_6(\mathbf{x}) \equiv x_3x_8 - x_3x_5 + 2500x_5 - 1250000 \geq 0 \\ & 100 \leq x_1 \leq 10000 \\ & 1000 \leq (x_2, x_3) \leq 10000 \\ & 10 \leq x_i \leq 1000, \quad i = 4, \cdots, 8 \end{aligned}$$

The optimal solution:

$$\mathbf{x}^* = (579.3167, 1359.943, 5110.071, 182.0174, 295.5985, 217.9799, 286.4162, 395.5979)$$
$$f^* = 7049.330923$$

f_4

$$\begin{aligned}
\min \quad & f(x) = 5.3578547x_3^2 + 0.8356891x_1x_5 + 37.293239x_1 - 40792.141 \\
s.t. \quad & g_1(x) \equiv 85.334407 + 0.0056858x_2x_5 + 0.0006262x_1x_4 - 0.0022053x_3x_5 \geq 0 \\
& g_2(x) \equiv 85.334407 + 0.0056858x_2x_5 + 0.0006262x_1x_4 - 0.0022053x_3x_5 \leq 92 \\
& g_3(x) \equiv 80.51249 + 0.0071317x_2x_5 + 0.0029955x_1x_2 + 0.0021813x_3^2 \geq 90 \\
& g_4(x) \equiv 80.51249 + 0.0071317x_2x_5 + 0.0029955x_1x_2 + 0.0021813x_3^2 \leq 110 \\
& g_5(x) \equiv 9.300961 + 0.0047026x_3x_5 + 0.0012547x_1x_3 + 0.0019085x_3x_4 \geq 20 \\
& g_5(x) \equiv 9.300961 + 0.0047026x_3x_5 + 0.0012547x_1x_3 + 0.0019085x_3x_4 \leq 25 \\
& 78 \leq x_1 \leq 102, \ 33 \leq x_2 \leq 45, \ 27 \leq x_i \leq 45, \ i = 3, 4, 5.
\end{aligned}$$

The optimal solution: $\mathbf{x}^* = (78, 33, 29.995, 45, 36.776)$, $f^* = -30665.5$.

f_5:

$$\begin{aligned}
\min \quad f(\mathbf{x}) = & \ x_1^2 + x_2^2 + x_1x_2 - 14x_1 - 16x_2 + (x_3 - 10)^2 + 4(x_4 - 5)^2 + (x_5 - 3)^2 \\
& + 2(x_6 - 1)^2 + 5x_7^2 + 7(x_8 - 11)^2 + 2(x_9 - 10)^2 + (x_{10} - 7)^2 + 45 \\
s.t. \quad & g_1(\mathbf{x}) = 105 - 4x_1 - 5x_2 + 3x_7 - 9x_8 \geq 0, \\
& g_2(\mathbf{x}) = -10x_1 + 8x_2 + 17x_7 - 2x_8 \geq 0, \\
& g_3(\mathbf{x}) = 8x_1 - 2x_2 - 5x_9 + 2x_{10} + 12 \geq 0, \\
& g_4(\mathbf{x}) = -3(x_1 - 2)^2 - 4(x_2 - 3)^2 - 2x_3^2 + 7x_4 + 120 \geq 0, \\
& g_5(\mathbf{x}) = -5x_1^2 - 8x_2 - (x_3 - 6)^2 + 2x_4 + 40 \geq 0, \\
& g_6(\mathbf{x}) = -x_1^2 - 2(x_2 - 2)^2 + 2x_1x_2 - 14x_5 + x_6 \geq 0, \\
& g_7(\mathbf{x}) = -0.5(x_1 - 8)^2 - 2(x_2 - 4)^2 - 3x_5^2 + x_6 + 30 \geq 0, \\
& g_8(\mathbf{x}) = 3x_1 - 6x_2 - 12(x_9 - 8)^2 + 7x_{10} \geq 0, \\
& -10 \leq x_i \leq 10, \quad i = 1, \cdots, 10.
\end{aligned}$$

The optimal solution:

$$\begin{aligned}
\mathbf{x}^* = & \ (2.171996, 2.363683, 8.773926, 5.095984, 0.9906548, 1.430574, \\
& 1.321644, 9.828726, 8.280092, 8.375927)
\end{aligned}$$
$$f(\mathbf{x}^*) = 24.3062$$

f_6:

$$\min \quad f(\mathbf{x}) = -\frac{\left|\sum_{i=1}^{n} \cos^4(x_i) - 2\prod_{i=1}^{n} \cos^2(x_i)\right|}{\sqrt{\sum_{i=1}^{n} i x_i}}$$

$$s.t. \quad \prod_{i=1}^{n} x_i \geq 0.75$$

$$\prod_{i=1}^{n} x_i \leq 7.5n$$

$$0 \leq x_i \leq 10,\ i = 1, \cdots, n$$

The current best solution: $f^* = -0.803553$.

References

1. Michalewicz, Z. 1994. *Genetic Algorithms + Data Structures = Evolution Programs*. Springer Science & Business Media.
2. Sivanandam, S, and S Deepa. 2008. *Genetic Algorithms*. In *Introduction to genetic algorithms*, 15–37. Springer.
3. Cukras, A.R., et al. 1999. Chess games: A model for RNA based computation. *Biosystems* 52 (1–3): 35–45.
4. Boneh, D., et al. 1996. On the computational power of DNA. *Discrete Applied Mathematics* 71 (1–3): 79–94.
5. Adleman, L.M. 1994. Molecular computation of solutions to combinatorial problems. *Science* 266 (1): 1021–1024.
6. Ding, Y, and L Ren. 2000. *DNA Genetic Algorithm For Design of the Generalized Membership-Type Takagi-Sugeno Fuzzy Control System*. In *2000 IEEE international conference on systems, man and cybernetics*. IEEE.
7. Yoshikawa, T, Furuhashi, and Y Uchikawa. 1997. *The effects of combination of DNA coding method with pseudo-bacterial GA*. In *Proceedings of 1997 IEEE international conference on evolutionary computation (ICEC'97)*. IEEE.
8. Tao, J.L., and N. Wang. 2007. Engineering, DNA computing based RNA genetic algorithm with applications in parameter estimation of chemical engineering processes. *Computers & Chemical Engineering* 31 (12): 1602–1618.
9. Yan, W., et al. 2006. A hybrid genetic algorithm-interior point method for optimal reactive power flow. *IEEE Transactions on Power Systems* 21 (3): 1163–1169.
10. Myung, H., and J.H. Kim. 1996. Hybrid evolutionary programming for heavily constrained problems. *Biosystems* 38 (1): 29–43.
11. Gudla, P.K., and R. Ganguli. 2005. An automated hybrid genetic-conjugate gradient algorithm for multimodal optimization problems. *Applied Mathematics Computation* 167 (2): 1457–1474.
12. Başokur, A.T., I. Akca, and N.W. Siyam. 2007. Hybrid genetic algorithms in view of the evolution theories with application for the electrical sounding method. *Geophysical Prospecting* 55 (3): 393–406.
13. Tao, J., and N. Wang. 2008. DNA double helix based hybrid GA for the gasoline blending recipe optimization problem. *Chemical Engineering Technology* 31 (3): 440–451.
14. Ponsich, A., et al. 2008. Constraint handling strategies in Genetic Algorithms application to optimal batch plant design. *Chemical Engineering and Processing* 47 (3): 420–434.
15. Rey Horn, J., N Nafpliotis, and D. E. Goldberg. 1994.*A niched Pareto genetic algorithm for multiobjective optimization*. In *Proceedings of the first IEEE conference on evolutionary computation, IEEE world congress on computational intelligence*. Citeseer.
16. Chakraborty, U.K., and C.Z. Janikow. 2003. An analysis of Gray versus binary encoding in genetic search. *Information Sciences* 156 (3): 253–269.

17. Tao, J., and Wang, N. 2007. DNA computing based RNA genetic algorithm with applications in parameter estimation of chemical engineering processes. *Computers and Chemical Engineering* 31 (12): 1602–1618.
18. Athreya, K.B., and P. Ney. 1978. A new approach to the limit theory of recurrent Markov chains. *Transactions of the American Mathematical Society* 245: 493–501.
19. Rosenthal, J.S. 1995. Minorization Conditions and Convergence Rates for Markov Chain Monte Carlo. *Publications of the American Statistical Association* 90 (430): 558–566.
20. Deb, K. 2000. An efficient constraint handling method for genetic algorithms. *Computer Methods in Applied Mechanics Engineering* 186 (2): 311–338.

Chapter 4
DNA Computing Based Multi-objective Genetic Algorithm

In this chapter, DNA computing based non-dominated sorting genetic algorithm is described for solving the multi-objective optimization problems. First, the inconsistent multi-objective functions are converted into Pareto rank value and density information of solution distribution. Then, the archive is introduced to keep the Pareto front individuals by Pareto sorting, and the maintaining scheme is executed to maintain the evenness of individual distribution in terms of individual crowding measuring. Finally, the gene-level operators of DNA computing are adopted to enhance the global searching capability of a multi-objective genetic algorithm (MOGA). The convergence speed is analyzed, and several suggestions on parameter setting are given based on the convergence analysis. Six multi-objective numeric functions are given, and the application results have shown the efficiency of DNA-MOGA in the evenness of population distribution and the convergence near the Pareto frontier.

4.1 Introduction

Multiple noncommensurable and simultaneously competing objectives are involved in many multi-objective optimization problems. If preference articulation of multiple objectives is aggregated into a scalar function with adequate weights, the multi-objective optimization problem can be transformed into a single-objective optimization problem, which can be solved by many methods [1]. However, various objectives often conflict with one another, and it is often unrealistic to obtain a single optimal solution for a multi-objective optimization problem. Hence, a group of compromise solutions will be derived through a multi-objective optimization algorithm, then, the decision maker will select one among those representative solutions.

Since Sckaffer first proposed Vector Evaluated Genetic Algorithm (VEGA) in 1985 [2], multi-objective evolution algorithms (MOEAs), such as PAES [3], SPEA2 [4], NSGA-II [5], etc., have gained significant attention from various fields. As the Pareto frontier is a set of solutions for multi-objective optimization problems, there

J. Tao et al., *DNA Computing Based Genetic Algorithm*,
https://doi.org/10.1007/978-981-15-5403-2_4

are two targets when applying the multi-objective optimization algorithm, that is, converge to the Pareto frontier and solution diversity preservation. Therefore, the individual fitness value according to the Pareto dominated relationship, as well as individual density information, is calculated by MOEAs. The corresponding individual maintaining and updating strategies are also studied in-depth. Usually, the individual updating strategy is implemented according to Pareto dominated relationship. In the cases that the individuals do not dominate each other, the individual density information considering the diversity index should be used. Though many schemes have considered the evenness degree of individual distribution, there are still short of the concrete measurement index. Some existing indexes are too complex to be applied easily in MOEA. Knowles et al. proposed an adaptive grid archiving strategy in PAES that provably maintained solutions in some "critical" regions of the Pareto front set once they were found [3]. Deb et al. computed the evenness of individual distribution using the cluster analysis method, however, the computing complexity was $O(N^3)$ with population size N [6].

As for theoretical research of MOEA, though most of MOEAs obtained satisfactory results of test functions, they lack theoretical analysis to guarantee the convergence of the solution distribution. Because of the importance of convergence analysis, theoretical research is carried out gradually. Rudolph proved that MOEA converged to the Pareto-optimal set with probability 1 [7]. The convergence of MOEA was usually analyzed by the Markov chain model without consideration of the evenness of individual distribution [8]. Laumanns et al. introduced the concept of ε-dominance and established MOEAs with the desired convergence and distribution properties, however, the convergence of MOEA was not analyzed [9]. Zitzler raised a theoretical analysis method to evaluate the performances of MOEA [10].

In this chapter, a DNA computing based non-dominated sorting multi-objective genetic algorithm (DNA-MOGA) is designed that guarantees both the progress towards the Pareto-optimal set and the evenness distribution of whole non-dominated solutions. The convergence of DNA-MOGA is analyzed based on Markov chain model to prove the effectiveness of the algorithm theoretically.

First, the inconsistent multi-objective fitness functions are converted into a single-objective function by Pareto sorting and individual crowding distance measuring.

Second, an external archive is introduced to keep the Pareto front individuals, and a maintaining scheme is used to maintain the evenness of individual distribution.

Third, the gene-level operators of DNA computing are adopted to enhance the global searching capability of DNA-MOGA.

Finally, six typical multi-objective test functions are applied to show an evenness distribution of Pareto frontier and a quick convergence to the true Pareto-optimal set.

4.2 Multi-objective Optimization Problems

Without loss of generality, multi-objective optimization algorithm seeks to optimize a vector of noncommensurable and competing objectives or cost functions, and the constrained multi-objective optimization problem is described as follows:

$$\begin{aligned} \min \quad & f(\mathbf{x}) = [f_1(\mathbf{x}), f_2(\mathbf{x}), \ldots, f_m(\mathbf{x})] \\ s.t. \quad & g_i(\mathbf{x}) \leq 0, \ i = 1, 2, \ldots, h, \ \mathbf{x} \in R^n \end{aligned} \tag{4.1}$$

where f_i $(1 \leq i \leq m)$ is the objective function with totally m objectives, $g_i(1 \leq i \leq h)$ is the constraint condition, and $\mathbf{x} = [x_1, x_2 \ldots x_n]$ is the solution vector to be solved with n variables. Solutions of the multi-objective optimization problem are a family of points known as the Pareto-optimal solution set, where each objective component of any point along with the Pareto frontier can only be improved by degrading at least one of the other objective components [6]. In the absence of preference information of the objectives, the Pareto ranking scheme is regarded as an appropriate approach to represent the strength information of each individual in a MOEA [7]. The vectors are compared according to the dominance relationship defined below [11].

Definition 4.1 (Dominance relationship) a vector $\mathbf{x}$ dominates another vector $\mathbf{y}$, denoted as $\mathbf{x} \prec \mathbf{y}$, if

$$f_i(\mathbf{x}) \leq f_i(\mathbf{y}), \ \forall i \in \{1, 2, \ldots, m\} \tag{4.2}$$

$$f_j(\mathbf{x}) < f_j(\mathbf{y}), \ \exists \ j \in \{1, 2, \ldots, m\} \tag{4.3}$$

Based on the concept of the dominance, the Pareto set can be defined.

Definition 4.2 (*Pareto set*) Let $F \subseteq R^n$ be a set of vectors, then the Pareto set F^* of F is defined as follows: F^* contains all vectors $x \in F$ that are not dominated by any other vector $f \in F$, i.e.

$$F^* := \{x \in F | \nexists f \in F : f \prec x\} \tag{4.4}$$

Vectors in F^* are called the Pareto vectors of F. All Pareto set of F is denoted as $P^*(F)$. Moreover, for a given set F, the Pareto set F^* is unique. Therefore, we have $P^*(F) = F^*$. Since the Pareto set F^* is of substantial size for multiple sets of F, the determination of F^* is quite difficult.

To illustrate the concept of Pareto-optimal solution set, the Pareto ranking scheme for a bi-objective minimization problem is shown in Fig. 4.1. As can be seen, it assigns the same smallest ranking value 1 for all non-dominated vectors, while the dominated ones are inversely ranked according to how many individuals in the population dominated them. The Pareto ranking was first introduced by Goldberg [12], and successfully applied by NSGA-II, which ensures that all the non-dominated individuals in the population will be assigned rank 1 and removed from a temporary

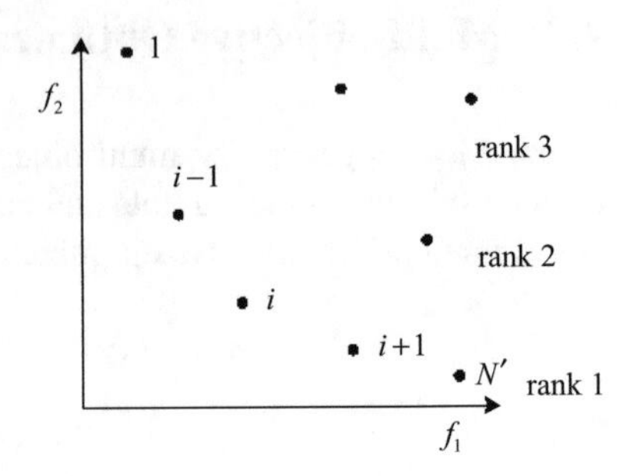

Fig. 4.1 Schematic illustration of Pareto ranking

assertion, then a new set of non-dominated individuals will be assigned 2, and so forth.

4.3 DNA Computing Based MOGA (DNA-MOGA)

Generally, the purpose of MOEA is to find or approximate the Pareto set and keep the diversity of Pareto-optimal solutions [1]. Hence, Pareto ranking and density values of the individual distribution are utilized as two important attributes to each individual. In this chapter, a fitness assignment strategy is designed and an external population is utilized in order to search for an approximated optimal Pareto frontier, and the maintaining scheme is also designed to keep the diversity of Pareto set. In addition, the DNA gene-level operators are introduced to improve the global searching capability of MOGA. Four crucial strategies to improve the performance of DNA-MOGA are discussed as follows.

4.3.1 *RNA Encoding*

RNA nucleotide base in Chap. 2 is also adopted, 0123 instead of AUGC is used to encode the variable to be solved, and the length of each variable is set as l. Since there are n variables in Eq. 4.1, the length of one chromosome becomes $L = nl$. Moreover, fitness values of multi-objective functions are calculated to analyze the ranking and density values of the chromosome. Thus, the structure of a chromosome is obtained as shown in Fig. 4.2. For a general multi-objective optimization problem, the quaternary encoding chromosome with the length of nl is generated randomly for n variables, Pareto sorting rank and density information are recorded in the $(nl + 1)$

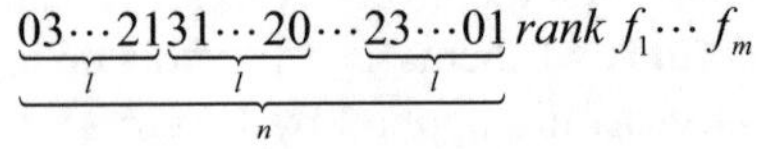

Fig. 4.2 The encoding and structure of one chromosome

$$\underbrace{\underbrace{03\cdots21}_{l}\underbrace{31\cdots20}_{l}\cdots\underbrace{23\cdots01}_{l}}_{n}\ rank\ f_1\cdots f_m$$

th location, and the fitness values of m objective functions are also recorded in the chromosome. The total length of a chromosome is then $nl+m+1$, where the former nl uses the quaternary encoding and the latter $m+1$ decimal numbers are obtained in terms of the former quaternary encoding. The decoding process of quaternary encoding is the same as that in Chap. 2.

$$\underbrace{\underbrace{03\cdots21}_{l}\,\underbrace{31\cdots20}_{l}\cdots\underbrace{23\cdots01}_{l}}_{n}\,rank\ f_1\cdots f_m$$

4.3.2 Pareto Sorting and Density Information

The rank value of Pareto sorting for the individuals has been described in Fig. 4.1. However, the ranking method may fail when most of the individuals do not dominate one another, i.e., all individuals have rank 1. Therefore, additional density information is incorporated to discriminate among individuals with identical rank value. Thus, any multi-objective optimization problem can be converted into a bi-objective optimization problem, i.e., minimizing the ranking value and maximizing the crowding distance [5], which can be further changed into a single-objective optimization problem through some modifications.

In Fig. 4.3, the size of individuals in the Pareto frontier with rank 1 is N' and the ith individual is alongside the Pareto frontier. The calculation of the density information for the ith individual for a bi-objective optimization problem is described as follows:

$$d_i' = 1\Big/\sum_{j=1}^{2}(f_j^{i+1} - f_j^{i-1}) \tag{4.5}$$

According to Eq. 4.5, the density value is inversely proportional to the perimeter of a rectangle composed of the dotted line, so the more crowding the individual distribution, the larger the individual density value. For the boundary individuals (1th and N'th), the rectangle is regarded as infinite and the density value is set to 0.

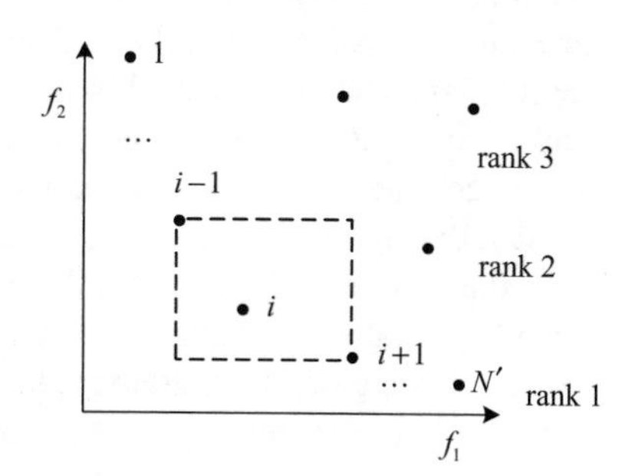

Fig. 4.3 Schematic illustration of individual density calculation

After a simple transformation, e.g., reciprocals operation in Eq. 4.5, the maximization crowding distance is changed into the minimization problem that is consistent with the rank value. As the minimum of Pareto sorting rank is 1, d_i' is normalized to $d_i = d_i' / d\text{max}$, where $d\text{max}$ is the maximum of the current d_i'. Hence, a single-objective fitness function can be derived:

$$F_{fit}(i) = i_{rank} + \lambda d_i \tag{4.6}$$

where i_{rank} is the Pareto sorting rank obtained by non-dominated sorting algorithm that requires $O(mN'^2)$ comparisons [5], and $\lambda = 0.99$ is a coefficient to guarantee $F_{fit}(i)$ less than $i_{rank} + 1$. Equation 4.6 changes the multi-objective optimization problem into a single-objective optimization problem by combining the rank value together with the density information. The calculation complexity of density value is dependent on the sorting algorithm for the current Pareto frontier, and the sorting algorithm requires $O(mN' \log N')$ computations. Hence, the overall computational complexity of the fitness calculation is $O(mN'^2)$.

Matlab Pseudocode of non-dominated sorting algorithm can be obtained from [5], and the code can be gained from their website. All the individuals in the first front are given rank 1, the second front individuals are assigned rank 2, and so on. After assigning the rank, the crowding in each front can be calculated in terms of Eq. 4.6.

4.3.3 Elitist Archiving and Maintaining Scheme

In terms of Eq. 4.6, the individuals satisfying $F_{\text{fit}} < 2$ are obviously the elitists, which will be kept at an archive. At each generation (g), a Pareto approximated set F will be produced and stored to the archive. Thus, the archive size will increase with the evolution process, and it may be far larger than N. Obviously, too many solutions cannot help decision maker deal with the problem. The desirable solution set is an approximation of F^* that dominates all elements of F and of bounded size. Hence, the maximal size of the elitist archive is limited to N. If the size of elitist archive (N') is larger than N, the maintaining scheme for the archive size limitation is implemented to keep the individual diversity and evenness distribution of the individuals. The following principles are abided by: if the new individual dominates partial individuals in the archive, then, the dominated ones are eliminated from the archive and the new individual is added, else, if the individuals do not dominate one another, the maintaining scheme is performed to make the archive size not larger than N and keep the evenness of individual distribution in the archive. Whether it can be added to the archive is depending on the Pareto rank. The computational complexity of the maintaining scheme is similar to the Pareto non-dominated sorting algorithm depending on the size of the elitist archive, i.e., $O(mNN')$. When all individuals in the archive are along the Pareto frontier, a modified adaptive cell density evaluation scheme is implemented that is originated from [13], as shown in Fig. 4.4.

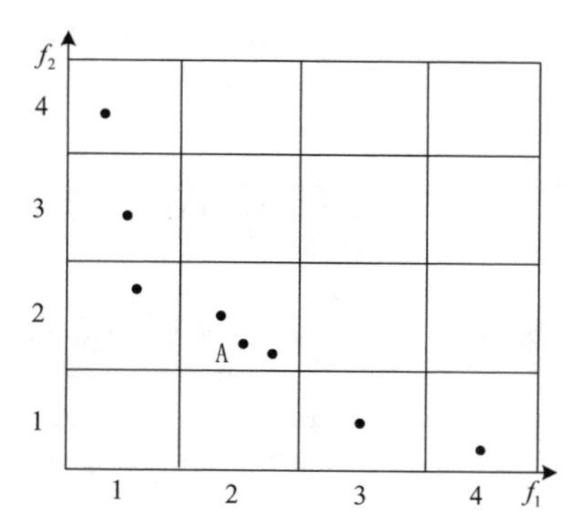

Fig. 4.4 Density map and density grid of the Pareto frontier

After elitist keeping, all individuals in the archive are non-dominated ones. If the archive size is still larger than N, the maintaining scheme is carried out to guarantee that at most one individual is kept at each cell. For example, there are three individuals at cell 'A' in Fig. 4.4, and two individuals will be removed by the maintaining scheme. The cell width in each dimension of objective functions can be calculated as

$$g_{wi} = [\max f_i(\mathbf{x}) - \min f_i(\mathbf{x})] / K_i \tag{4.7}$$

where g_{wi} is the cell width in the ith dimension at generation (g), K_i denotes the number of cells designated for the ith dimension, e.g., in Fig. 4.4, and there exist 16 cells if $K_1 = K_2 = 4$. Since the maximal and minimal fitness values in the objective space will change with the process of evolution, the cell size will vary from generation to generation.

At every generation (g), the cell information is first calculated in terms of Eq. 4.7, and the cells with more than 1 individual will be maintained. The main procedure of the maintaining scheme is given as follows [14].

(1) Calculate the width of the cell (g_{wi}) and add ζ to g_{wi} in order to guarantee that all individuals locate in the cell but are not at the boundary line of the cell.
(2) Repeat to find the position of each individual and judge whether there exist other individuals in the cell; if so, compare the fitness value obtained by Eq. 4.6 and keep the best individual, else the individual is directly added to the archive.
Through the above procedure, there is at most one individual in a cell. The calculation complexity of the maintaining scheme is $O(N'^2)$ to locate the position of the individuals, where N' is the archive size, $N' > N$.
By using the elitist archive and its maintaining scheme, the diversity of the population is kept, and the archive size (N') satisfies $K_j < N' < \sum_{j=1}^{m} K_j$. When there are two or more individuals in a cell, the excessive individuals will be deleted. Matlab code of the maintenance function is given as follows:

```
function CompressedData=maintainf( RawData,ColumnIndex, Count)
% RawData is individuals in the archive, ColumnIndex is nl+M+V, Count is
Population size
[m,n]=size(RawData);
TempPoint=zeros(1,n);
for i=m:-1:2
    for j=1:1:i-1
        if RawData(j,ColumnIndex)>RawData(j+1,ColumnIndex)
            TempPoint=RawData(j,:);
            RawData(j,:)=RawData(j+1,:);
            RawData(j+1,:)=TempPoint;
        end
    end
end
% Locate the cell position
maxX=max(RawData(:,ColumnIndex));
minX=min(RawData(:,ColumnIndex));
maxY=max(RawData(:,ColumnIndex+1));
minY=min(RawData(:,ColumnIndex+1));
 % obtain the length of cell
unitX=(maxX-minX)/Count+1e-4;
unitY=(maxY-minY)/Count+1e-4;
SelectedDataFirst=[zeros(1,ColumnIndex-1),100,100,0,0];
for i=1:m
    areaX=ceil((RawData(i,ColumnIndex)-minX)/unitX+1e-8);
    areaY=ceil((RawData(i,ColumnIndex+1)-minY)/unitY+1e-8);
    [mm,nn]=size(SelectedDataFirst);
    flag=0;
    for j=1:mm
        if SelectedDataFirst(j,nn-1)==areaX &&
SelectedDataFirst(j,nn)==areaY
            flag=1;
              if SelectedDataFirst(j,ColumnIndex)>RawData(i,ColumnIndex)
                  &&
                  SelectedDataFirst(j,ColumnIndex+1)>RawData(i,ColumnIn
                  dex+1)
SelectedDataFirst(j,1:ColumnIndex+1)=RawData(i,:); %COPY the X
information
            end
        end
    end
    if flag == 0
SelectedDataFirst=[SelectedDataFirst;[RawData(i,:),areaX,areaY]];
    end
end
[mm,nn]=size(SelectedDataFirst);
SelectedDataFirst=SelectedDataFirst(2:1:mm,:);
CompressedData=SelectedDataFirst(:,1:ColumnIndex+1);
```

4.3.4 DNA Computing Based Crossover and Mutation Operators

When solving the single-objective optimization problem, various crossover and mutation operators are proposed and the shortcomings of SGA can be alleviated greatly [15]. Recently, DNA computing based gene-level operators have made some significant achievements [16, 17]. However, most of MOEAs utilize the traditional crossover and mutation operators. In this chapter, DNA computing based crossover and mutation operators are utilized to improve the global searching capability of MOGA.

(1) Selection operator

Because the population in the archive is composed of the non-dominated individuals, they are directly selected as the parents of the genetic operators. If the number of the non-dominated individuals N' is less than N, the random selection for a better individual with the value of Eq. 4.6, is adopted to choose the rest $(N - N')$ of the individuals. The Matlab code is given as follows: Note that the proportional selection and Roulette wheel selection can also be used here.

```
function parent_chromosome=chooserest(BestS,SizeBestS,chromosome,pop,BsJi)
% BestS is the individuals in the archive, SizeBestS is N', pop is
population size
if SizeBestS<=pop
      parent_chromosome(1:SizeBestS,:)=BestS;
      kk=SizeBestS+1;
   while kk<=pop
       choose_1=kk;
       choose_2=ceil(rand*pop); % randomly selected individual
       while choose_1==choose_2
           choose_2=ceil(rand*pop);
       end

       if BsJi(choose_1)<=BsJi(choose_2) %better individual is selected.
            parent_chromosome(kk,:)=chromosome(choose_1,:);
            kk=kk+1;
       else
            parent_chromosome(kk,:)=chromosome(choose_2,:);
            kk=kk+1;
       end
   end
 else
     parent_chromosome(1:pop,:)=BestS(1:pop,:);
 end
```

(2) Crossover and mutation operators

Since DNA sequence is made up of four nucleotide bases, i.e., Adenine (A), Uracil (U), Guanine (G), and Cytosine (C). The quadruple encode is also selected as the encoding of chromosomes. Three operations on a single RNA sequence: translocation, transformation, and permutation are adopted as the crossover operators [18]. While three operations for mutation of nucleotide base: reversal, transition, and exchange are utilized as the mutation operators [18]. The corresponding description about RNA crossover and mutation operators, and their Matlab code can refer to Chap. 2.

4.3.5 The Procedure of DNA-MOGA

In DNA-MOGA, a fitness function using Pareto sorting rank with density information is given in Sect. 3.2. The elitist individuals are then kept in the archive. To keep the evenness distribution of the elitist, a maintaining scheme is designed based on the adaptive cell. DNA computing based crossover and mutation operators are introduced to improve the global searching capability of MOGA. The whole procedure of DNA-MOGA is described in the following steps.

Step 1: Initialize the population size N, the number of cells for the ith dimension K_i, and the maximum generation G.

Step 2: Generate the N quaternary encoding chromosomes randomly in the search space, decode, and calculate the fitness value in terms of Eq. 4.6.

Step 3: Keep the elitists in the archive and maintain the archive when the number of archive individuals N' is larger than N.

Step 4: Select the archive population as the parents of the genetic operators. If $N' < N$, use the random selection operation for choosing the rest of the individuals.

Step 5: Carry out the crossover operator in the best $N/2$ individuals in terms of the value of the single-objective function Eq. 4.6, where the permutation operator is implemented with probability 1 and the transformation operator is implemented with probability 0.5. If the transformation operator is not performed, the translocation operator is then implemented. N offspring are generated by the crossover operator.

Step 6: Implement the mutation operator in the left $N/2$ individuals. The nucleotide base is replaced by one of the three integers when the mutation operator is performed.

Step 8: Repeat steps 3–6 until the termination criterion is met, that is, the set maximal generation or the set minimal distance to the true Pareto frontier is satisfied.

4.3.6 Convergence Analysis of DNA-MOGA

At present, the convergence analysis of MOEA mainly lies in the infinite Pareto solutions [9]. In this chapter, the elitists are kept in the archive with bounded size, and the convergence is analyzed according to the above definition of Pareto non-dominated relationship and the running procedure of DNA-MOGA.

Theorem 4.1 Let $F^{(g)} = \cup_{j=1}^{g} f_i^{(j)}, 1 \leq f_i^{(j)} \leq N', i \in \{1, \ldots, m\}$ be the set of elitists in the archive and maintained by the maintaining scheme. Then $A^{(g)}$ is an approximated Pareto set of $F^{(g)}$ with bounded size $\left|A^{(g)}\right|$, i.e., (1) $A^{(g)} \in P^*(F^{(g)})$, (2) $\left|A^{(g)}\right| \leq \sum_{i=1}^{m} K_i$.

Proof

(1) Suppose $A^{(g)} \notin P^*(F^{(g)})$ is at generation g. According to Definition 4.2, the case occurs only if f is not dominated by any individual of $A^{(g)}$ or not in $A^{(g)}$.

For f that is not in $A^{(g)}$, i.e., f is not an elitist or f is an elitist but removed later on. Removal, however, only takes place when some new f^p enters the archive which dominates f. The case contradicts the assumption that f is not dominated by any individuals of $A^{(g)}$. Likewise, removal takes place in the maintaining scheme when f^p and f are located in the same cell, i.e., both of them are individuals in the Pareto frontier, and the fitness value of f^p is superior to f's, which contradicts with the assumption that f is not in $A^{(g)}$.

For f in $A^{(g)}$, however, f does not belong to the Pareto-optimal set. Hence, there exists $f^p \in F^{*(g)}$ that $f^p \prec f$. If this is the case, f will be eliminated by elitist maintaining scheme, i.e., $f \notin A^{(g)}$, which contradicts the assumption.

(2) In terms of the maintaining scheme, the objective spaces are divided into $\prod_{i=1}^{m} K_i$ cells. For $A^{(g)} \in P^*(F^{(g)})$, only those with rank 1 is maintained, and at most one individual can be kept at each cell. Therefore, the maximum number of individuals distributed at the cells is $\sum_{i=1}^{m} K_i$, i.e., $\left|A^{(g)}\right| \leq \sum_{i=1}^{m} K_i$.

Theorem 4.2 If $\left|A^{(g)}\right| \leq N$, DNA-MOGA converges to the Pareto-optimal set with probability 1.

Proof In terms of theorem 4.1, the size of Pareto-optimal set $\left|A^{(g)}\right|$ is the boundary. If the transition probability matrix (P) is the finite state homogeneous Markov chain, then the state of the population is ergodic. Because the elitist maintaining scheme is adopted in DNA-MOGA, the Pareto-optimal individual can be kept and the dominated ones are eliminated finally.

Since the population size is finite and the crossover and mutation operators with elitist are adopted in DNA-MOGA, the transition probability matrix has been proved as the finite states homogeneous Markov chain in [19], i.e., the whole state of population can be reached by DNA-MOGA. Obviously, $\left|A^{(g)}\right| \leq N$ is the premise of the ergodicity.

In terms of theorem 4.1 and 4.2, DNA-MOGA can not only converge to the Pareto-optimal set with probability 1 but also keep the evenness of the population distribution using maintaining scheme. Though $\left|A^{(g)}\right| \leq N$ is required by the convergence analysis, $\left|A^{(g)}\right| = \sum_{i=1}^{k} K_i$ is an extreme number obtained by DNA-MOGA. Hence, it is not necessary for N to be larger than $\sum_{i=1}^{m} K_i$, however, N should be at least greater than $\max(K_i)$.

4.4 Simulations on Test Functions by DNA-MOGA

4.4.1 Test Functions and Performance Metrics

Six test problems are used to examine the performances of DNA-MOGA as listed in the Appendix, where ZDT3, ZDT4, and ZDT6 were designed by Zitzler, as can be found in [4] and [20], and other test functions include DEB, FON, and KUR problems. These problems have characteristics that are suitable for examining the effectiveness of multi-objective optimization approaches in terms of maintaining the population diversity, as well as converging to the Pareto frontier. Many researchers, such as Zitzler et al. [4], Deb et al. [5], Knowles and Corne [13], and Tan et al. [21], have used these test problems in their research on MOEAs.

Three different quantitative performance metrics for multi-objective optimization approaches are utilized, which are capable of evaluating non-dominated individuals and widely used in the different MOEAs [4, 5, 21].

(1) Generational Distance (*GD*): The metric of generational distance represents how "far" the known Pareto front (PF_{known}) is from the true Pareto front (PF_{true}), which is defined as

$$GD = \left(\frac{1}{n_{PF}} \sum_{i=1}^{n_{PF}} d_i^2 \right)^{1/2} \tag{4.8}$$

where n_{PF} is the number of individuals in PF_{known}, and d_i is the Euclidean distance in the objective domain between the individuals in PF_{known} and its nearest individual in PF_{true}. The smaller the value of *GD*, the better the approximation of the Pareto-optimal set becomes.

(2) Evenness Spacing (*ES*): The metric of spacing measures how "even" the individuals in PF_{known} are distributed. It is defined as

$$ES = \frac{\left[\frac{1}{n_{PF}} \sum_{i=1}^{n_{PF}} (d_i' - \bar{d}')^2 \right]^{1/2}}{\bar{d}'}, \quad \bar{d}' = \frac{1}{n_{PF}} \sum_{i=1}^{n_{PF}} d_i' \tag{4.9}$$

where d_i' is the Euclidean distance in the objective domain between the *i*th individual in PF_{known} and its nearest one. The smaller the value of *ES*, the more evenness the distribution of Pareto frontier will be.

(3) Maximum Spread (*MS*): The metric of maximum spread measures how 'well' the PF_{true} is covered by the PF_{known} through hyper-boxes formed by the extreme function values observed in the PF_{true} and PF_{known}. In order to normalize the metric, it is described as

$$MS = \sqrt{\frac{1}{m}\sum_{i=1}^{m}\left\{\frac{[\min(f_i^{\max}, F_i^{\max}) - \max(f_i^{\min}, F_i^{\min})]}{F_i^{\max} - F_i^{\min}}\right\}^2} \tag{4.10}$$

where m is the number of objectives, $f_i^{\max}$ and $f_i^{\min}$ are the maximum and minimum values of the ith objective function in PF_{known}; $F_i^{\max}$ and $F_i^{\min}$ are the maximum and minimum values of the ith objective in PF_{true}, respectively. The maximum of MS is 1, which shows that the extremum of each objective function is covered by the solutions in the Pareto frontier.

4.4.2 Calculation Results

When DNA-MOGA is used, quaternary encoding is adopted to simulate four nucleotide bases, the individual length for each variable is set to 10, initial population size N is set to 60, the maximum generation number is limited to 10000, and the cell number for each objective dimension K_i is set to 50. The runs of DNA-MOGA terminate when $SD < 10^{-3}$ is satisfied or the maximum generation is performed. NSGA-II is also used to optimize the same multi-objective optimization problems and compared with DNA-MOGA. NSGA-III has obtained a great improvement in solving the multi-objective problem with more than two objectives [22], here, it is not adopted to solve the bi-objective optimization functions.

Two MOEAs are run 50 times, respectively, and the best results with Pareto frontier are selected as the final results. The simulation results are illustrated in Figs. 4.5, 4.6, 4.7, 4.8, 4.9, 4.10, and listed in Table 4.1. It can be seen that the DNA-MOGA guarantees both progress towards the Pareto-optimal frontier and covering the whole range of non-dominated solutions evenly. For DEB problem and ZDT4

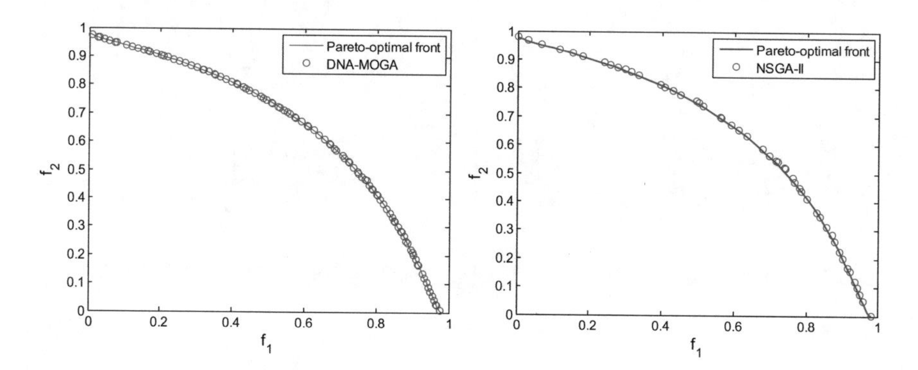

Fig. 4.5 Pareto-optimal solutions to FON problem found by DNA-MOGA and NSGA-II

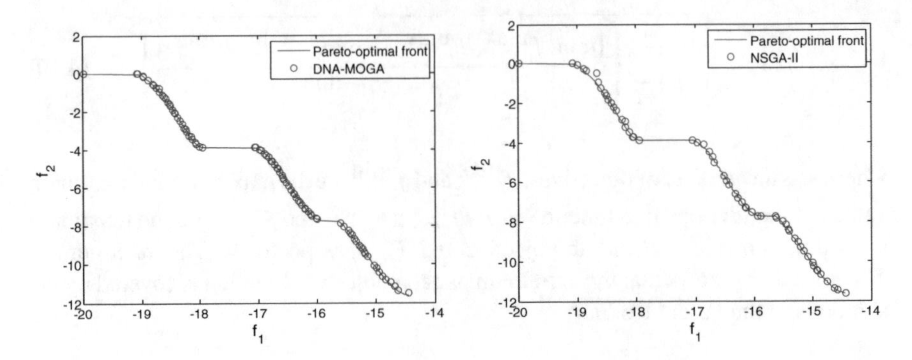

Fig. 4.6 Pareto-optimal solutions to KUR problem found by DNA-MOGA and NSGA-II

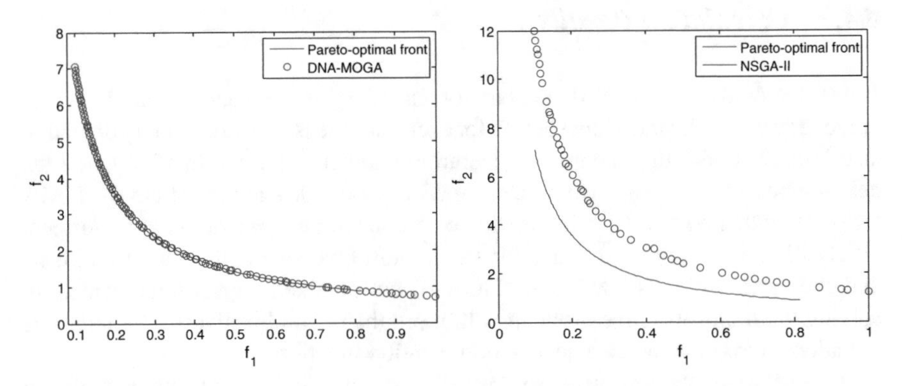

Fig. 4.7 Pareto-optimal solutions to DEB problem found by DNA-MOGA and NSGA-II

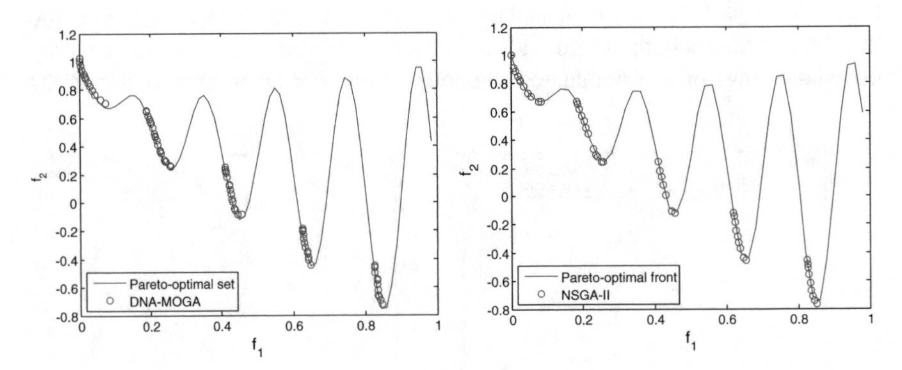

Fig. 4.8 Pareto-optimal solutions to ZDT3 problem found by DNA-MOGA and NSGA-II

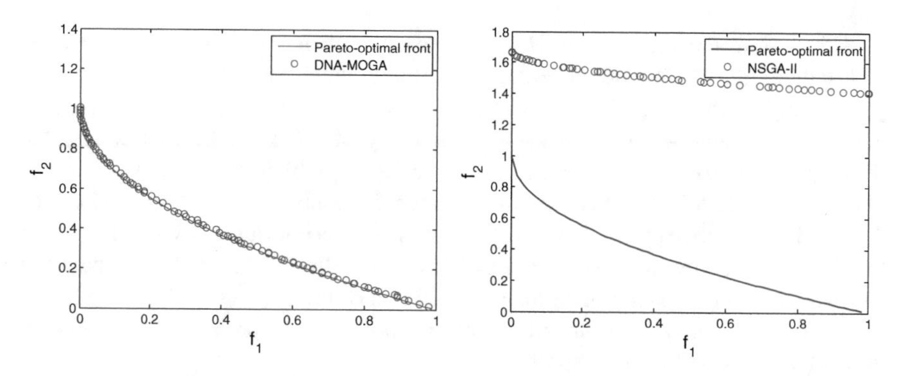

Fig. 4.9 Pareto-optimal solutions to ZDT4 problem found by DNA-MOGA and NSGA-II

Table 4.1 The comparison results for test problems over 50 runs

Test problems	NSGA-II			DNA-MOGA		
	SD	ES	MS	SD	ES	MS
FON	0.0014	0.6236	1.0000	0.0012	0.3032	0.9948
KUR	0.0056	0.6717	1.0000	0.0045	0.9995	0.9935
DEB	0.0723	0.7114	0.7181	0.0056	0.4403	0.9981
ZDT3	0.0069	0.6374	0.9333	0.0058	0.4620	0.9311
ZDT4	0.0956	0.7425	0.7279	0.0021	0.4589	0.9999
ZDT6	0.0069	0.5811	1.0000	0.0049	0.2304	1.0000

problem with many local minima and fraudulence characteristics, the advantages of DNA-MOGA are especially obvious from Figs. 4.7 and 4.9, and the measure of *SD* in Table 4.1 also shows the improvement of DNA-MOGA in approximating to the Pareto-optimal solution set. The metrics of *ES* indicate the superiority of DNA-MOGA in the evenness of population distribution except for the KUR problem.

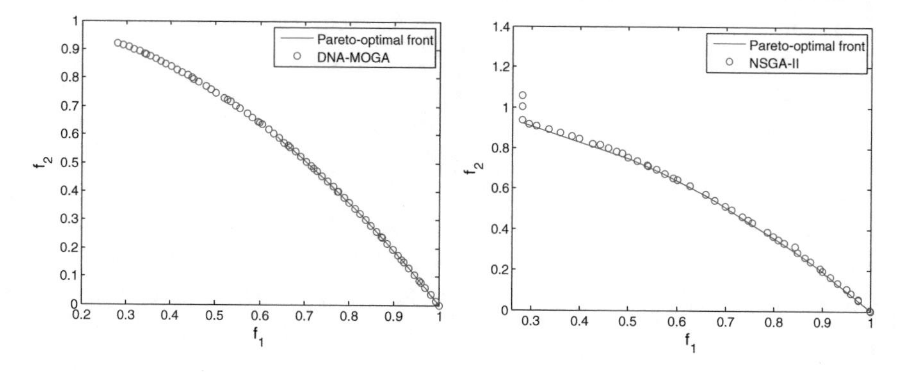

Fig. 4.10 Pareto-optimal solutions to ZDT6 problem found by DNA-MOGA and NSGA-II

Concerning the measure of *MS*, because one cell is required to keep at most one individual, the individual with larger fitness value but better single-objective value will be removed, which may affect the performance of *MS*. The comparison results from Table 4.1 illuminate this deficiency of DNA-MOGA, especially the KUR problem. Since the Pareto front of the KUR problem is noncontinuous and the distance between 2 noncontinuous frontiers is relatively large, the drawback of DNA-MOGA is then prominent and the metrics of *ES* is inferior to NSGA-II's, which can be improved by increasing the number of the cells. However, the computation burden will become heavy with the increasement of the cell number. Since *SD* and *ES* are the main indexes of the multi-objective optimization problem, DNA-MOGA is superior to NSGA-II as a whole.

4.5 Summary

The DNA computing based non-dominated sorting genetic algorithm is suggested for multi-objective optimization problems. The inconsistent multi-objective fitness functions are converted into a single-objective function by Pareto sorting and individual crowding measuring. And an external archive is introduced to keep the individuals in the Pareto frontier, and the maintaining scheme is designed to keep the evenness of individual distribution. The gene-level operators of DNA computing are also adopted to enhance the global searching capability of MOGA. Convergence analysis and simulation results on typical multi-objective test problems show the improvement of DNA-MOGA in the spread of solutions and the convergence near the true Pareto-optimal set.

Appendix

Typical multi-objective test problems

Problems	The minimum objective functions	Optimal solutions	Comments
DEB	$f_1(\mathbf{x}) = x_1$ $f_2(\mathbf{x}) = g/\, f_1(\mathbf{x})$ $g = 2 - \exp\left\{-\left(\frac{x_2 - 0.2}{0.004}\right)^2\right\} - 0.8\exp\left\{-\left(\frac{x_2 - 0.6}{0.4}\right)^2\right\}$	$x_i \in [0.1, 1]$ $i = 1, 2$	Non-convex
FON	$f_1(\mathbf{x}) = 1 - \exp(-\sum_{i=1}^{3} (x_i - 1/\sqrt{3})^2)$ $f_2(\mathbf{x}) = 1 - \exp(-\sum_{i=1}^{3} (x_i + 1/\sqrt{3})^2)$	$x_i \in [-4, 4]$ $i = 1, 2, 3$	Non-convex
KUR	$f_1(\mathbf{x}) = \sum_{i=1}^{2} [1 - 10\exp(-0.2\sqrt{x_i^2 + x_{i+1}^2})]$ $f_2(\mathbf{x}) = \sum_{i=1}^{3} [\lvert x_i\rvert^{0.8} + 5\sin(x_i^3)]$	$x_i \in [-5, 5]$ $i = 1, 2, 3$	Non-convex Noncontinuous
ZDT3	$f_1(\mathbf{x}) = x_1$ $f_1(\mathbf{x}) = x_1$ $f_2(\mathbf{x}) = g[1 - (x_1/g)^{1/2} - x_1\sin(10\pi x_1)/g]$ $f_2(\mathbf{x}) = g[1 - (x_1/g)^{1/2} - x_1\sin(10\pi x_1)/g]$ $g = 1 + 9\left(\sum_{i=2}^{n} x_i/(n-1)\right)$ $g = 1 + 9\left(\sum_{i=2}^{n} x_i/(n-1)\right)$	$x_i \in [0, 1]$ $i = 1, 2, \ldots, 30$	Convex, disconnected
ZDT4	$f_1(\mathbf{x}) = x_1$ $f_2(\mathbf{x}) = g[1 - (x_1/g)^{1/2}]$ $g = 1 + 10(n-1) + \sum_{i=2}^{n} [x_i^2 - 10\cos(4\pi x_i)]$	$x_1 \in [0, 1]$ $x_i \in [-5, 5]$ $i = 2, \cdots, 10$	Non-convex
ZDT6	$f_1(\mathbf{x}) = 1 - \exp(-4x_1)\sin^6(6\pi x_1)$ $f_2(\mathbf{x}) = g[1 - (x_1/g)^2]$ $g = 1 + 9\left(\sum_{i=2}^{n} x_i/(n-1)\right)^{0.25}$	$x_i \in [0, 1]$ $i = 1, 2, \ldots, 10$	Non-convex non-uniformly spaced

Matlab code of the main function for improved DNA-MOGA to solve the multi-objective function

```
clc;
clear all;
close all;

global howlong G
fig=0;
G=10000;                                        % Maximal generation
pop=60;                                         % Population size
V=2;  fmax=0;   M=2;    howlong=8;   num=V;
CodeL=V*howlong;
Rep=1;

amax=ones(1,V);
amin=0.1*ones(1,V);
max_range=amax;min_range=amin;
 for rep=1:1:Rep
   t0=clock;  SizeBestS1=0;deltkg=1;

  chromosome = initialize_variablesRNADEB(pop,M, V, min_range,
max_range);
  chromosome= non_domination_sort_modRNA(chromosome, M, V);
  BsJi=chromosome(:,CodeL+1);
  a=find(BsJi<2);%%%find the non-dominated individuals
  ma=size(a,1);
  BestS=[];
  for i=1:ma
      BestS=[BestS;chromosome(a(i),:)];
  end
kkk=1;
for kg=1:1:G
   time(kg)=kg;
  BsJi=chromosome(:,CodeL+1);
  a=find(BsJi<2);
  ma=size(a,1);
  for i=1:ma
     BestS=[BestS;chromosome(a(i),:)];
  end
  SizeBestS=size(BestS,1);
  tempBest= non_domination_sort_modRNA1(BestS, M, V);
  tempBsJi=tempBest(:,CodeL+1);
     a=find(tempBsJi<2);
     ma=size(a,1); BestS=[];
      for i=1:ma
           BestS=[BestS;tempBest(a(i),:)];
      end
```

```
        SizeBestS=size(BestS,1);
        if SizeBestS>pop
             BestS=SelectComData(BestS(:,1:M+CodeL+1),CodeL+M,60);
            SizeBestS=size(BestS,1);
            deltsize(deltkg)=SizeBestS-SizeBestS1;
             deltkg=deltkg+1;
            SizeBestS1=SizeBestS;
        end
          if deltkg>1201
            sumdeltsize=sum(deltsize(deltkg-1200:deltkg-1));
            if sumdeltsize<1
                break;
            end

       end

    parent_chromosome=chooserest(BestS,SizeBestS,chromosome,pop,BsJi);
     [temp_chromosome1,indexp]=sort(parent_chromosome(:,CodeL+1));
     for i=1:pop
         parent_chromosome1(i,:)=parent_chromosome(indexp(i),:);
     end
     offspring_chromosome = ...
         genetic_operatorRNADEB(parent_chromosome1,i,pop,...
         M, V, min_range, max_range);
     chromosome=non_domination_sort_modRNA(offspring_chromosome, M, V);
end
wholetime=etime(clock,t0)
bestjj(rep,:,:)=BestS;
rep
end
    plot(BestS(:,CodeL + 2),BestS(:,CodeL + 3),'o');
    xlabel('f_1','fontsize',12);
    ylabel('f_2','fontsize',12);
```

References

1. Deb, K. 2001. Multi-objective optimization using evolutionary algorithms. John Wiley & Sons.
2. Schaffer, J.D. 1985. Multiple Optimization with vector evaluated genetic algorithms. In *International conference on genetic algorithms.*
3. Knowles, J.D., and D.W. Corne. 2000. Approximating the nondominated front using the pareto archived evolution strategy, vol. 8.
4. Zitzler, E., K. Deb, and Thiele, L. 2000. Comparison of multiobjective evolutionary algorithms: empirical results. *Evolutionary Computation* 8 (2): 173–195.
5. Deb, K., and D. Kalyanmoy. 2002. A fast and elitist multiobjective genetic algorithm: NSGA-II. *IEEE Transactions on Evolutionary Computation* 6 (2): 182–197.

6. Konak, A., Coit D.W., and Smith, A.E. 2006. Multi-objective optimization using genetic algorithms: A tutorial. *Reliability Engineering & System Safety* 91 (9): 992–1007.
7. Rudolph, G. 1998. On a multi-objective evolutionary algorithm and its convergence to the Pareto set. In *IEEE World Congress on IEEE International Conference on Evolutionary Computation.*
8. Rudolph, G.N., and A. Agapie. 2000. Convergence properties of some multi-objective evolutionary algorithms. In *Congress on Evolutionary Computation.*
9. Laumanns, M., et al. 2002. Combining convergence and diversity in evolutionary multiobjective optimization. *Evolutionary Computation* 10 (3): 263–282.
10. Zitzler, E., et al. 2003. Performance assessment of multiobjective optimizers: An analysis and review. *IEEE Transactions on evolutionary computation* 7 (2): 117–132.
11. Fonseca, C.M., and P.J.J.E.C. Fleming. 2014. An overview of evolutionary algorithms in multiobjective optimization. 3 (1): 1–16.
12. Goldberg, D.E. 1989. Genetic algorithms in search, optimization and machine learning. Addison-Wesley, Boston, MA.
13. Knowles, J., and D. Corne. 1999. The pareto archived evolution strategy: A new baseline algorithm for pareto multiobjective optimisation. In *Proceedings of Congress on Evolutionary Computation.*
14. Tao, J., Q. Fan., and Chen X, et al. 2012. Constraint multi-objective automated synthesis for CMOS operational amplifier.*Neurocomputing* 98: 108–113.
15. Michalewicz Z. 2013. Genetic algorithms + data structures = evolution programs[M]. Springer Science & Business Media.
16. Ren, L., et al. 2010. Emergence of self-learning fuzzy systems by a new virus DNA–based evolutionary algorithm. *International Journal of Intelligent Systems* 18 (3): 339–354.
17. Jan, H.Y., C.L. Lin, and Hwang, T.S. 2006. Self-organized PID control design using DNA computing approach. *Journal of the Chinese Institute of Engineers 29* (2): 251–261.
18. Tao, J., and N. Wang. 2007. DNA computing based RNA genetic algorithm with applications in parameter estimation of chemical engineering processes. *Computers Chemical Engineering*31 (12): 1602–1618.
19. Liepins, G.E. 1992. Global convergence of genetic algorithms. Proceedings of SPIE - The International Society for Optical Engineering 1766: 61–65.
20. Fonseca, C.M. and P.J. Fleming 2002. Multiobjective optimization and multiple constraint handling with evolutionary algorithms. II. Application example. *IEEE Transactions on Systems, Man Cybernetics, Part A 28* (1): 38–47.
21. Tan, K.C., Y.J. Yang, and T.H. Lee. 2006. A distributed cooperative coevolutionary algorithm for multiobjective optimization. *IEEE Transactions on Evolutionary Computation* 10 (5): 527–549.
22. Mkaouer, W., M, Kessentini., and Shaout A, et al. 2015. Many-objective software remodularization using NSGA-III. *ACM Transactions on Software Engineering and Methodology (TOSEM)* 24 (3): 1–45.

Chapter 5
Parameter Identification and Optimization of Chemical Processes

Because of the complex nonlinear characteristics of chemical processes, traditional numerical optimization algorithms generally cannot be used to solve the modeling and optimization problems. In this chapter, the estimation of model parameters for heavy oil thermal cracking is firstly solved by RNA-GA. Then, we use DNA-DHGA to solve the recipe optimization problem of gasoline blending with heavy nonlinear inequality constraints. DNA computing based GAs are efficient in solving the optimization problems in chemical processes.

5.1 Introduction

Complex chemical processes are often nonlinear with serious coupling of multiple inputs and multiple outputs. The precise mathematical models of chemical processes are required with higher production requirements of modern industry [1]. The estimation problem of the model parameters is actually approximated to the real industrial processes by using the sampling data and minimizing the modeling error. The heavy oil cracking chemical processes is chosen as an example of the chemical process modeling [5]. The structure of three lumping models has been obtained, and there are totally eight parameters to be estimated. Due to the nonlinear characteristics of the process, traditional parameter estimation algorithms, such as the least squares algorithm [6], maximum-likelihood method [7], etc., are not used to solve such parameter estimation problems. Similarly, the parameter estimation of the FCCU main fractionator was not easy to be solved by the traditional parameter estimation method, due to the severe coupling between multiple variables [8]. In Chap. 2, an RNA-GA is proposed based on DNA computing, which overcomes some shortcomings of SGA. In this chapter, RNA-GA is applied to solve the parameter estimation problem of the above two chemical processes [9]. Simulation results are illustrated to show the effectiveness and practicability of RNA-GA in chemical process parameter estimation.

J. Tao et al., *DNA Computing Based Genetic Algorithm*,
https://doi.org/10.1007/978-981-15-5403-2_5

Except for the system identification of chemical processes, the system parameter optimization of the chemical process, such as gasoline blending recipe optimization [10], is also quite challenging. The reformulated gasoline has strict regulations on the content of benzene and oxygen, Reid vapor pressure (RVP), olefin, and aromatics. Refineries have to reduce production costs while meeting the gasoline quality specifications for a higher profit. On the one hand, the refinery uses advanced equipment to improve the production efficiency and gasoline quality. On the other hand, the blending technology is utilized to blend the multi-component oil, especially low-grade gasoline according to a certain formula to produce high quality gasoline with the lowest cost and qualified quality.

Gasoline blending is the last step of the refinery, and the blending efficiency plays an important role in the economic benefits of production enterprises. The tank blending is usually applied in traditional gasoline blender. The octane number in different component gasolines should be known in advance, the proportion of each component gasoline can then be determined. Because of the complex blending effect of gasoline components, it is difficult to predict the octane number of the blended gasoline [11]. Therefore, the octane number of blended gasoline is usually much higher than the required index, which means that refineries have to bear unnecessary and considerable economic losses every year. In order to obtain the target octane number, the gasoline blending was simplified to a linear system, and GA was used to solve the gasoline blending formula [12]. The basic particle swarm optimization (PSO) algorithm and its improved algorithms were also used to solve the gasoline blending optimization problem [13, 14]. However, the octane number was still higher than the required index using the above methods.

Recently, many companies have done a lot of theoretical research and technology development work. For example, the BOSS (Blending Optimization Scheduling System) of Haverly Systems, can automatically calculate and optimize the blending formula, reduce the surplus of various indicators, and eliminate the re-adjustment. The software has been used in 16 refineries around the world. G-Spare software of Honeywell Profimatics is another blending monitoring system, which has been used in 12 blenders of 3 refineries. ABB Simcon has developed a blending control software, which combines online optimization control with off-line formula calculation. They provide automatic calculations of economically optimal gasoline recipes according to price analysis and refinery conditions [15, 16]. In China, many enterprises have developed gasoline blending systems, most of them are still at the level of off-line optimization [17].

The product oil indexes, such as octane number and vapor pressure, etc., are nonlinear functions of the gasoline components blending formula. Moreover, material balance constraints and index constraints have to be satisfied in the blending processes. Therefore, the calculation of the oil blending formula is a complex constrained nonlinear optimization problem. Conventional optimization algorithms are difficult to obtain satisfactory optimization solutions. The DNA-DHGA [20] in Chap. 3 is then applied to solve the problem.

5.2 Problem Description of System Identification

Due to the superior performances of RNA-GA, it is applied for model parameter estimation problem, which is given as follows:

$$y(t) = g(\mathbf{u}(t), \boldsymbol{\theta}) \tag{5.1}$$

where $y(t)$ is system output, $\mathbf{u}(t)$ is the system input vector, and $\boldsymbol{\theta} = [\theta_1, \theta_2, \ldots, \theta_k]^{\mathrm{T}}$ are the parameters to be estimated. The modeling error is a function of the system output $y(t)$ and the model output $\hat{y}(t)$ when the system input $\mathbf{u}(t)$ is given. Suppose the structure of the model g is known, the parameters $\boldsymbol{\theta} = [\theta_1, \theta_2, \ldots, \theta_k]^{\mathrm{T}}$ can then be estimated by minimizing the modeling error. Obviously, the parameter estimation problem is essentially an optimization problem, where the objective function can be defined as follows:

$$f(\boldsymbol{\theta}) = \sum_{t=0}^{n_s-1} |y(t) - \hat{y}(t)| \tag{5.2}$$

where n_s is the number of samples. The framework to solve the model parameter estimation problem can be illustrated by Fig. 5.1, where the RNA-GA algorithm is used to identify the estimated parameters with the objective function. Because of the nonlinearity and complication of chemical processes, traditional numerical optimization methods for parameter estimation are ineligible and not applied here. Here, the RNA-GA is used to solve the parameter estimation and optimization problems. The implementation process of RNA-GA can be referred to in Chap. 2. It's worth noting that n_s is crucial for the parameter estimation, because the larger n_s is, the more precise parameter estimation can be obtained, while the less n_s is, the more sensitive time delay becomes. Hence, n_s should be selected according to the specific issue.

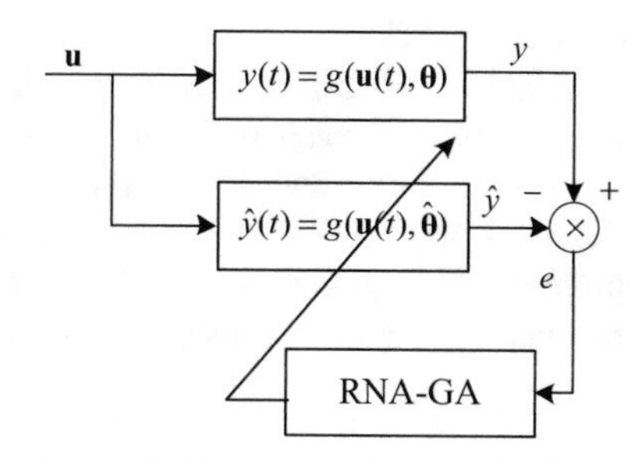

Fig. 5.1 Framework of solving the model parameter estimation

5.2.1 Lumping Models for a Heavy Oil Thermal Cracking Process

A heavy oil thermal cracking three lumping model has been described as follows [5]:

$$x_L = \frac{K_{\mathrm{LP0}}\mathrm{e}^{-E_{\mathrm{LP}}/T}}{n_L}[1-(1-z)^{n_L}] + \frac{K_{\mathrm{WP0}}K_{\mathrm{WLP0}}e^{-(E_{\mathrm{WP}}+E_{\mathrm{WLP}})/T}}{n_W - K_{\mathrm{WLP0}}e^{-E_{\mathrm{WLP}}/T}}\left\{\frac{1-(1-z)^{K_{\mathrm{WLP0}}e^{-E_{\mathrm{WLP}}/T}}}{K_{\mathrm{WLP0}}e^{-E_{\mathrm{WLP}}/T}} + \frac{1}{n_W}\left[(1-z)^{n_W}-1\right]\right\} \tag{5.3}$$

where z and T are input variables, and x_L is the system output. RNA-GA is used to estimate parameters in the above model$K_{\mathrm{LP0}}, K_{\mathrm{WP0}}, K_{\mathrm{WLP0}}, E_{\mathrm{LP}}, E_{\mathrm{WP}}, E_{\mathrm{WLP}}, n_L, n_W$. We chose 20 groups of data from [5], to estimate the unknown parameters. Here, the optimization objective Eq. (5.2) is changed as follows:

$$f = \sum_{i=0}^{n_s-1} |x_L(i) - \hat{x}_L(i)| \tag{5.4}$$

where $\tilde{x}_L$ is obtained by using Eq. (5.1). The bounds of the parameters to be estimated are set as follows: $K_{\mathrm{LP0}} \in (0, 10)$, $K_{\mathrm{WP0}} \in (0, 10)$, $K_{\mathrm{WLP0}} \in (0, 10)$, $E_{\mathrm{LP}} \in (800, 1500)$, $E_{\mathrm{WP}} \in (1500, 4000)$, $E_{\mathrm{WLP}} \in (1500, 4500)$, $n_L \in (0, 5)$, $n_W \in (0, 5)$. The maximum evolution generation of RNA-GA is set as 2000, and the other parameters of RNA-GA are the same as those in Chap. 2. Run the RNA-GA independently 50 times for the above model, and the best results are listed in Table 5.1, where the comparisons of the results between RNA-GA and SGA methods are also listed. The best fitness value (F_b) of SGA is also calculated with the same data as RNA-GA. We also compare the model outputs obtained by SGA and RNA-GA with the same training and testing data. The corresponding outputs are shown in Figs. 5.2 and 5.3.

All of the 56 groups of data provided by [5], are used as the test samples in order to verify the efficiency of the above model parameters. The model outputs are shown in Fig. 5.3. The value of the objective function of Eq. (5.4), is 1.2116 when using SGA, while it is 0.8759 when using RNA-GA. From Figs. 5.2 and 5.3, and their corresponding performance index listed in Table 5.1, we can find that the modeling precision of RNA-GA is superior to that of SGA.

Table 5.1 Results of parameter estimation by SGA and RNA-GA

Methods	K_{LP0}	K_{WP0}	K_{WLP0}	E_{LP}	E_{WP}	E_{WLP}	n_L	n_W	F_b	f (test data)
SGA	4.680	5.155	4.197	1257	1850	3776	1.191	1.488	0.4883	1.2116
RNA-GA	3.221	9.333	3.385	998.1	2779	3662	1.439	3.308	0.2815	0.8738

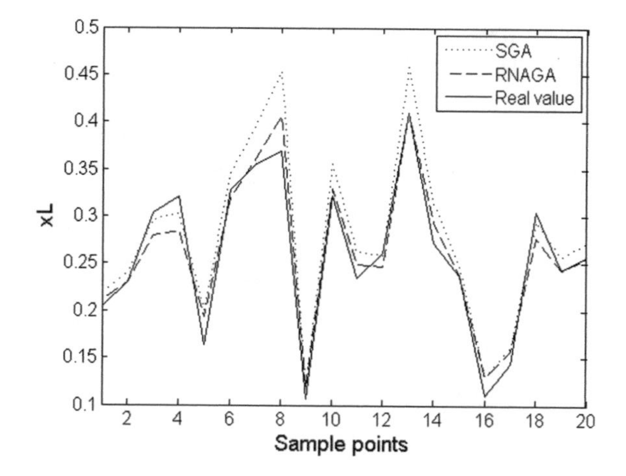

Fig. 5.2 Comparison of model outputs obtained using training data

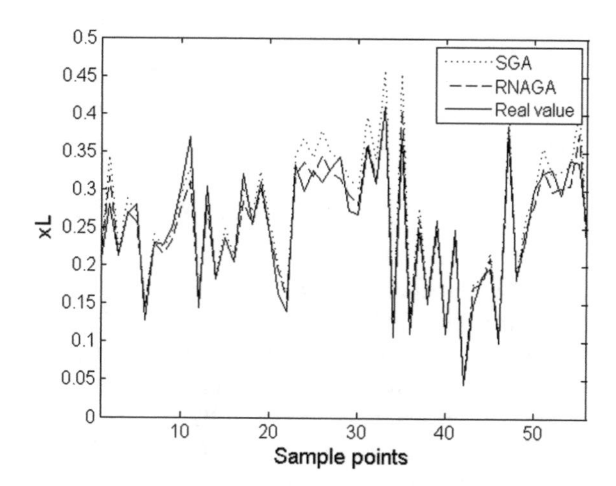

Fig. 5.3 Comparison of model outputs using testing data

5.2.2 *Parameter Estimation of FCC Unit Main Fractionator*

5.2.2.1 FCC Unit Description

Fluid catalytic cracking (FCC) is an important oil refinery process, which converts high molecular weight oils into lighter hydrocarbon products. It consists of the reactor–regenerator, the riser reactor, the main fractionator, the absorber-stripper-stabilizer, the main air blower, the wet gas compressor, etc. Among them, the main fractionator is the most important one to realize the advanced control of the FCC unit. The main fractionator configuration of a 140 wt FCC unit in an oil refinery factory is given in Fig. 5.4.

The configuration is shown as follows: FIC2207, top circulation flow controller; FIC2202, mid circulation flow controller 1. FIC2208, mid circulation flow controller

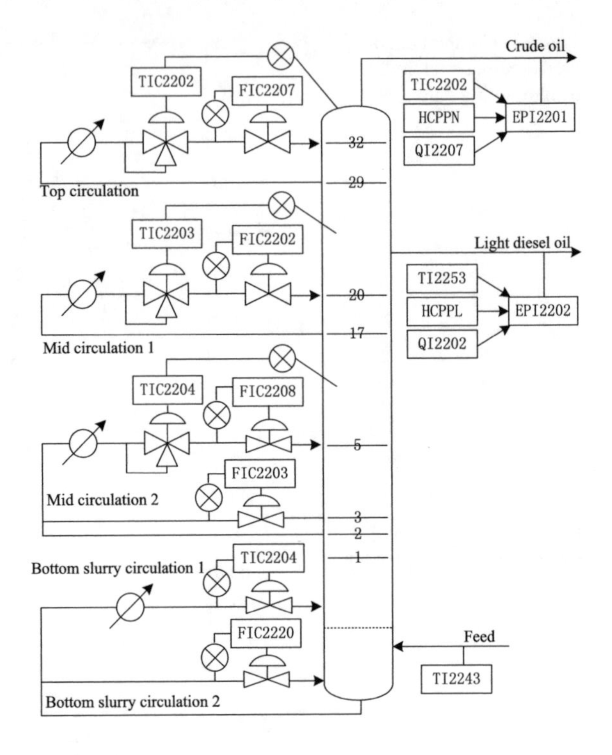

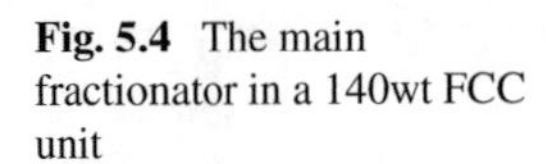
Fig. 5.4 The main fractionator in a 140wt FCC unit

2. FIC2203, mid circulation flow controller 3. FIC2204, bottom slurry circulation flow controller 1. FIC2220-bottom slurry circulation flow controller 2. TIC2202, top temperature controller; TIC2203, mid circulation temperature controller 1. TIC2204, mid circulation temperature controller 2. TI2243, feeding temperature; EPI2201, soft sensing for end point of crude oil; EPI2202, soft sensing for pour point of light diesel oil; QI2207, top circulation heat quantity; HCPPN, crude oil gas pressure; TI2253, tray 20 vapor temperature; QI2202, mid circulation heat quantity; HCPPL, light diesel oil pressure.

The oil quality is conventionally controlled by the top temperature and tray 20 vapor temperature (TI2253). However, the product quality cannot be precisely reflected by the controlled temperature. It mainly depends on the parting accuracy in the boiling range, such as the end point of crude oil and the pour point of light diesel oil. Therefore, the quality index of the main fractionator is not the top temperature or tray 20 vapor temperature but the end point of naphtha. An online soft sensing for the end point of naphtha has been successfully established by the members in our laboratory [8].

The 400 °C oil gas from the reactor is fed into the bottom of the main fractionator at tray 1 after heat removal. Once it is in contact with the 275 °C counter flow of slurry from top circulation, the oil gas is cooled down, and separated into gas, crude oil, light diesel oil, cycle oil, and slurry. In order to provide enough inner reflux and make the load distribution uniform, the fractionator contains four heat circulation systems, i.e., top heat removal circulation, the first mid heat removal circulation, the second mid heat removal circulation, and slurry heat removal circulation. In the slurry heat removal system, the slurry is extracted from the tower bottom and synchronously exchanges heat quantity with fuel oil. The slurry is then separated into two parts: one is the cycle slurry and the other is discharged from the fractionator. In the second heat removal circulation system, there exist three parts: the first part returns to tray 2 as an inner circulation; the second part returns to tray 5; the third part is extracted as the cycle oil. The first heat removal circulation system locates between tray 17 and tray 20, the light diesel oil is drawn out from tray 20. In the top heat removal circulation system, the oil gas is abstracted from tray 29 and returns to tray 32 when the temperature cools down to 80 °C. When the oil gas enters the tower top, the vapor phase and liquid phase (crude oil) can then be obtained.

5.2.2.2 Process Modeling

From the above analysis, the influencing factors of the end point are top temperature, top pressure, top heat removal, etc. The top heat removal is the main method to adjust the end point. The factors affecting the pour point are mainly top load changes, top pressure changes, first and second mid heat removal. Since most of the heat in the second mid circulation acts as the thermal resource of the bottom reboiler, only a small amount of heat is used to adjust the temperature of tray 1, and the first mid heat removal becomes the most important adjustment method for pour point. The change of heat removal is implemented by the change of flow or by the change of temperature. Therefore, the top circulation flow, the first mid flow, and the second mid flow are chosen as the manipulated variables, denoted as MV_1–MV_3; the top temperature, the end point of crude oil, and the pour point of light diesel oil are selected as controlled variables, denoted as CV_1–CV_3. To implement the advanced control of product quality for the main fractionator, the dynamic model is necessary to be established. As it is designed to be applied in industrial implementation, the model structure cannot be too complicated. It is supposed as the following discrete representation: $\frac{a(1)+a(2)z^{-1}}{1-bz^{-1}}z^{-d}$. Because the coupling mainly exists between CV_1 and CV_2, as well as CV_2 and CV_3, the MIMO process models of the main fractionator of the FCC unit can be simplified as shown in Table 5.2.

The multiple inputs multiple outputs (MIMO) process model is then obtained as follows:

Table 5.2 FCC unit main fractionator process models

	CV_1	CV_2	CV_3
MV_1	$\frac{a_{11}(1)+a_{11}(2)z^{-1}}{1-b_{11}z^{-1}}z^{-d_{11}}$	0	0
MV_2	$\frac{a_{21}(1)+a_{21}(2)z^{-1}}{1-b_{21}z^{-1}}z^{-d_{21}}$	$\frac{a_{22}(1)+a_{22}(2)z^{-1}}{1-b_{22}z^{-1}}z^{-d_{22}}$	0
MV_3	0	$\frac{a_{32}(1)+a_{32}(2)z^{-1}}{1-b_{32}z^{-1}}z^{-d_{32}}$	$\frac{a_{33}(1)+a_{33}(2)z^{-1}}{1-b_{33}z^{-1}}z^{-d_{33}}$

$$\mathrm{CV}_1(z^{-1}) = \frac{a_{11}(1) + a_{11}(2)z^{-1}}{1 + b_{11}z^{-1}} z^{-d_{11}} \mathrm{MV}_1(z^{-1}) + \frac{a_{21}(1) + a_{21}(2)z^{-1}}{1 + b_{21}z^{-1}} z^{-d_{21}} \mathrm{MV}_2(z^{-1}) \tag{5.5}$$

$$\mathrm{CV}_2(z^{-1}) = \frac{a_{22}(1) + a_{22}(2)z^{-1}}{1 + b_{22}z^{-1}} z^{-d_{22}} \mathrm{MV}_2(z^{-1}) + \frac{a_{32}(1) + a_{32}(2)z^{-1}}{1 + b_{32}z^{-1}} z^{-d_{32}} \mathrm{MV}_3(z^{-1}) \tag{5.6}$$

$$\mathrm{CV}_3(z^{-1}) = \frac{a_{33}(1) + a_{33}(2)z^{-1}}{1 + b_{33}z^{-1}} z^{-d_{33}} \mathrm{MV}_3(z^{-1}) \tag{5.7}$$

Since there exists the coupling of the estimated parameters in 5.5–5.7, these parameters are difficult to estimate. RNA-GA in Chap. 2 can be used to solve the complicated parameter estimation problem. The model of the main fractionator in a FCC unit was established according to the typical field data by the members of our laboratory [21]. Therefore, the input–output data were produced by the model provided in [22]. Moreover, the noise satisfying the practical conditions is added into the original data based on the knowledge of the refinery process unit.

5.2.2.3 System Parameter Estimation by RNA-GA

The objective functions for the above three models are listed as follows:

$$f_i = \sum_{k=1}^{n_s} |CV_i(k) - \hat{C}V_i(k)| \quad i = 1, 2, 3 \tag{5.8}$$

where $\hat{C}V_i$ is the model output. The outputs are generated by a group of step signals using the given model with a maximum divergence of $\pm 10\%$. The inputs and outputs are then normalized between 0 and 1. The parameter domain of the fraction equations is set as [−1, 1], and as it is a stable process, the range of denominator coefficient is reduced to [−1,0). The range of time delay is supposed as [1, 10].

All parameters of RNA-GA are kept unchanged as those in parameter estimation of the heavy oil thermal cracking three lumping model. RNA-GA for each f_i is also implemented for 50 independent runs. The results with the best values of the objective function are selected as the ultimate estimated parameters, they are listed in Table 5.3.

The model outputs of CV_1, CV_2, and CV_3, as well as the corresponding values of the real process, are shown in Figs. 5.5, 5.6, and 5.7. To verify the efficiency of the obtained model, another group of test data is selected, which is shown in Figs. 5.8, 5.9, 5.10. We can find that using RNA-GA to estimate the parameter is applicable in real processes, and the established model can make a good description of the

Table 5.3 Results of estimated parameters for main fractionator by RNA-GA

Models	Numerator	Denominator	Delay	F_b
CV_1MV_1	$a_{11} = [0.0904, -0.0045]$	$b_{11} = -0.9138$	$d_{11} = 6.0000$	1.1913
CV_1MV_2	$a_{21} = [0.0626, -0.0789]$	$b_{21} = -0.0012$	$d_{21} = 3.0000$	
CV_2MV_2	$a_{22} = [0.0626, -0.0196]$	$b_{22} = -0.4165$	$d_{22} = 8.0000$	0.8636
CV_2MV_3	$a_{32} = [0.4284, -0.3602]$	$b_{32} = -0.9254$	$d_{32} = 1.0000$	
CV_3MV_3	$a_{33} = [0.5171, -0.3801]$	$b_{33} = -0.8599$	$d_{33} = 3.0000$	0.5439

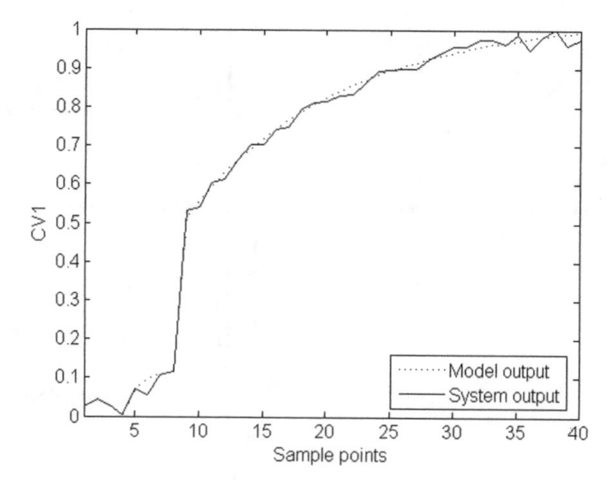

Fig. 5.5 Comparisons of model prediction and real value for CV_1

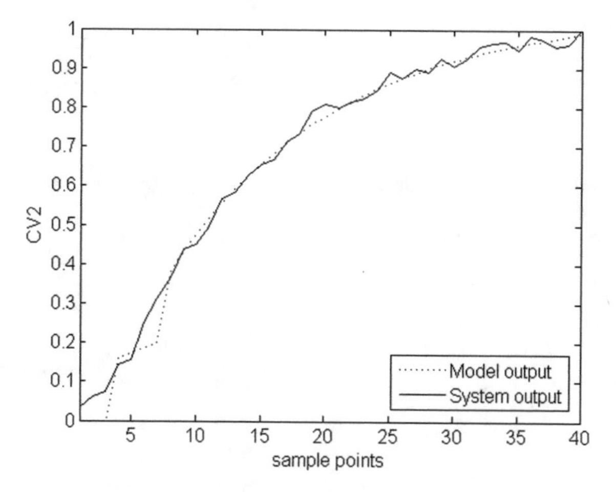

Fig. 5.6 Comparisons of model prediction and real value for CV_2

dynamic characteristic between manipulated variables and controlled variables. We can conclude that RNA-GA should be an effective and efficient approach for model parameter estimation of this kind of chemical process.

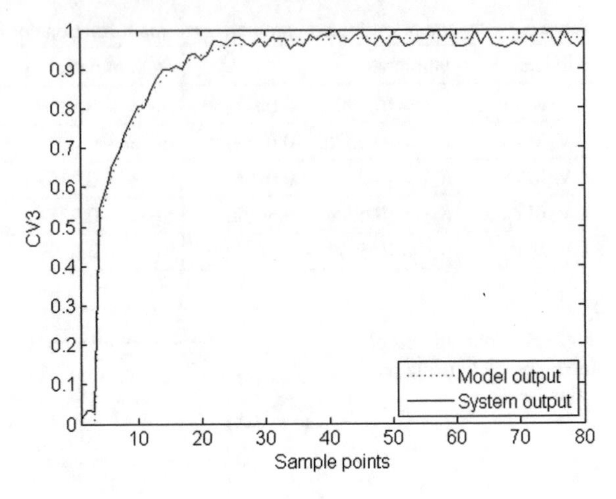

Fig. 5.7 Comparisons of model prediction and real value for CV_3

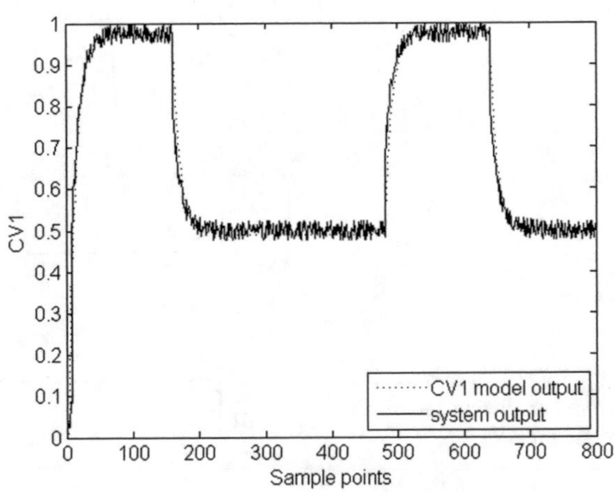

Fig. 5.8 Comparisons of model prediction and real value for CV_1 by test data

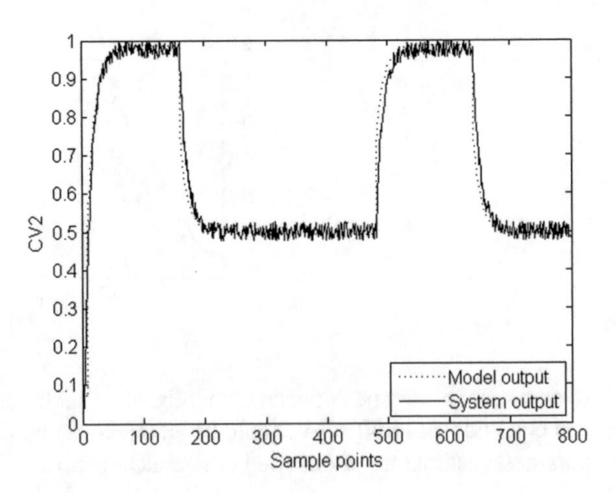

Fig. 5.9 Comparisons of model prediction and real value for CV_2 by test data

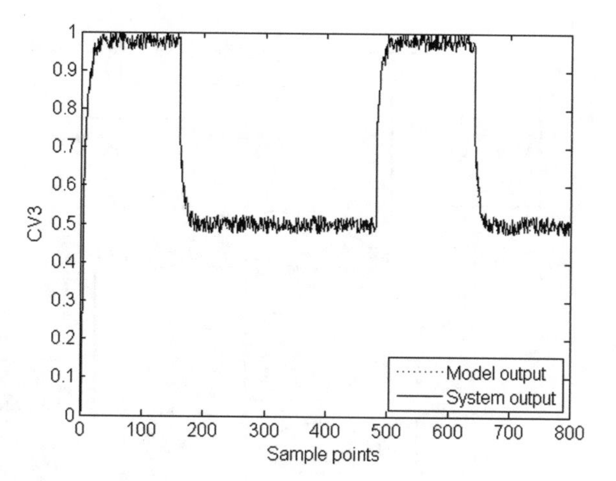

Fig. 5.10 Comparisons of model prediction and real value for CV_3 by test data

5.3 Gasoline Blending Recipe Optimization

In Chap. 3, a DNA-DHGA is described in detail. When the feasible domains are located, the sequential quadratic programming (SQP) method is applied to efficiently find the local optimum and improve the solution accuracy. The DNA-DHGA effectively alleviates the premature convergence and improves the weak exploitation capability of GA. The solving of gasoline blending recipe optimization problem will further demonstrate how to apply DNA-DHGA to solve the constrained industrial optimization problem efficiently.

5.3.1 Formulation of Gasoline Blending Scheduling

Various component streams are mixed in the gasoline blending process to produce an automotive gasoline product stream meeting certain quality specifications [13], as shown in Fig. 5.11.

The usual objective function of blending scheduling is to maximize the product profit given as follows:

$$\max \sum_{t=1}^{T} \sum_{n=1}^{N_p} (Cp_{n,t} Vp_{n,t} - \sum_{m=1}^{Nc_n} Cc_{m,t} Vc_{n,m,t}) \tag{5.9}$$

where the first part is the gasoline production value, the second part is the spent cost of blending components, T is the time scale of the blending scheduling, Nc_n is the component categories for product n, N_p is the product categories, $Vc_{n,m,t}$ is the volume of the component m in product n, $Vp_{n,t}$ is the blending volume in product n, $Cc_{m,t}$ is the cost of component m, and $Cp_{n,t}$ is the price of product n.

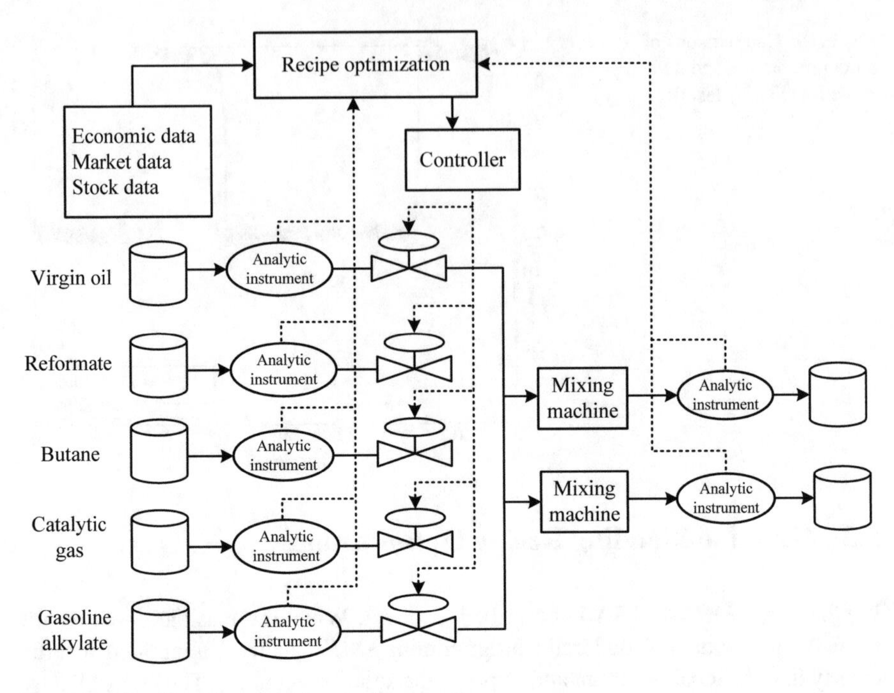

Fig. 5.11 Simplified flowsheet for gasoline blending

Specifications on gasoline qualities include octane number, volatility, sulphur content, aromatics content, RVP, and viscosity, etc. As the most important quality index of gasoline products, research octane number (RON), motor octane number (MON), and Reid vapor pressure (RVP) are constraints with nonlinear property in the process of blending scheduling. In contrast with other methods of predicting components of octane properties, the ethyl RT-70 models exhibit the best combination of predictive accuracy and parsimony for octane numbers [23]. Hence, the ethyl RT-70 models have been used to represent the mixing rules for octane numbers, which is shown as follows:

$$P_{\mathrm{RON}} = \mathbf{r}^T\mathbf{x} + \alpha_1\left(\mathbf{r}^T\mathrm{diag}(\mathbf{s})\mathbf{x} - \frac{(\mathbf{r}^T\mathbf{x})(\mathbf{s}^T\mathbf{x})}{\mathbf{e}^T\mathbf{x}}\right) + \alpha_2\left(\mathbf{o}_s^T\mathbf{x} - \frac{(\mathbf{o}^T\mathbf{x})^2}{\mathbf{e}^T\mathbf{x}}\right) + \alpha_3\left(\mathbf{a}_s^T - \frac{(\mathbf{a}^T\mathbf{x})^2}{\mathbf{e}^T\mathbf{x}}\right) \tag{5.10}$$

$$P_{\mathrm{MON}} = \mathbf{m}^T\mathbf{x} + \alpha_4\left(\mathbf{m}^T\mathrm{diag}(\mathbf{s})\mathbf{x} - \frac{(\mathbf{m}^T\mathbf{x})(\mathbf{s}^T\mathbf{x})}{e^T\mathbf{x}}\right) + \alpha_5\left(\mathbf{o}_s^T\mathbf{x} - \frac{(\mathbf{o}^T\mathbf{x})^2}{\mathbf{e}^T\mathbf{x}}\right) + \frac{\alpha_6}{10000(\mathbf{e}^T\mathbf{x})}\left(a_s^T - \frac{(\mathbf{a}^T\mathbf{x})^2}{\mathbf{e}^T\mathbf{x}}\right)^2 \tag{5.11}$$

where $\mathbf{x}$ is the feedstock flow rate, r is the RON of each component, m is the MON of each feedstock, $\mathbf{s} = \mathbf{r} - \mathbf{m}$, $\mathbf{o}$ is the olefin content of each feedstock, $\mathbf{o_s}$ is the square of the olefin content, $\mathbf{a}$ is the aromatics content of each feedstock, $\mathbf{a_s}$ is the square of the aromatics content, P_{RON} is the blended RON, P_{MON} is the blended MON, and a_i are the coefficients in the above model, where $a_1 = 0.03224$, $a_2 = 0.00101$, $a_3 = 0$, $a_4 = 0.0445$, $a_5 = 0.00081$, $a_6 = -0.00645$.

The RVP model to represent the blending process is the blending index approach which has the following form [24]:

$$P_{RVP} = \left(\sum_{i=1}^{n} u_i (\mathrm{RVP}_i)^{1.25} \right)^{0.8} \tag{5.12}$$

where n is the number of components in the blend, P_{RVP} is the blended RVP, u_i is the volume fraction of component i.

5.3.2 *Optimization Results for Gasoline Blending Scheduling*

To demonstrate the application process of the DNA-DHGA scheme to a large extent, we shall further apply it for the optimization of a short-term gasoline blending recipes in [13]. The blending process obtains two products using five components as shown in Fig. 5.11. The regular gasoline is derived using Reformate, LSR naphtha, n-Butane, and Catalytic gas, while the premium gasoline is produced using Reformate, LSR naphtha, n-Butane, Catalytic gas, and Alkylate. The blending product requires to have the following specifications: a minimum RON, a minimum MON, and a maximum RVP. The optimization objective is to maximize the profit of the gasoline blend recipe on the premise that the quality and quantity of their products are satisfied. The market demand for the production is listed in Table 5.4. The parameters of the refined oil and five components are listed in Tables 5.5 and 5.6, respectively. Equation 5.9, is then accordingly rewritten as

Table 5.4 Production requirement

Demand	Regular	Premium
Demand in the first day(ton)	3000	3500
Demand in the second day(ton)	3300	3300
Demand in the third day(ton)	2500	3300

Table 5.5 Refined oil parameters

Types	Cost(¥/ton)	Minimum RON	Minimum MON	Maximum RVP
Regular	1900	88.5	77.0	10.8
Premium	2200	91.5	80.0	10.8

Table 5.6 Component data

Feedstock	Reformate	LSR naphtha	n-Butane	Catalytic gas	Alkylate
RON	94.1	70.7	93.8	92.9	95.0
MON	80.5	68.7	90.0	80.8	91.7
Olefin (%)	1.0	1.8	0	48.8	0
Aromatics (%)	58.0	2.7	0	22.8	0
RVP/psi	3.8	12.0	138	5.3	6.6
Available (ton/day)	1700	700	300	4000	200
Cost (¥/ton)	1960	1500	600	1800	2100

$$\max\ f(\mathbf{x}) = 1900x_r + 2200x_h - 1960(x_{r1} + x_{h1}) - 1500(x_{r2} + x_{h2}) \\ - 600(x_{r3} - x_{h3}) - 1800(x_{r4} + x_{h4}) - 2100x_{h5} \quad (5.13)$$

where x_r is the yield of regular gasoline, x_h is the yield of premium gasoline, x_{ri} is the quantity of components used in the blending process for regular gasoline, x_{hi} is the quantity of components used in the blending process for premium gasoline. The equality constraints of material balance are shown as follows:

$$x_r = x_{r1} + x_{r2} + x_{r3} + x_{r4} \\ x_h = x_{h1} + x_{h2} + x_{h3} + x_{h4} + x_{h5} \quad (5.14)$$

Accordingly, the inequality constraints of material balance in Table 5.6 are described as follows:

$$0 \le x_{r1} + x_{h1} \le 1700 \\ 0 \le x_{r2} + x_{h2} \le 700 \\ 0 \le x_{r3} + x_{h3} \le 300 \\ 0 \le x_{r4} + x_{h4} \le 4000 \\ 0 \le x_{r5} \le 200 \quad (5.15)$$

According to Table 5.5, the inequality constraints for quality control are given as follows:

$$88.5 \le P_{r\mathrm{RON}} \le 90 \\ 77.0 \le P_{r\mathrm{MON}} \le 80 \\ 10.8 \le P_{r\mathrm{RVP}} \le 11 \\ 88.5 \le P_{h\mathrm{RON}} \le 90 \\ 91.5 \le P_{h\mathrm{MON}} \le 92 \\ 10.8 \le P_{h\mathrm{RVP}} \le 11 \quad (5.16)$$

It is noteworthy that to reduce the difficulty of solving the optimization problem, the recipe for premium gasoline with higher profit is first solved and the recipe of regular gasoline is optimized using the left components. Moreover, the left components are added to the components the next day. When DNA-DHGA is applied, the maximum evolution generation G is set as 1000, the population size N is initialized to 60, and the individual length is set to 40. The calculated results for three days are listed in Tables 5.7 and 5.8. The profit using DNA-DHGA is ¥5,317,630. The comparisons with the PSO algorithm were listed in Table 5.9, and the profit using PSO was ¥5,289,908 [24]. Obviously, the profit obtained by DNA-DHGA is more than PSO's. However, there still exists larger redundancy of the quality indicator for

Table 5.7 Component requirement calculated by DNA-DHGA

Time	Product	Reformate	LSR naphtha	*n*-Butane	Catalytic gas	Alkylate
First day	Regular	1044.0	105.0	63.7	1254.1	0.0
	Premium	34.1	592.4	67.0	2738.4	67.3
Second day	Regular	1158.8	93.8	74.6	1522.8	0.0
	Premium	75.9	598.6	55.4	2483.1	86.8
Third day	Regular	350.3	81.0	49.8	1576.1	0
	Premium	14.5	627.5	57.6	2417.6	134.1
	Remaining components	2422.4	1.7	531.9	7.9	311.8

Table 5.8 Quality indicators and profit of refined oil calculated by DNA-DHGA

Time	Product	RON	MON	RVP	Profit(¥)
First day	Premium	92.0224	80.0622	10.7155	1,819,730
	Regular	92.3884	80.9026	10.7615	
Second day	Premium	91.7619	80.0170	10.2946	1,748,840
	Regular	92.5109	80.9802	10.7995	
Third day	Premium	91.5117	80.0433	10.5995	1,749,060
	Regular	92.9466	80.9197	10.7208	

Table 5.9 Component requirement calculated by PSO algorithm

Time	Product	Reformate	LSR naphtha	n-Butane	Catalytic gas	Alkylate
First day	Regular	1652.9	644.7	64.4	637.7	0.3
	Premium	13.7	22	86.2	3362.2	15.9
Second day	Regular	1666.7	629.7	71.7	817.7	114.2
	Premium	0	36.9	80.7	3182.4	0
Third day	Regular	787.1	540	50.2	1122.7	0
	Premium	254.3	87.7	80.7	2877.3	0

the regular gasoline. The production of LSR naphtha and Catalytic gas is insufficient to meet the market demand according to the optimization result listed in Table 5.7, which is the key issue to improve the profit of the refinery factory.

5.4 Summary

In this chapter, RNA-GA is applied to the model parameter estimation of complex chemical processes with given model structures. In the parameter estimation of three lumped models, the errors of this model are less than those of SGA. In the model parameter estimation for the catalytic cracking main fractionation tower, satisfactory modeling precision is also obtained. However, the successful application of the above model parameter estimation problem mainly depends on the choice of the model structure. If the model structure is not appropriate, any algorithm can obtain satisfactory results. In the next chapter, we will apply neural networks to solve the modeling problem of chemical processes in the case of an unknown model structure.

Aiming at the difficulty of dealing with the nonlinear constrained problem of gasoline blending with traditional algorithms, DNA-DHGA in Chap. 3, is applied to solve the constraint optimization problem of oil blending formula. The optimization results illustrate that the algorithm can achieve the quality index of the refined oil, which can effectively avoid repeated blending.

Appendix

Modeling data of heavy oil thermal cracking

NO.	T/K	X	XL	NO.	T/K	X	XL
1	673	0.3122	0.2034	29	723	0.4371	0.2802
2	673	0.3138	0.2136	30	723	0.3637	0.2674
3	673	0.3879	0.2806	31	673	0.2075	0.1282
4	683	0.3348	0.2305	32	683	0.311	0.2276
5	683	0.3384	0.2497	33	703	0.3927	0.3049
6	683	0.4684	0.3693	34	683	0.222	0.1447
7	693	0.405	0.3041	35	683	0.2599	0.1822
8	695	0.3355	0.2369	36	698	0.2846	0.2045
9	703	0.4034	0.3209	37	723	0.342	0.2559
10	708	0.4266	0.3107	38	713	0.3306	0.2426
11	698	0.2674	0.1638	39	713	0.2062	0.1397
12	713	0.4514	0.3326	40	733	0.4585	0.2977

(continued)

(continued)

NO.	T/K	X	XL	NO.	T/K	X	XL
13	708	0.4534	0.326	41	733	0.472	0.3099
14	713	0.4552	0.3282	42	713	0.4447	0.3441
15	715	0.4075	0.2724	43	733	0.3729	0.2688
16	723	0.51	0.3555	44	733	0.4234	0.3083
17	725	0.5918	0.4101	45	663	0.1739	0.1064
18	733	0.574	0.3685	46	663	0.1893	0.1104
19	708	0.3637	0.2449	47	673	0.2169	0.1541
20	689	0.3562	0.2609	48	673	0.1619	0.1103
21	693	0.3397	0.2482	49	673	0.0649	0.0439
22	693	0.2381	0.1483	50	683	0.2512	0.1867
23	684	0.3031	0.1976	51	683	0.141	0.0983
24	703	0.5436	0.3831	52	693	0.2532	0.1833
25	703	0.3478	0.2346	53	703	0.3973	0.3018
26	708	0.4719	0.3222	54	713	0.4162	0.3266
27	713	0.4232	0.2945	55	723	0.4203	0.339
28	719	0.5367	0.3354	56	723	0.3011	0.251

References

1. Turton, R., R.C. Bailie, W.B. Whiting, et al. 2008. *Analysis, synthesis and design of chemical processes*. Pearson Education.
2. Thompson, M.L., and M.A. Kramer. 1994. Modeling chemical processes using prior knowledge and neural networks. *AIChE Journal* 40 (8): 1328–1340.
3. Li, C., C. Yang, and H. Shan. 2007. Maximizing propylene yield by two-stage riser catalytic cracking of heavy oil. *Industrial and Engineering Chemistry Research* 46 (14): 4914–4920.
4. He, Y.L., and Q.X. Zhu. 2016. A novel robust regression model based on functional link least square (FLLS) and its application to modeling complex chemical processes. *Chemical Engineering Science* 153: 117–128.
5. Song, X., et al. 2003. Eugenic evolution strategy genetic algorithms for estimating parameters of heavy oil thermal cracking model. *Journal of Chemical Engineering of Chinese Universities* 17 (4): 411–417.
6. Björck, Å. 1996. *Numerical methods for least squares problems*. Society for Industrial and Applied Mathematics.
7. Murshudov, G.N., A.A. Vagin, and E.J. Dodson. 1997. Refinement of macromolecular structures by the maximum-likelihood method. *Acta Crystallographica. Section D, Biological Crystallography* 53 (3): 240–255.
8. Zhong, X., and S. Wang. 1998. On-line soft sensing for end point of naphtha based on neural network. *Journal of Chemical Industry & Engineering* 49 (2): 251–255.
9. Tao, J., and N. Wang. 2007. DNA computing based RNA genetic algorithm with applications in parameter estimation of chemical engineering processes. *Computers & Chemical Engineering* 31 (12): 1602–1618.

10. Li, J., I.A. Karimi, and R. Srinivasan. 2010. Recipe determination and scheduling of gasoline blending operations. *AIChE Journal* 56 (2): 441–465.
11. Pasadakis, N., V. Gaganis, and C. Foteinopoulos. 2006. Octane number prediction for gasoline blends. *Fuel Processing Technology* 87 (6): 505–509.
12. Litvinenko, V.I., et al. 2002. Application of genetic algorithm for optimization gasoline fractions blending compounding. In *IEEE international conference on artificial intelligence systems.*
13. Zhao, X. 2010. *Blending scheduling based on particle swarm optimization algorithm.* In *2010 Chinese control and decision conference.* IEEE.
14. Pan, H., and L. Wang, 2006. Blending scheduling under uncertainty based on particle swarm optimization with hypothesis test. In 2006 *international conference on intelligent computing,* 109–120. Springer.
15. Kirgina, M.V., et al. 2014. Computer program for optimizing compounding of high-octane gasoline. *Chemistry Technology of Fuels Oils* 50 (1): 17–27.
16. Sakhnevitch, B., et al. 2014. Complex system for gasoline blending maintenance. *Procedia Chemistry* 10: 289–296.
17. Shang, B., L. Jiao, and Z. Shang. 2007. The optimum control design for gasoline piping automatic blending. *Automation in Petro-Chemical Industry* 1: 22–24.
18. Li, W., L. Shi, and C. Liang. 2009. Forecasting model of research octane number based on PSO-VB-LSSVM. *Chinese Journal of Scientific Instrument* 30 (2): 335–339.
19. Zhang, J., X. Luo, and L. Yang. 2006. Dvelopment of online optimization system of gasoline blending. *Computer Engineering and Applications* 42 (33): 216–218.
20. Tao, J., and N. Wang. 2008. DNA double helix based hybrid GA for the gasoline blending recipe optimization problem. *Chemical Engineering and Technology* 31 (3): 440–451.
21. Zhong, X., Q. Zhang, and S. Wang. 2001. Multivariable constrained generalized predictive control strategy for the FCCU main fractionator. *Control Theory & Applications* 18: 134–140.
22. Liu, Q. and Q. Kang. 1997. The DCS implementation of FCCU main fractionator tray 20 temperature feed forward control system. *Automation in Refined and Chemical Industry* 5: 29–32. (In Chinese).
23. Zhang, Y., D. Monder, and J.F. Forbes. 2017. Real-time optimization under parametric uncertainty: a probability constrained approach. *Journal of Process Control* 12 (3): 373–389.
24. Zhao, X., Research on Refinery Production Scheduling Problems. 2005, Zhejiang University.

Chapter 6
GA-Based RBF Neural Network for Nonlinear SISO System

Radial basis function (RBF) neural network is efficient to model nonlinear systems with its simpler network structure and faster learning capability. The temperature and pressure modeling of the coke furnace in an industrial coke equipment is not very easy due to disturbances, nonlinearity, and switches of coke towers. To construct the temperature and pressure models in a coke furnace, RBF neural network is utilized to improve the modeling precision. Moreover, the shortcoming of RBF neural network, such as over-fitting is overcome. Moreover, the improved RNA-GA, MOGA, and PCA-based NSGA-II are utilized to optimize both the structure and parameters of the RBF network. Encoding/decoding, genetic operations, and fitness functions are designed to obtain satisfying modeling performances. The industrial data sets in the industrial coke furnace are utilized to construct the RBF neural network model by using three modeling optimization strategies.

6.1 Introduction

Since Broomhead and Lowe, proposed0 Radial basis function (RBF) neural networks in 1988 [1], RBF networks have attracted a lot of interests to application research in various fields because of the partial response character of the neuron, better approximation capability, simpler network structure, and faster learning capability than other artificial neural networks (ANNs) [2–4]. However, how to design radial basis functions remains a critical issue for RBF networks. The number and parameters of radial basis functions control the structure complexity and the generalization capability of RBF networks. A RBF network with too few radial basis functions gives poor generalization on new data because of the limited flexibility, while a RBF network with too many radial basis functions yields poor generalization since it is too flexible and may fit the noise in the training data. The best generalization performance can be obtained via a compromise between the conflicting requirements of reducing prediction error while simultaneously decreasing model complexity [5, 6]. This trade-off

J. Tao et al., *DNA Computing Based Genetic Algorithm*,
https://doi.org/10.1007/978-981-15-5403-2_6

highlights the importance of optimizing the structure complexity of the RBF network to improve its generalization capability.

More specifically, the network structure of the RBF network needs to be given before training other parameters in the neural network. The procedures usually proceed in two steps: First, the centers of radial basis functions are determined by a clustering method; second, the final-layer weights are calculated by the least square method. Usually, an unsupervised method that is separated from the actual objective of minimizing the modeling error will be executed in the first stage. The structure optimization in the construction of the network is desirable, however, it is a rather difficult problem and cannot be easily solved by the standard optimization method [7].

An interesting alternative for solving this complicated problem can be offered by the recently developed evolutionary algorithms. Perhaps the most popular and successful strategies are the genetic algorithms (GAs), which have succeeded in the structure selection of several kinds of neural networks, such as, Back propagation (BP) neural networks [8, 9] and recurrent neural networks [10, 11], etc. As for RBF neural networks, Vesin et al. used a GA to solve the whole optimization problem of the RBF network, but the centers of the potential nodes were restricted among the training data set [12]. Esposito et al. employed a GA-based technique for the determination of the widths of Gaussian radial basis functions [13], while Sarimveis et al., utilized a GA approach for optimizing RBF network only based on prediction errors [14].

When RBF networks are used to model the nonlinear system, the learning algorithm of the RBF network to determine its structure and parameters is critical, because different learning algorithms have a great influence on the performances of the derived RBF-based models.

Studies on parameter learning algorithm and the network structure optimization have been developed in-depth. Huang et al., proposed a simple sequential learning algorithm for RBF neural networks, which is referred to as the RBF growing and pruning algorithm [15]. Du et al., proposed a multi-output fast recursive algorithm (MFRA) that formulates the construction of an RBF network as a linear parameter optimization problem [16]. Han et al., presented a flexible structural radial basis function (FS-RBF) neural network, which changed its structure dynamically in order to maintain the prediction accuracy [17]. Most previous algorithms will become inefficient with too large search space and trap into the local minimum.

Although GA is a global searching algorithm, it is challenged by its weak local-search capability and premature convergence. As such, some biological operations at the gene level are effectively adopted in SGA, and the global searching speed can be largely improved [18, 19]. Moreover, the pruning operation is introduced to simplify the structure of the RBF neural network. In addition, the fewest process variables for accurate modeling are often of great interest by means of the most relevant variables selection, thus, the modeling, control, optimization, and monitoring issues for quality improvement of industrial production will be much easier [20–22]. Hence, it is anticipated that prediction accuracy can be improved by variable selection techniques, which will reduce the model complexity and capture the nature of industrial processes better [23–26].

Research on various variable selection methods for ANNs has been developed continuously. Huang et al., utilized the least absolute shrinkage operator for the input variables selection of a multilayer perceptron neural network in nonlinear industrial processes [27]. A sequential backward multiplayer perceptron (SBS-MLP) was proposed to perform feature selection [28]. Souza et al., have considerably reduced the computational cost and improved the model accuracy by variable selection comparing with SBS-MLP [29]. Estévez et al., proposed an improved variable selection method by introducing the average normalized mutual information for the measurement of redundancy [30].

The variable selection using principal component analysis (PCA) has also been studied in recent years [31–33]. However, the principal components are obtained by the linear combination of all variables, which makes the interpretation of principal component variable quite difficult. Therefore, a variety of criterion functions, such as Similarity indices, RM criterion, RV criterion, Generalized Coefficient of Determination (GCD) criterion have been proposed for subset selection [34]. In addition, the heuristics algorithm [35], simulated annealing [36], stochastic approximation iteration [37], genetic algorithm [38], etc., have also been applied to select the variables. Though some variable selection methods are efficient in the literatures [31–35, 37–39], most of them are not included in system modelling, while the variable selection in neural network only considered modeling accuracy [27].

Coking is an important process to improve economic benefits and has been widely used for refineries [40, 41]. A coke unit usually consists of coke furnaces, fractionating towers and coke towers. The temperature control of the coke furnaces is one of the operation goals in the unit, due to the coke furnaces, fractionating towers and coke towers in a completed process stream with their dynamic characteristics interacting with one another, the tasks are complicated. For example, the temperature affects the coking rate in the tubes of coke towers, which in turn has an impact on the temperature in the furnace [42]. Modeling is very important for advanced controller design but is even more difficult in term of the nonlinear characteristics, time delay and other various disturbances, such as feeding quantity, feeding temperature, fuel amount, etc. One of the most serious disturbances is the switches of coke towers, which disturb the temperature periodically and cause severe temperature fluctuations.

In this chapter, the structure optimization is included, and the fitness value of each chromosome is calculated based on the prediction error and the structure complexity criterion. In order to simplify the optimization of the RBF network, thin-plate-spline function can be chosen as the radial basis function, which is not required to determine its widths. However, the Gaussian function may obtain better performance with suitable centers. Generally, the RBF centers are determined based on a self-organizing clustering process, such as k-means clustering, the nearest neighbor clustering. The application of the above algorithms requires the network structure to be selected through trial and error, and only the input data is considered. Herein, several RBF neural network optimization methods are given as follows:

First, a pruning operator is designed to simplify the RBFNN structure, and a RNA-GA is first developed to optimize the RBF neural network structure and its corresponding parameters of radial basis functions to improve the approximation and generalization performance of RBFNN for temperature modeling in a coke furnace [7].

Second, the structure of the input and hidden layers, the parameters of the Gaussian basis functions are encoded in a chromosome. The local search operator and the prolong operator are proposed to obtain multiple RBF neural network structure. And an improved MOEA is then designed for the RBFNN modeling of the chamber pressure [43].

Finally, PCA variable selection is combined with ANN for nonlinear system modeling, and an RV criterion function of PCA is used to select the effective variables. Since both RV criterion and modeling accuracy are considered, the multi-objective evolution algorithm (MOEA) is adopted. Among MOEAs, NSGA-II is adopted due to its popularity and efficiency in solving ANN optimization and modeling problems [44, 45]. Here, it is also used to solve the variable selection and ANN modeling problem.

6.2 The Coke Unit

The whole process flow is shown in Fig. 6.1. It consists of such equipment as one fractionating tower (T102), three coke furnaces (F101/1, 2, 3), and six coke towers (T101/1, 2, 3, 4, 5, 6). The detailed flow of each part of the unit is shown in Fig. 6.2, its main job is to coke residual oil. Take furnace (F101/3) as an example, the process flow is as follows: The flow of residual oil is divided into two branches (FRC8103, FRC8105) and sent into the convection chamber of the furnace (F101/3) to be heated to about 330 °C, then the two branches are combined and flow out of the radiation chamber of the furnace and go to the fractionating tower (T102) for heat exchange with gas oil from the coke towers (T101/5, 6). After heat exchange, the heavy part of both residual oil and the gas oil join together, which is called circulating oil. The circulating oil is then divided into two branches (FRC8107, FRC8108) by pumps (102/1, 2, 3) and returned to enter the radiation chamber of the furnace (F101/3) to be heated to about 495 °C. Finally, the two branches join together and go to the coke towers (T101/5, 6) to remove coke. This process is called the coking of residues. The flows of the other two furnaces are the same as that of the furnace (F101/3), but the corresponding coke towers are different. The coke towers (T101/1, 2) are for the furnace (F101/1) and (T101/3, 4) for the furnace (F101/2). Each time, only one of each pair of coke towers works for its corresponding furnace, and when it is full, the other one replaces it. This replacement is called the switch of coke towers and the procedure recycles. The switch time of three pairs of coke towers is different. The heat exchange with gas oil from the coke towers poses a continuous disturbance on the outlet temperature because of the volume of the gas oil from the coke towers. During the switch of the coke towers, the outlet temperature of the furnace often

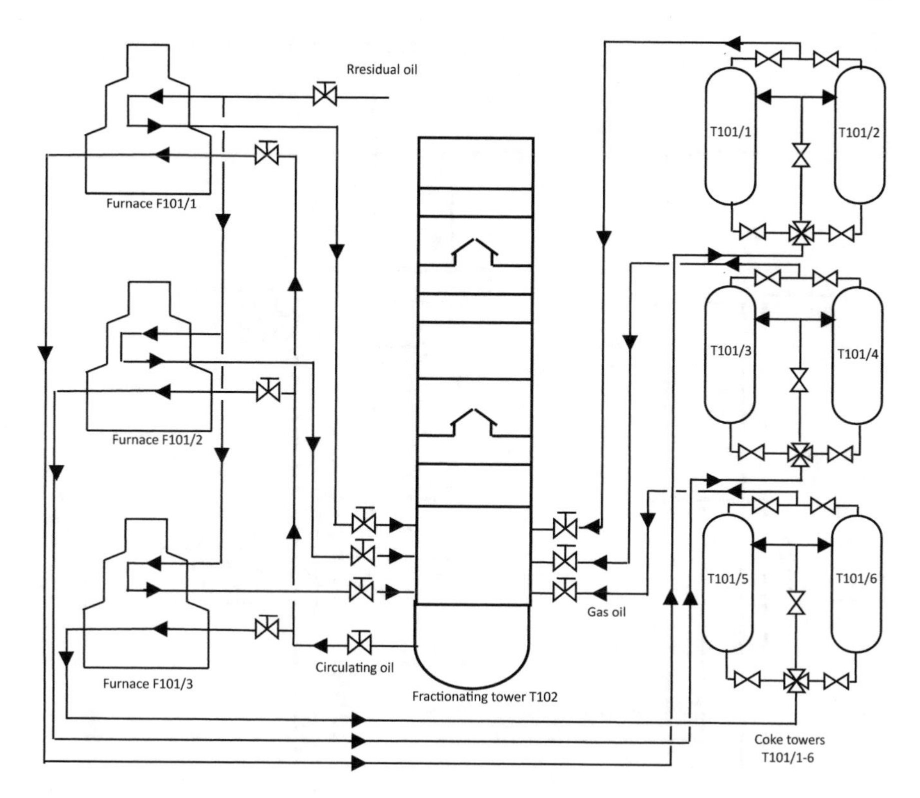

Fig. 6.1 Overall flow of coke unit

drops and rises sharply because some of the oil in it will become gas oil, and part of the inlet gas oil flowing into it will be used for the heating of coke towers. What's more, the random switch time of three pairs of coke towers adds to this serious problem.

The outlet pressure, temperature, and relevant variables are sampled using the experimental equipment CENTUM CS3000 Distributed Control System (DCS), as shown in Fig. 6.3. The DCS has a database, namely PAI database, for process data acquisition. To ensure the modeling precision and make the administrator conveniently analyze the process data, the sampling period 0.5 s is set in the PAI database, and 2 digits after the decimal point are retained in the sampling dataset.

6.3 RBF Neural Network

A schematic of the RBF network with n inputs and a scalar output is shown in Fig. 6.4.

In the RBF neural network (RBFNN), the function form $\varphi(\cdot)$ and the centers $\mathbf{c}_i$ are assumed fixed. Here we denote a set of the inputs $\mathbf{x}(k)$ as $\mathbf{x}(k) =$

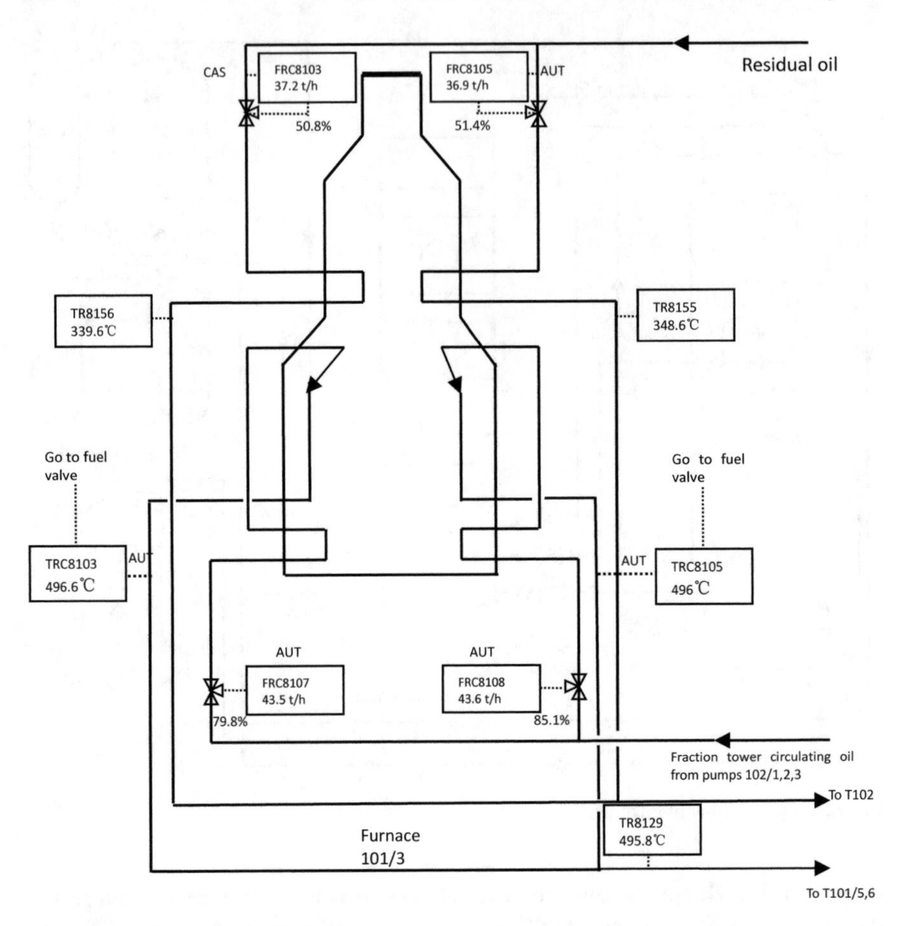

Fig. 6.2 Overall flow of coke furnace

Fig. 6.3 Data acquisition configuration for system outputs

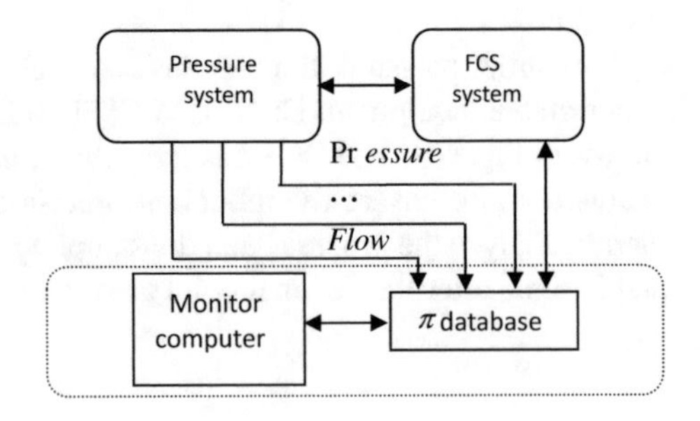

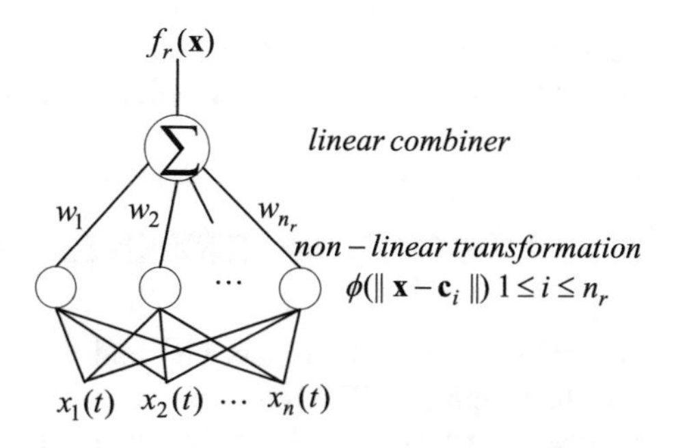

Fig. 6.4 Schematic of RBF network

$[y(k-1), \ldots, y(k-n), \mathbf{u}(k-1), \ldots, \mathbf{u}(k-m)]$, $\mathbf{u}(k)$ as the selected manipulated variables evaluated by RV criterion of PCA, $\hat{y}$ as the output of RBFNN, and $\boldsymbol{\omega} = [\omega_1, \ldots, \omega_{n_r}]$ as the weights between the hidden layer and the output layer, where n_r is the number of the nodes in the hidden layer. The choices of $\varphi(\cdot)$ and $\mathbf{c}_i$ must be carefully considered for the RBF neural network to obtain both the approximation capability and generalization performance. The thin-plate-spline function and the Gaussian function are two typical choices, both of them have obtained good approximation capabilities according to the fitting result of RBF networks [44].

Here, $\phi_i(\mathbf{x})$ is the ith neuron output in the hidden layer, which is selected as the Gaussian function or thin-plate-spline function:

$$\phi_i(\|x\|) = \exp\left(-\frac{\|x - c_i\|}{\sigma_i^2}\right) \text{ or } \phi_i(\|x\|) = \|x - c_i\|^2 \log\|x - c_i\|,\ i = 1,\ 2, \ldots,\ n_r \tag{6.1}$$

where $\|\mathbf{x} - \mathbf{c}_i\|$ is the Euclidean distance between $\mathbf{x}$ and $\mathbf{c}_i$, $c_i \in \Re^{n+m}$ is the center vector and $\sigma_i \in \Re$ represents the spread of radial basis function, respectively.

The prediction of RBFNN, $\hat{y}(k)$, can be expressed as a linear weighted sum of n_r hidden functions

$$y(\mathbf{x}(k)) = \sum_{i=1}^{n_r} \omega_i \phi_i(\|\mathbf{x}(k)\|) = \omega \Phi(k) \tag{6.2}$$

where $\mathbf{\Phi} = [\phi_1, \cdots, \phi_{n_r}]^T$. Given N_1 samples of training data, $\mathbf{Y}_1 = [y_1(1), \cdots, y_1(N_1)]$ and $\mathbf{U} = [\mathbf{u}(1), \cdots, \mathbf{u}(N_1)]$, the weight coefficients can be calculated by recursive least squares (RLS) method [20]

$$\begin{cases} \boldsymbol{\omega}(k) = \boldsymbol{\omega}(k-1) + \mathbf{K}(k)[f_i(k) - \mathbf{\Phi}^T(k)\boldsymbol{\omega}_i(k-1)] \\ \mathbf{K}(k) = \mathbf{P}(k-1)\mathbf{\Phi}(k)[\mathbf{\Phi}^T(k)\mathbf{P}(k-1)\mathbf{\Phi}(k) + \mu]^{-1} \\ \mathbf{P}(k) = 1/\mu[\mathbf{I} - \mathbf{K}(k)\mathbf{\Phi}^T(k)]\mathbf{P}(k-1) \end{cases} \tag{6.3}$$

where $0 < \mu < 1$ is the forgetting factor, $\boldsymbol{P}(k)$ is a positive definite covariance matrix, $\boldsymbol{P}(0) = \alpha^2\boldsymbol{I}$, and $\boldsymbol{I}$ is an $(n+m) \times (n+m)$ identity matrix, α is a sufficiently large real number set to 10^5 and $\boldsymbol{\omega}(0) = \boldsymbol{\varepsilon}$, and $\boldsymbol{\varepsilon}$ is a sufficiently small $n + m$ real vector as set to 10^{-3}, $\mathbf{K}(k)$ is a weight matrix.

6.4 RNA-GA Based RBFNN for Temperature Modeling

By giving a set of the inputs $\mathbf{x}(t)$ and the corresponding output $y(t)$ for $t = 1$ to N_1, the weights of RBFNN can be derived using RLS method in 0.3. However, the number of neuron nodes in the input and hidden layer will determine the structure complexity of RBFNN, and the parameter selection of radial basis function is quite important in order to obtain a good approximation capability. The better modeling capability with a simpler structure was tried to be obtained by an improved RNA-GA. Since the encoding/decoding method and the genetic operations will affect the efficiency of GA, this section is focused on the optimization of the RBF network by the RNA-GA.

6.4.1 Encoding and Decoding

Select the Gaussian function as the radial basis function, σ_i, $\mathbf{c}_i$, and the number of hidden nodes of RBFNN of the lth chromosome is shown as follows:

$$\mathbf{C}_l = \begin{bmatrix} c^l_{1,1} & c^l_{1,2} & \cdots & c^l_{1,n} & \sigma_1 \\ c^l_{2,1} & c^l_{2,2} & \cdots & c^l_{2,n} & \sigma_2 \\ \vdots & \vdots & \ddots & \vdots & \vdots \\ c^l_{n_r,1} & c^l_{n_r,2} & \cdots & c^l_{n_r,n} & \sigma_{n_r} \\ 0 & 0 & 0 & 0 & 0 \\ \vdots & \vdots & \vdots & \vdots & \vdots \\ 0 & 0 & 0 & 0 & 0 \end{bmatrix} \tag{6.4}$$

where $l = 1, 2, \ldots N$, N is the population size, n_r is generated randomly between 1 and D, D is the maximal number of hidden neurons. The rows between $[n_r + 1, D]$ are set to zeros and do not correspond to a center. The number of neurons in the input layer (n) is generated randomly between 2 and 5. Since the choice of the input layer is limited, it then uses the enumerated method during the optimization process. There are entirely $D \times (n + 1)$ real parameters to be optimized in the RBFNN, which means one chromosome should represent $D \times (n + 1)$ real number. The elements of $\mathbf{C}_l$ are then encoded by 0123/CUAG as shown in Fig. 6.5.

The parameters of the chromosome can be decoded by using the following equations:

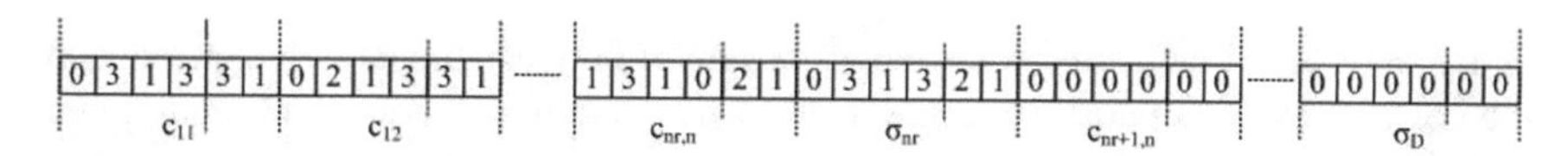

Fig. 6.5 Quanternary encoding for $\mathbf{C}_l$

$$c_{ij} = x_{j,\min} + \frac{x}{4^L - 1} \cdot \left(x_{j,\max} - x_{j,\min}\right), \quad 1 \le i \le n_r, \quad 1 \le j \le n \tag{6.5}$$

$$\sigma_j = \frac{x}{4^L - 1} w_{\max} \tag{6.6}$$

where x is the integer decoded by quaternary encoding with the encode length L, $x_{j,\min}$ and $x_{j,\max}$ are the minimum and maximum values of input variables given in the problem, $w_{\max}$ is the maximum width of Gaussian basis function.

6.4.2 Fitness Function

The training procedures using the improved RNA-GA (IRNA-GA) are processed in two steps. First, the network structure and the parameters of radial basis functions are determined by the chromosomes in an individual. Second, the final-layer weights are calculated by the RLS method, because IRNA-GA is a random search algorithm, and the constructed RBFNN system maybe ill-conditioned. Hence, the ordinary least square method cannot be applied in the optimization procedure. At each generation of IRNA-GA, the calculation of the weights in the output layer completes the formulation of N RBFNN, which can be expressed by the pairs $(\mathbf{C}_1, \ \mathbf{w}_1), \ \ (\mathbf{C}_2, \ \mathbf{w}_2), \ldots (\mathbf{C}_N, \ \mathbf{w}_N)$.

To obtain good generalization capability of RBF network, the sampled data set is divided into 3 groups, where one group of data subset $(\mathbf{X}_1, \mathbf{Y}_1)$ are used to calculate the weights of the final layer, the second group $(\mathbf{C}_1, \boldsymbol{\omega}_1)$ are utilized to evaluate the modeling performance of RBFNN at each generation, and the third group $\cdots(\mathbf{C}_N, \ \boldsymbol{\omega}_N)$ is used to verify the modeling performance of the optimal RBF network. This scheme incorporates a testing procedure into the training process and ensures good generalization performance of RBFNN. However, to obtain a better approximation capability with a simpler structure and avoid neural network overfitting, the objective function considering both the approximation capability and structure complexity is shown as follows:

$$J(\mathbf{C}_i, \mathbf{w}_i) = \sum_{t=1}^{N_1} |Y_1(t) - \hat{Y}_1(t)|^2 + \sum_{t=1}^{N_2} |Y_2(t) - \hat{Y}_2(t)|^2 + \lambda(n_r + n) \tag{6.7}$$

It can be seen that a compromise has been made between the modeling errors and the complexity of network structure. Here, λ is a coefficient between 0 and 1, and the bigger λ is, the more complicated the structure of RBFNN.

6.4.3 Operators of RBFNN Optimization

Li et al. has summarized various operations of DNA computing, such as elongation operation, deletion operation, absent operation, insertion operation, translocation operation, transformation operation, and permutation operation, etc. [46]. In addition to selection, crossover, and mutation operators, other appropriate operations of DNA computing can also be adopted to improve the performances of RBFNN modeling.

(1) Selection operator

A set of individuals from the previous population must be selected for reproduction depending on their fitness values. Individuals with bigger fitness value have more probability to survive. There exist several types of selection operators, and Roulette wheel method is applied to produce the parents of crossover and mutation operators. The probability of an individual being selected, $P(\mathbf{C}_i)$, is given by

$$P(\mathbf{C}_i) = \frac{f(\mathbf{C}_i)}{\sum_{l=1}^{N} f(\mathbf{C}_i)} \tag{6.8}$$

where $f\mathbf{C}_i$ is the fitness value of an individual $\mathbf{C}_i$ by using reciprocals of Eq. (6.7), i.e., $1/j(\mathbf{C}_i\mathbf{W}_i)$. The roulette wheel is placed with N equally spaced pointers. A single spin of the roulette wheel will simultaneously pick N individuals of the next population.

(2) Crossover operator

The crossover operator is executed with the crossover probability p_c among the selected individuals, and generates new structure and the parameters of RBFNN. If the randomly generated number is less than p_c, crossover operation is carried out between the current chosen individual $\mathbf{C}_l$ and the next individual $\mathbf{C}_{l+1}$, and yields the offspring chromosomes $\mathbf{C}'_l, \mathbf{C}'_{l+1}$. Since the number of input neurons n is fixed during an optimization process, the procedure is illustrated with an example presented in Fig. 6.6, which includes a scheme of the multi-point crossover operation, where the crossover points are generated randomly between 1 and L. The operator is prone to generate more hidden neurons, e.g., after the crossover of $c_{\mathrm{nr}+1,\mathrm{n}}$ of $\mathbf{C}_l$ and $c_{\mathrm{nr}+1,\mathrm{n}}$ of $\mathbf{C}_{l+1}$, the new nonzero chromosomes are generated, and the number of the hidden nodes in $\mathbf{C}'_l$ becomes $n_r + 1$.

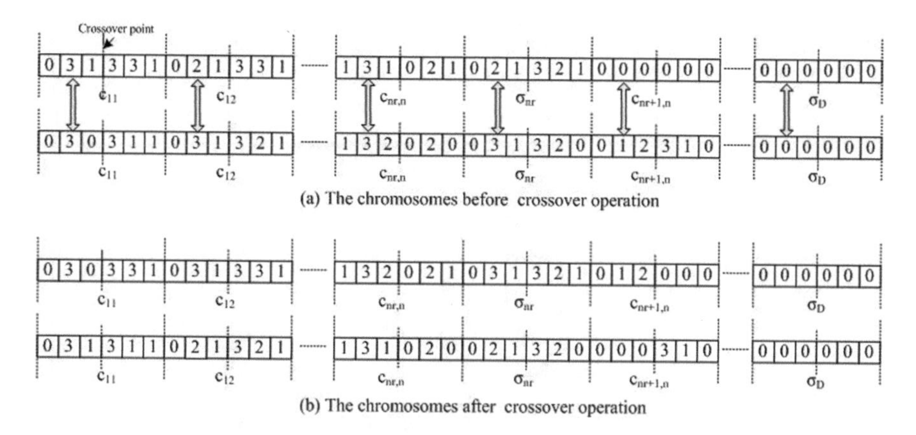

Fig. 6.6 Example of the crossover operation

(3) Mutation operator

To have a better exploration of the search space, the mutation operator is implemented. Because there exist four elements (0123/CUAG) in RNA sequence, the mutation of the nucleotide base is relatively complex. Three mutation operations on a single RNA sequence, i.e., reversal, transition, and exchange operations are adopted. The reversal operator makes $0 \leftrightarrow 2$, $1 \leftrightarrow 3$, transition operator makes $0 \leftrightarrow 1$, $2 \leftrightarrow 3$, and exchange operator makes $2 \leftrightarrow 1$, $0 \leftrightarrow 3$. When the element of an individual is mutated with a probability p_m, three mutation operators are executed simultaneously. This will generate more than N individuals after mutation operators, but the population size still remains invariant after selection operator.

The mutation probability is critical and generally small since too large mutation probability makes RNA-GA become a random search algorithm. At the beginning stage of the evolution process, larger probability of mutation is assigned to explore the larger feasible region. When the region of the global optimum is found, the mutation probabilities are reduced to prevent better solutions from disruption. Therefore, the dynamic mutation probability p_m is described as follows:

$$p_m = a_0 + \frac{b_0}{1 + e^{aa(g - g_0)}} \tag{6.9}$$

where a_0 denotes the initial mutation probability of p_m, b_0 is the variation range of mutation probability, g is the evolution generation, g_0 decides the generation where a great change of mutation probability occurs, and aa denotes the speed of change. The coefficients of Eq.(6.9) are selected as follows: $a_0 = 0.02$,$b_0 = 0.2$, $g_0 = G/2$, $aa = 20/G$. Let G be 1000, the probability curve $\mathbf{C}_l$ changing with evolution generation is shown in Fig. 6.7.

After calculating the mutation probability in terms of 0.9, $L \times N$ decimal fractions between 0 and 1 are produced compared with the above dynamic mutation probability.

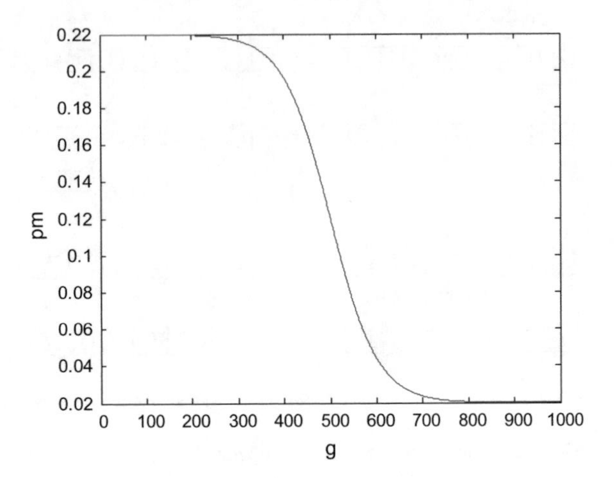

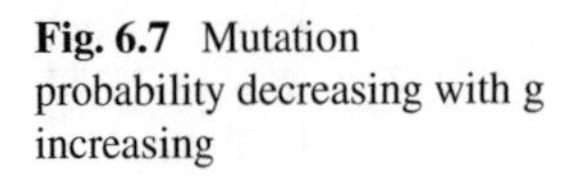
Fig. 6.7 Mutation probability decreasing with g increasing

If the decimal fraction is less than the corresponding probability in Fig. 6.7, 3 RNA mutation operators are executed meanwhile, and 3 new individuals will be produced.

(4) Pruning operator

Since the chromosomes are generated randomly, the effectiveness of every hidden neuron is evaluated in terms of the active firing (AF) of the hidden neurons, which is described as follows [17]:

$$Af_i = \rho e^{-\|\mathbf{x}-\mathbf{c}_i\|} \frac{\phi_i(\mathbf{x})}{\sum_{i=1}^{n_r} \phi_i(\mathbf{x})}, i = 1, \ldots, n_r \tag{6.10}$$

where Af_i is the active firing of the ith hidden neuron, $\phi_i(\mathbf{x})$ is the output of the ith hidden neuron, $\rho > 1$ is a positive constant, which is set as 100. When Af_i is less than the activity threshold Af_o ($0.05 < Af_o < 0.3$), the hidden neuron i is regarded as an inactive neuron. The number of the hidden neurons (n_r) will be decreased and the corresponding $\boldsymbol{c}_i$ is moved to the last location of $\boldsymbol{c}_{nr}$, its values of chromosomes are then set to zeros.

6.4.4 Procedure of the Algorithm

The fitness function evaluation, selection, crossover, mutation, and pruning operators are described for RNA-GA to be appropriate to optimize RBFNN, the running procedure is given in the following steps.

Step 1: Generate input layer with n inputs, $\boldsymbol{x}$(t), which consists of $\langle n/2 \rangle$ system input (u) and $n-\langle n/2 \rangle$ previous values of system output (y). Here $\langle \cdot \rangle$ is to round the elements to the nearest integer. As an example of 3 inputs, 2 inputs $u(k)$, $u(k-1)$ and 1 previous system output $y(k-1)$ are produced, that is, $\boldsymbol{x}(t) = [u(k), u(k\text{-}1), y(k\text{-}1)]$.

Step 2: Generate randomly N quaternary encoding chromosomes with a length of $D \times L$ in the search space, where N is the population size.

Step 3: Decode and compute the performance J of each individual.

Step 4: Select the chromosomes to generate N new chromosomes as the parents of the next generation by tournament selection operator. Before selection operator, the best $\langle 3\ N/4\rangle$ individuals and the worst $\langle N/4\rangle$ individuals are derived to make up of N individuals to keep population diversity.

Step 5: Judge if the crossover probability is satisfied, if yes, select one point randomly in l quaternary genes, and totally Dn points are generated as shown in Fig. 6.6, and exchange the codes of $\mathbf{C}_l$ and the next individual $\mathbf{C}_{l+1}$. Repeat this for all the $p_c \times N/2$ pairs of parents produced at step 4.

Step 6: For effective mutation, execute 3 RNA mutation operators once the random number is less than the dynamic mutation probability in 0.7, and this step may generate the individuals more than N.

Step 7: If the number of individuals is greater than N, the pruning operator is performed to improve the quality of RBFNN, else the pruning operator is not carried out.

Step 8: Repeat steps 3–7 until a termination criterion is met, that is, the maximal evolution generation (G). Moreover, elitism, the inclusion of the best individual in the next population is used throughout the optimization procedure.

Step 9: Increase the number of the input nodes and repeat steps 2–8. Choose the best RBFNN in terms of the value of an objective function using the test data set $(\mathbf{X}_3, \mathbf{Y}_3)$.

6.4.5 *Temperature Modeling in a Coke Furnace*

Advanced temperature control is critical for the coke unit and the first important issue to advanced controller design is system modeling. In this section, RBFNN optimized by the IRNA-GA is used to construct the north and south sides of the temperature models and the main disturbances in the coking furnace.

The experimental data are collected from the industrial coking unit of a refinery controlled by CENTUM CS3000, which is described in Sect. 6.2. The temperature is measured by thermocouple with the measuring precision ± 1.5°C. The flow rate is measured by the mass flowmeter. There are totally 1350 data sampled from PAI database of the control system. Each measurement sample includes four inputs and four outputs, that is, the north side primary channel model of the outlet temperature (TRC8105) and the input fuel flow (FRC8105), its disturbance channel model of the perturbation of FRC8105 and its corresponding temperature perturbation of TRC8105, the south side primary channel model of the outlet temperature (TRC8103) and the input fuel flow (FRC8103), and its disturbance channel model of the perturbation of FRC8103. Four groups of input and output data are plotted from Fig. 6.8a–d,

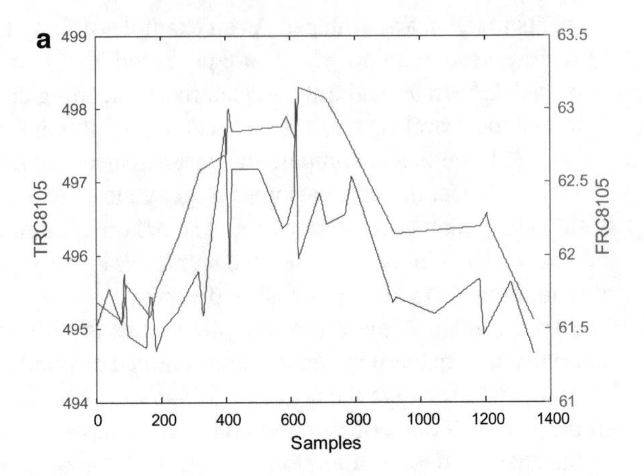

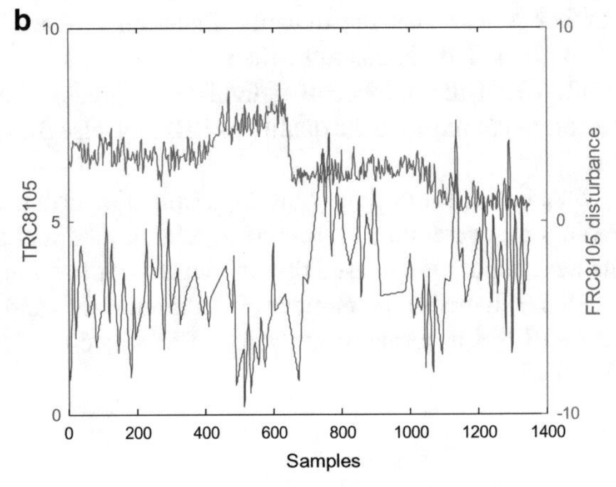

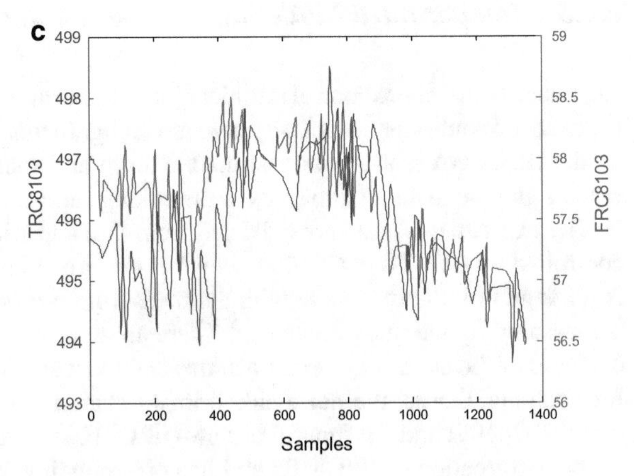

Fig. 6.8 **a** Input FRC8105 and output TRC8105 for north side primary channel modeling. **b** Input FRC8105 perturbation and output TRC8105 for north side disturbances modeling. **c** Input FRC8103 and output TRC8103 for south side primary channel modeling. **d** Input FRC8103 perturbation and output TRC8103 for south side disturbances modeling

Fig. 6.8 (continued)

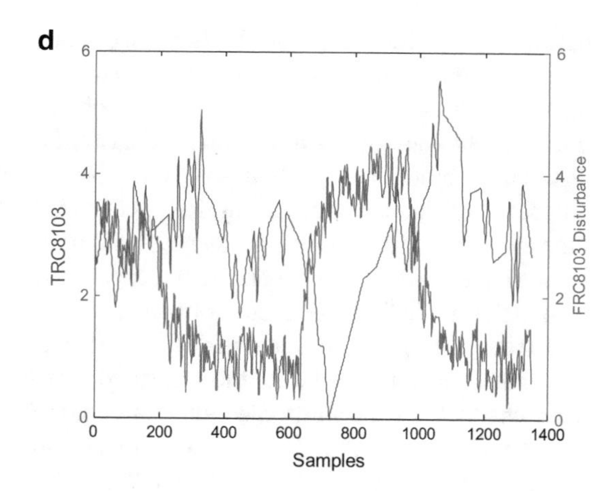

where the x-axis is the number of samples, y-axis labeling on the left is the system output, and the right one is the system input.

All collected 1350 samples are divided into three groups. The first group of 450 samples is selected as the training set, and the intermediate 450 samples are used to verify the generalization capability of RBFNN, the remaining 450 samples are used as the final testing set. Based on the three sets of data, the IRNA-GA is employed to optimize the structure and parameters of RBFNN by minimizing 0.7. Here, the parameters of the IRNA-GA are set as follows: the population size N is 60, the maximal evolution generation G is 1000, the individual length L is $3 \times D$, the probability of crossover operator p_c is 0.6, the mutation probability p_m is dynamically changed according to Eq. (6.9), the activity threshold Af_o is 0.1, and λ is 0.3. To examine the generalization capability of the constructed model, the trained RBFNN is used to predict the coke temperature yield of the testing samples, which are not included in the training data. In addition, for validation of the effectiveness of the random optimization algorithm, RBFNN is trained for 10 times. At each time, the parameters of IRNA-GA and data set are kept invariant. The best results are listed in Table 6.1, where e_1 is Root Mean Squared Error (RMSE) of the testing data.

The IRNA-GA is compared with the k-means method, which is used to train the centers of the RBF network. The pruning operator is also applied and final-layer weights are derived using the RLS method, and the number of the input nodes is the same as the optimized RBFNN. The maximal number of hidden neuron nodes is set to 38, which is also obtained based on the maximal number of hidden nodes

Table 6.1 The simulation results comparison with 2 methods

Methods	TRC8105			TRC8105 disturbances			TRC8103			TRC8103 disturbances		
	n_1	n_2	e_1	n_1	n_2	e_1	n_1	n_2	e_1	n_1	n_2	e_1
IRNA-GA	4	32	0.0094	3	38	0.0439	4	31	0.0305	3	28	0.0813
k-means	4	38	0.0584	3	38	0.3245	4	38	0.2707	3	38	0.0866

optimized by IRNA-GA. From Table 6.1, it can be seen that the best results of IRNA-GA can obtain better prediction precision than using the k-means method in terms of e_1. Moreover, RBFNN using IRNA-GA can obtain smaller errors with fewer hidden nodes for the four groups of the testing dataset. Though the RMSE of the TRC8103 disturbance model using IRNA-GA is similar to that of the k-means method, the number of the hidden nodes using IRNA-GA is reduced greatly. All the results in 10 runs are superior to those of the k-means method, because the RBFNN with fewer hidden nodes gains better generalization capability. The simpler structure of RBFNN with higher modeling precision is obtained after running IRNA-GA.

To reflect the prediction accuracy of the established RBFNN model, the predicted temperature is compared with the measured temperature on the testing set for the main channels of the north side and south side (TRC8105, TRC8103) and their disturbance channels, the comparison results are given in Figs. 6.9, 6.10, 6.11, 6.12, 6.13, 6.14, 6.15, 6.16, 6.17, 6.18, 6.19, 6.20, 6.21, 6.22, 6.23, and 6.24, respectively.

Figure 6.9 shows the predicted yields comparing with measured outputs on the testing set by IRNA-GA, while the corresponding prediction error is plotted in Fig. 6.10. Figure 6.12 shows the fitting curve of the prediction outputs and the measured outputs using the k-means method, and the estimation errors are given in Fig. 6.11. Comparing Figs. 6.9 with 6.11, it can be seen that the maximal modeling error obtained by the k-means method is several times larger than that obtained by IRNA-GA. Similar results can be observed by comparing with the modeling error of TRC8103 and their disturbance models, which are shown Figs. 6.14, 6.16, 6.18, 6.20, 6.22, and 6.24, respectively, it can be seen from Figs. 6.9, 6.10, 6.11, 6.12, 6.13, 6.14, 6.15, 6.16, 6.17, 6.18, 6.19, 6.20, 6.21, 6.22, 6.23, and 6.24 that the IRNA-GA optimal RBFNN modeling approach has obtained considerably smaller modeling error with simpler network structure.

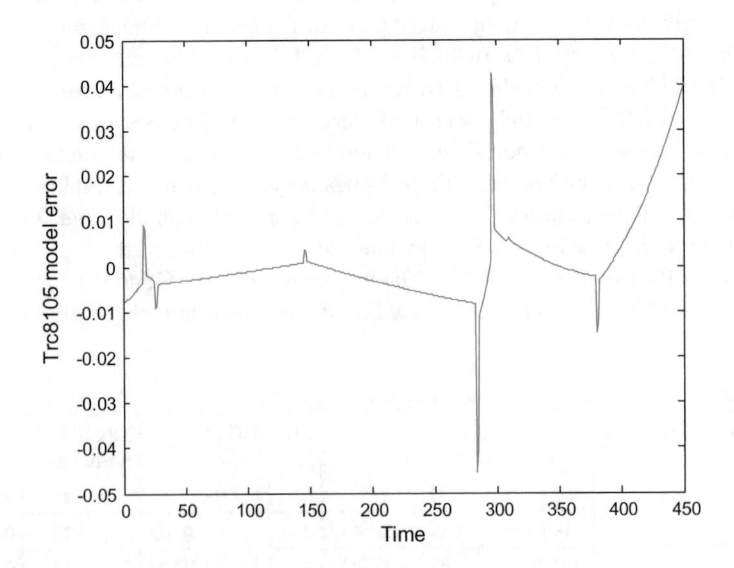

Fig. 6.9 Modeling error of TRC8105 using IRNA-GA

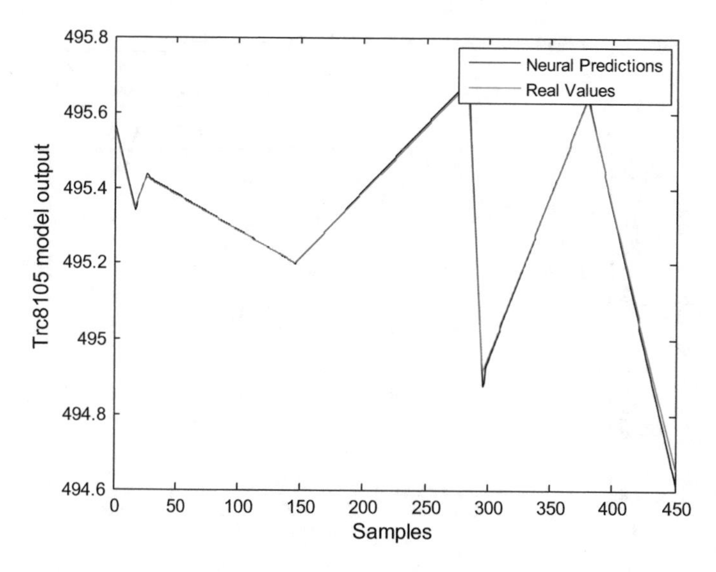

Fig. 6.10 RBFNN model output for TRC8105 using IRNA-GA

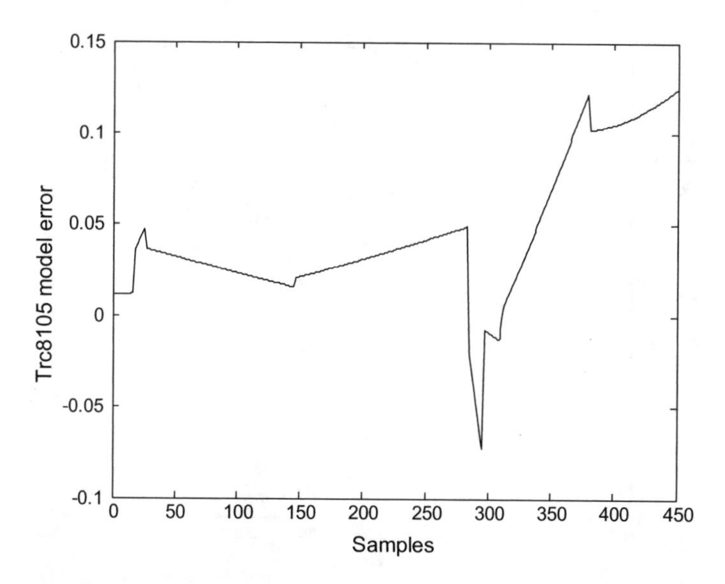

Fig. 6.11 Modeling error of TRC8105 using k-means method

6.5 Improved MOEA Based RBF Neural Network for Chamber Pressure

In Sect. 6.4, RBFNN is optimized by the weighted-sum method in (6.7). In this section, RBFNN is to be optimized by an improved MOEA (IMOEA) considering

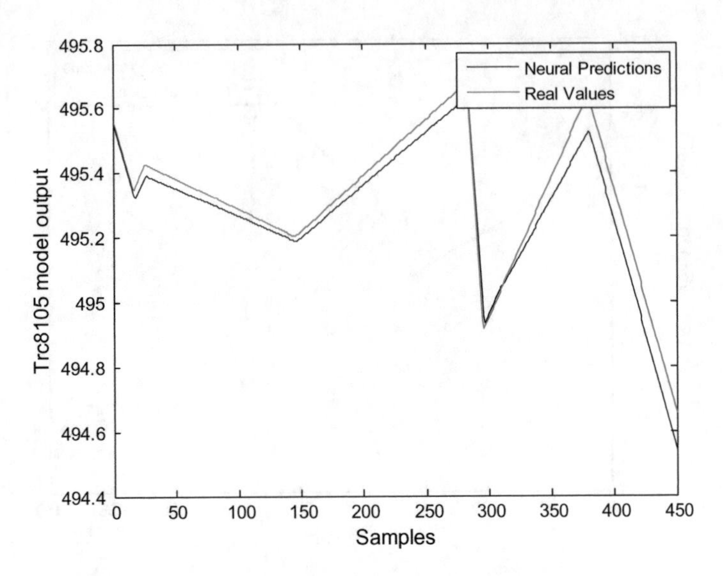

Fig. 6.12 RBFNN model output for TRC8105 using k-means method

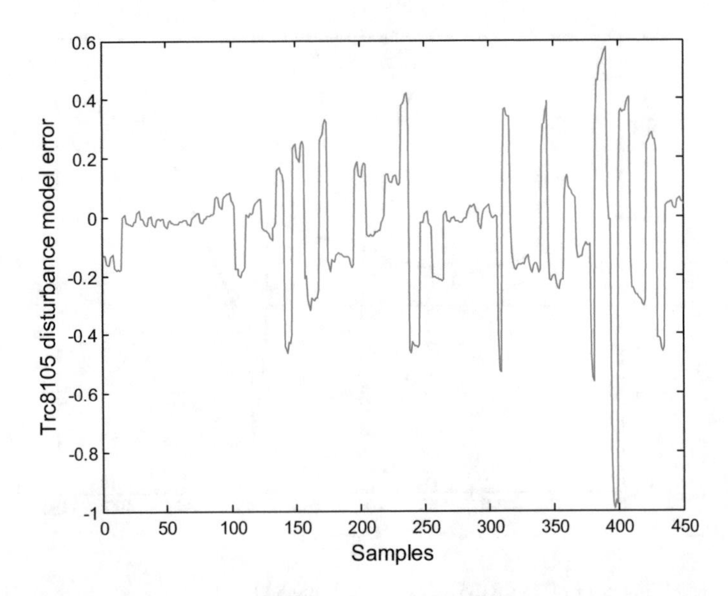

Fig. 6.13 Modeling error for TRC8105 disturbances using IRNA-GA

two objectives: the smallest modeling error and the simplest structure. The encoding method and various operators for the RBFNN structure and parameter optimization are also designed to solve the bi-objective optimization problem.

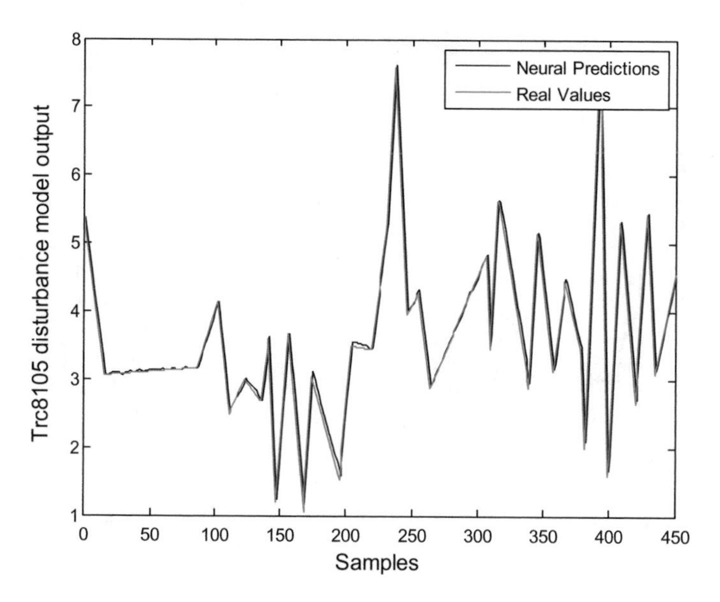

Fig. 6.14 RBFNN model output for TRC8105 disturbances using IRNA-GA

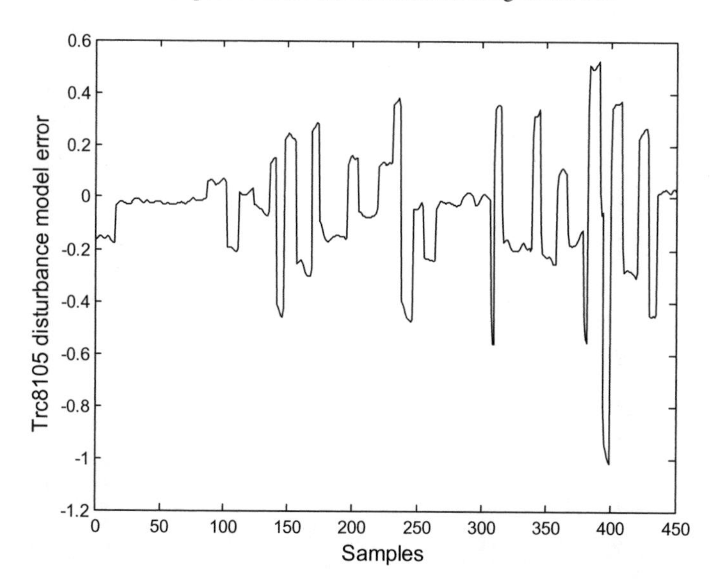

Fig. 6.15 Modeling error of TRC8105 disturbances using k-means method

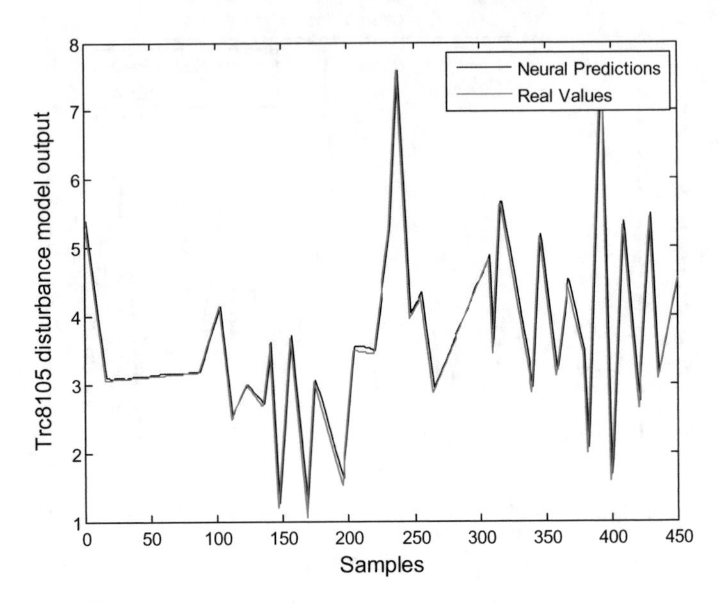

Fig. 6.16 RBFNN model output for TRC8105 disturbances using k-means method

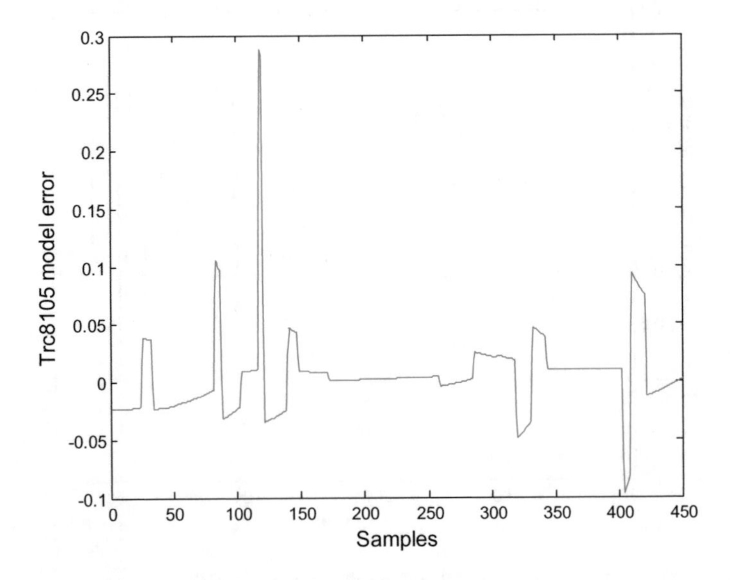

Fig. 6.17 Modeling error of TRC8103 using IRNA-GA

6.5.1 Encoding of IMOEA

Herein, m and n in the input layer, the number of the neurons in the hidden layer n_r and the parameters of the Gaussian functions $\mathbf{c}_i, \sigma_i, i = 1, \ldots, n_r$ are optimized

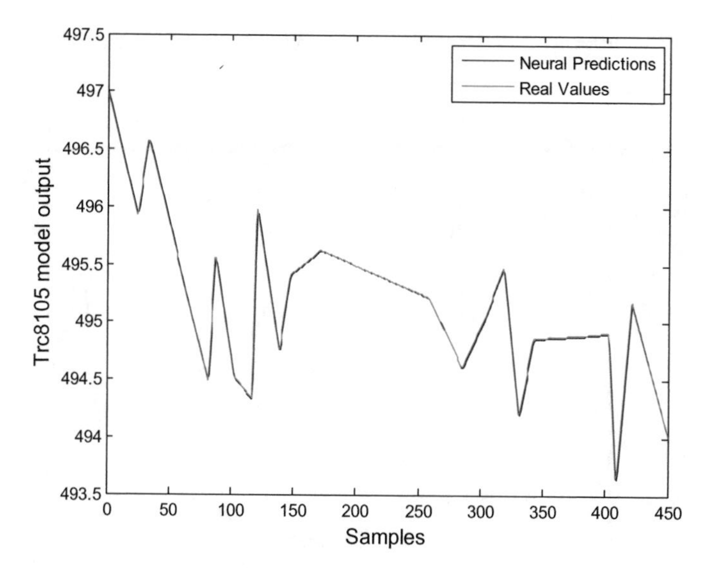

Fig. 6.18 RBFNN model output for TRC8103 using IRNA-GA

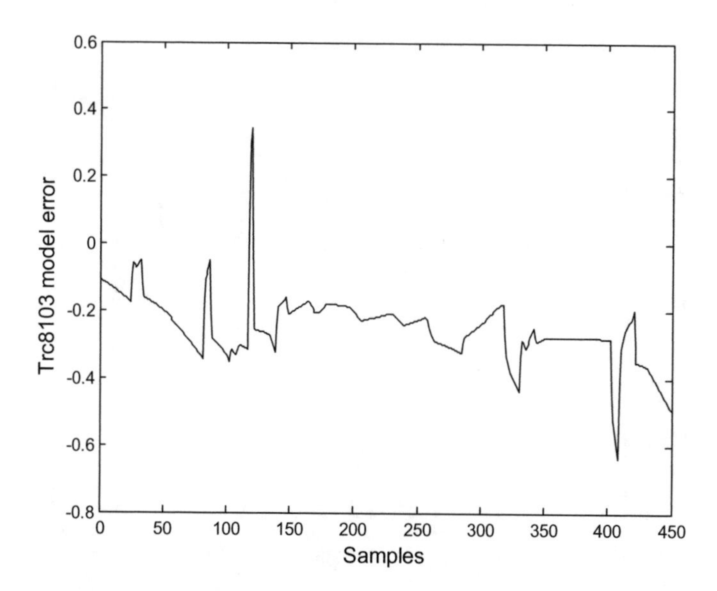

Fig. 6.19 Modeling error of TRC8103 using k-means method

simultaneously. The encoding for all the parameters is designed similarly to Eq. (6.4), and the lth chromosome is given as follows:

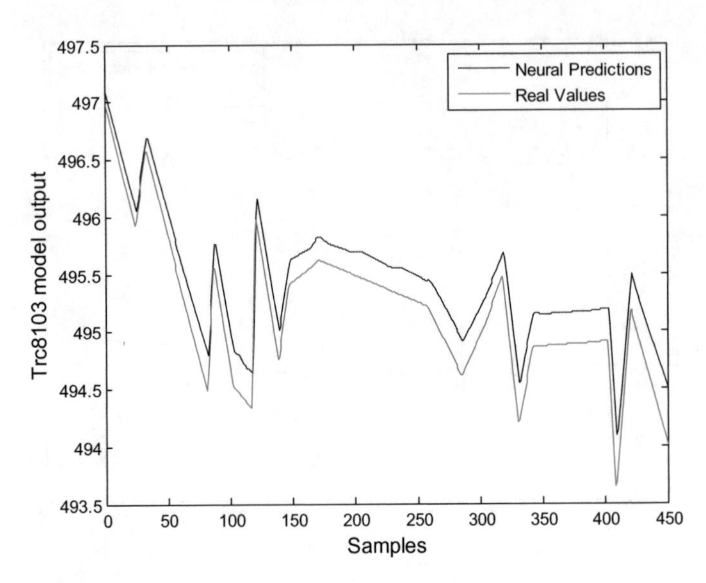

Fig. 6.20 RBFNN model output for TRC8103 using k-means method

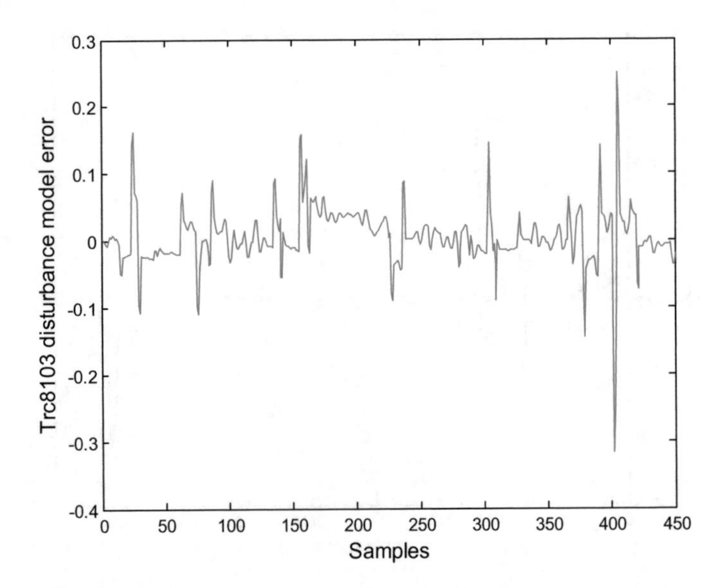

Fig. 6.21 Modeling error of TRC8103 disturbances using IRNA-GA

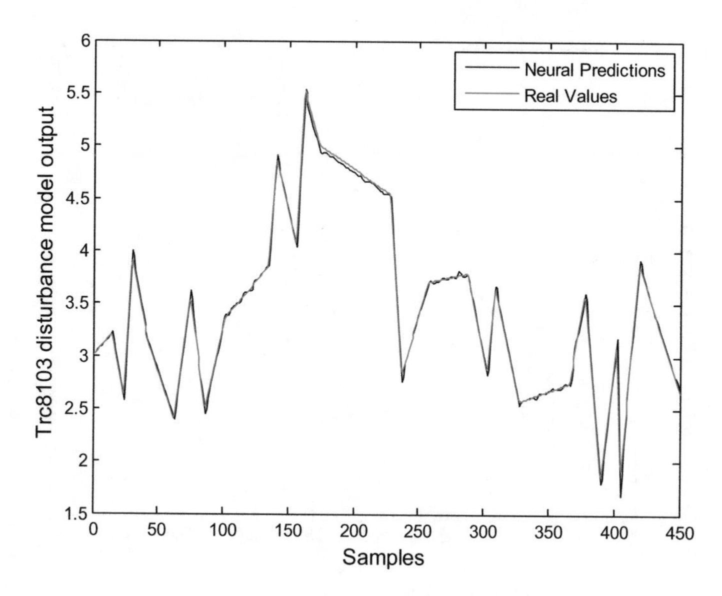

Fig. 6.22 RBFNN model output for TRC8103 disturbances using IRNA-GA

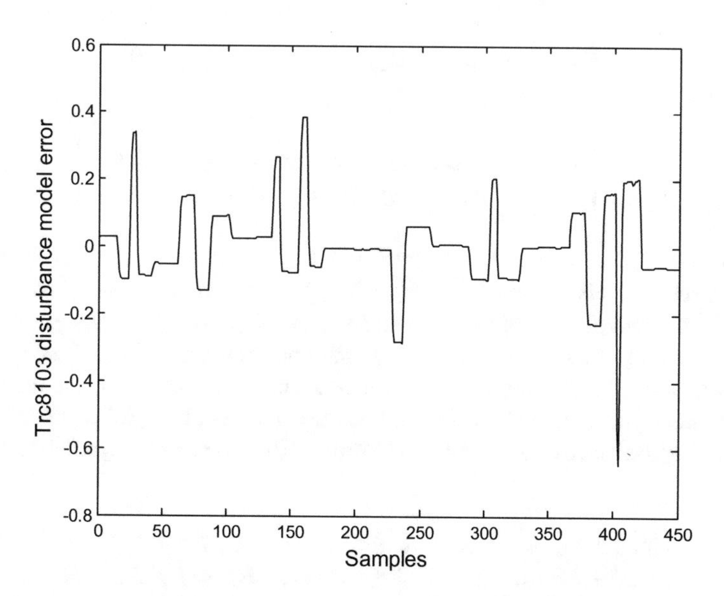

Fig. 6.23 Modeling error of TRC8103 disturbances using k-means method

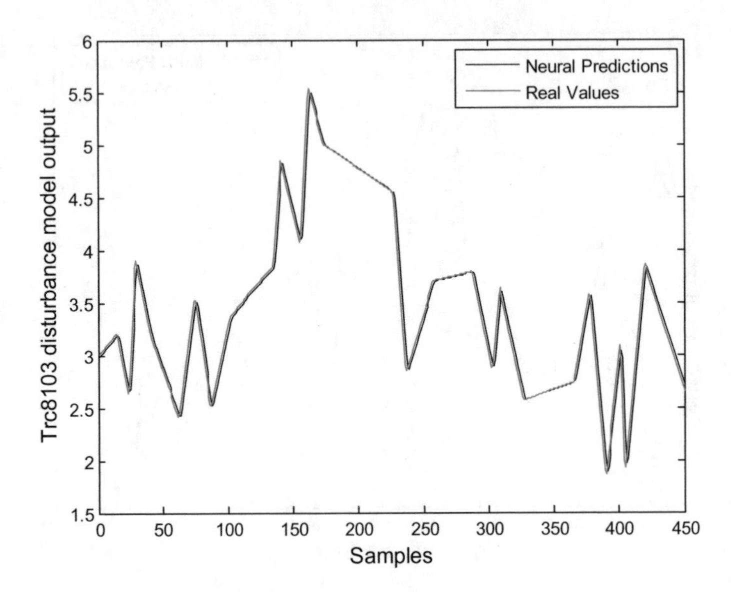

Fig. 6.24 RBFNN model output for TRC8103 disturbances using k-means method

$$\mathbf{C}_l = \begin{bmatrix} c_{1,1} & \cdots & c_{1,n} & 0 & c_{1,N_n+1} & \cdots & c_{1,N_n+m+1} & 0 & \sigma_1 \\ & & \vdots & \vdots & \vdots & \vdots & \vdots & \vdots & \vdots \\ c_{n_h,1} & \cdots & c_{n_h,n} & 0 & c_{n_h,N_n+1} & \cdots & c_{n_h,N_n+m+1} & 0 & \sigma_{n_h} \\ & 0 & \cdots & 0 & 0 & 0 & \cdots & 0 & 0\ 0 \end{bmatrix} \tag{6.11}$$

where $l = 1, 2, \ldots N$, m, n, and n_r are limited to $1 \leq m \leq N_m$, $1 \leq n \leq N_n$,$1 \leq n_r \leq D$, respectively. $\boldsymbol{C}_l$ is a $D \times (N_m+N_n+1)$ matrix and the rows below n_r are set to zeros. The columns of $(n, N_n]$ and $(m + N_n, N_m + N_n]$ are also set to zeros. Hence, there are actually $n_r \cdot (m + n + 1)$ parameters to be optimized. m, n, and n_r are first generated randomly among the given range. In Sect. 6.4.1, the number of the input nodes is obtained by using enumeration method, here, it is encoded in Eq. (6.11) and optimized by the evolution algorithm. The elements in $\boldsymbol{C}_l$ can be obtained as follows:

$$c_{ij} = \begin{cases} y_{\min} + r(y_{\max} - y_{\min}) & 1 \leq i \leq n_r, \quad 1 \leq j \leq n \\ u_{\min} + r(u_{\max} - u_{\min}) & 1 \leq i \leq n_r, \quad N_n < j \leq N_n + m \end{cases} \tag{6.12}$$

$$\sigma_i = r w_{\max} \quad 1 \leq i \leq n_h \tag{6.13}$$

where r is randomly generated between [0.01, 1], $u_{\min}$ and $u_{\max}$ are the minimal and maximal values of the system inputs, and $y_{\min}$ and $y_{\max}$ are the minimal and maximal values of the system outputs. $w_{\max}$ is the maximal width of the Gaussian basis function that is set to $\max(u_{\max}, y_{\max})$.

Once C_l is generated randomly, the structure and parameters of the RBFNN are determined and the connecting weight vector can be derived by the RLS algorithm by using the training data. N RBFNNs can then be obtained, denoted as $(\mathbf{C}_1, \boldsymbol{\omega}_1), \ldots (\mathbf{C}_N, \boldsymbol{\omega}_N)$.

6.5.2 Optimization Objectives of RBFNN Model

Two objectives considering the structure complexity and modeling accuracy of RBFNN are expressed as follows:

$$\text{Min}\begin{cases} f_1 = \sqrt{\sum_{k=1}^{N_1} \left|y_1(k) - \hat{y}_1(k)\right|^2} + \sqrt{\sum_{k=1}^{N_2} \left|y_2(k) - \hat{y}_2(k)\right|^2} \\ f_2 = (m+n)n_r \end{cases} \tag{6.14}$$

We denote $\mathbf{Y}_1 = [y_1(1), \ldots y_1(N_1)]$ as the training data set used to calculate the weight vector $\boldsymbol{\omega}$, $\mathbf{Y}_2 = [y_2(1), \ldots y_2(N_2)]$ as the testing data set, $\hat{\mathbf{Y}}_1 = [\hat{y}_1(1), \ldots \hat{y}_1(N_1)]$ and $\hat{\mathbf{Y}}_2 = [\hat{y}_2(1), \ldots, \hat{y}_2(N_2)]$ as the prediction outputs of the RBFNN. Here f_1 is the modeling accuracy by using the sum of the root of square errors (RSE) for Y_1 and Y_2, in which the generalization capability of the RBFNN is involved. f_2 is the structure complexity of the RBFNN by using the product of the number of neurons in the input layer and the hidden layer. Y_3 is used to choose the best RBFNN among the Pareto frontier.

6.5.3 Operators of IMOEA for RBFNN

After the Roulette wheel selection of the parents from individuals in terms of the top $N/2\,f_1$ and the top $N/2\,f_2$, respectively, the crossover and mutation operators are then implemented to generate the offspring.

(1) Crossover and mutation operators

The crossover operation is performed with probability p_c between individuals $\mathbf{C}_l$ and $\mathbf{C}_{l+1}$, and the offspring $\mathbf{C}_l^{'}$ and $\mathbf{C}_{l+1}^{'}$ are produced. The crossover position is generated between $[1, n_r]$ randomly. The number of the input nodes $m + n$ and the corresponding parameters of the radial basis functions are changed dynamically with the evolution processes going on. However, the number of the hidden nodes cannot be changed by crossover operation. In Fig. 6.25, an example of the crossover operation is given in the genes surrounded by a dotted line, all genes in the dotted line are exchanged, and obviously, this is a multi-point crossover operator in nature.

For a better exploration, a mutation operator is also designed with the probability p_m. When the mutation operator is implemented, m, n, and n_r are first produced in

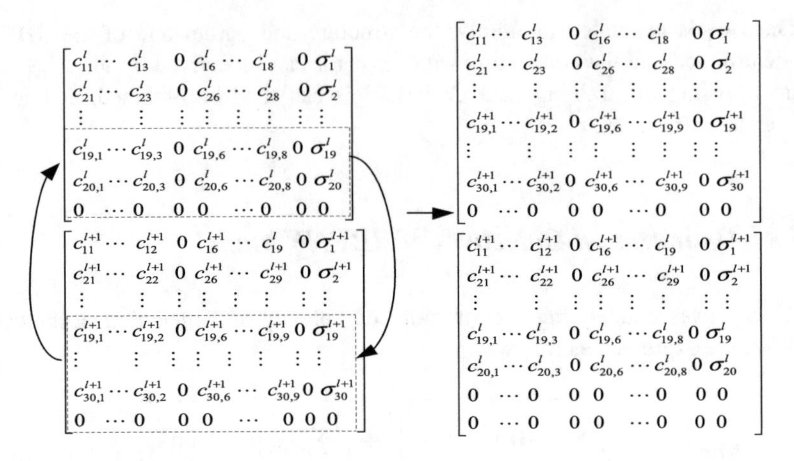

Fig. 6.25 Example of the crossover operator

random among the given ranges in Sect. 6.5.1. and the elements of the mutation individual are replicated according to the Eqs. (6.12)–(6.13). The new structure of the RBFNN is thus generated.

In addition to the crossover and mutation operators, the local search operator, prolong and pruning operators are designed to improve the search capability of MOEA and guarantee the rationality of the RBFNN.

(2) Local search operator

The local search operator is given as follows:

$$\mathbf{C} = \alpha_1 \mathbf{C}_l + (1 - \alpha_1)\mathbf{C}_l^{'} \tag{6.15}$$

$$\mathbf{C} = \mathbf{C}_l + \alpha_2 \mathbf{C}_l \tag{6.16}$$

where C_l is selected randomly from the former $N/2$ parents, $\mathbf{C}'_l$ is chosen randomly from the latter $N/2$ parents, and α_1 is randomly generated between (0,1) that is to generate excellent offspring inheriting the gene information of C_l and $\mathbf{C}'_l$. If C_l is equal to $\mathbf{C}'_l$, 0.146 is utilized to generate the offspring and α_2 is generated randomly between (−1, 1). In order to keep the population diversity and avoid running into the local optima as the evolution goes on, a similar dynamical probability of 0.9 is adopted for local search operator. The difference is that the probability is increased from p_{l0} to p_{lG} with the generation increasing from 1 to G by using minus aa.

(3) Prolong and pruning operators

Since the crossover operator cannot generate new structures of the hidden layer and the probability of mutation is low, the prolong operator is designed. That is, the number of the hidden nodes is reproduced randomly between n_r and D with probability p_p. The elements of the newly added node can be calculated according to Eqs. (6.12)–(6.13).

Because the chromosomes are randomly generated, the crossover and mutation operations are also of randomness and there may be inactive structures in the population. The neuron with $\boldsymbol{c}_i = 0$, $i < n_r$ will first be deleted, and each hidden neuron is evaluated in terms of the active firing (AF) in Sect. 6.4.3, using the same value of ρ. Here the upper threshold Af_o is also selected from [0.05, 0.3]. When the hidden neuron is judged as inactive, the corresponding hidden neuron is deleted.

(4) Elitism maintaining scheme

The fast non-dominated sorting scheme is adopted and all non-dominated individuals in the population are regarded as the elitists, which will be found and stored to an archive. Because the size of the archive will increase with the evolution going on, the maximum size of the elitist archive is set as N_e. If the current size of the elitist archive is larger than N_e, the maintaining scheme will be performed to keep the evenness of the elitist population. The fast non-dominated sorting algorithm is implemented and the dominated individuals will be removed from the archive. If the archive size becomes less than N_e, the maintaining procedure will not be carried out, otherwise, a modified adaptive cell density maintaining scheme will be implemented by dividing the objective spaces into $\prod_{i=1}^{2} k_1$ cells, and at most one individual can be kept at each cell [45]. Matlab code of the maintaining scheme has been given in Chap. 4. When the maximal number of the individuals distributed at the Pareto frontier is set as, and $\sum_{i=1}^{2} k_i - 1$, and $N_e < \sum_{i-1}^{2} K_i$ must be satisfied to keep the evenness of the population distribution which can refer to the analysis of MOEA in Chap. 4.

6.5.4 The Procedure of IMOEA

The whole procedure of IMOEA can be run by using the following steps.

Step 1: Initialize the population size N, the maximum generations G, the operator probabilities $p_c, p_m, p_{l0}, p_{lG}\ p_p$, and Af_o, the RBFNN N_m, N_n, D, the number of cells for the ith objective function K_i, and the maximal archive size N_e, then generate randomly N chromosomes using 0.11.

Step 2: Calculate the two fitness functions based on 0.14.

Step 3: Implement the non-dominated sorting algorithm in NSGA-II and keep the elitists in the archive. Execute the elitist maintaining scheme when the size of the elitist archive is larger than N_e.

Step 4: Select the parent individuals using the Roulette wheel method in terms of f_1 and f_2, and Pareto Elite individuals are also selected as the parents to produce the offspring by genetic operators. To keep the individual diversity, N_e is set as not larger than $N/2$.

Step 5: Execute the crossover and mutation operators with probability p_c and p_m, respectively, then implement the local search operator with dynamic probability, prolong operator with probability p_p, and pruning operator with probability 1 to produce the offspring.

Step 6: Repeat steps 2–5 until the maximum generation G is met.

Step 7: Calculate RMSE of an unused data set for determining the final solution, and the RBFNN model with a minimal value of f_1 is selected as the final optimal one.

6.5.5 The Chamber Pressure Modeling in a Coke Furnace

This section describes the application of the IMOEA to optimize the RBFNN model for the chamber pressure in the industrial coke furnace in Sect. 6.2. Herein, the parameters of the IMOEA are set as $N = 60$, $G = 1000$, $p_c = 0.9$, $p_m = 0.1$, $p_{l0} = 0.02$, $p_{lG} = 0.22$, and $Af_o = 0.1$. Since the prolong operator is used to increase the number of the hidden nodes, the probability p_p is set relatively small as 0.1. The pruning operator is designed to keep the rationality of the RBFNN structure and its probability is set as 1. K_i, $i = 1, 2$, is set to 20 and the archive size N_e is set to 30 to satisfy $N_e \leq N/2$ and $N_e < \sum_{i-1}^{2} K_i \cdot N_m$, N_n and D are directly related to the model complexity and can be selected among the following ranges: N_m, $N_n \in [3, 10]$, $D \in [10, 60]$, where a little of prior knowledge is required to set suitable values for these parameters. Note that the simpler the modeled system, the smaller the value is to be set, which may speed up the convergence of the algorithm. N_m, N_n, and D in this section are set as 5, 5, 60, respectively.

Two pressure branches, i.e., the main channel and its coupling disturbance should be modeled here. Then, several sets of step tests are performed for system analysis and modeling. The input step signal is the set point of the originally designed PID controller and this signal also poses disturbances on the other side of the chamber pressure. The experimental data are collected from the same industrial coke unit equipped with a distributed control system CENTUM DCS3000. And all data are filtered to reduce the impact of measurement noise. There are totally 4 groups of 1200 input/output samples as plotted in Figs. 6.26 and 6.27. For the main channel of the chamber pressure PRC8112A/PRC8112B, the input is the valve opening given by the PID controller and the output is the chamber pressure PRC8112A/PRC8112B. The output responses of PRC8112A and its coupling disturbance on PRC8112B are shown in Fig. 6.26 when the set point is set as −0.029 kPa, −0.024 kPa, −0.019 kPa, and −0.024 kPa, respectively. Figure 6.27 shows the output responses of PRC8112B and its coupling disturbance on PRC8112A when the set point is set as −0.022 kPa, −0.016 kPa, −0.02 kPa, −0.026 kPa, and −0.02 kPa, respectively. The sampled dataset is equally divided into three groups, where the former 1/3 data are selected as the training data $\mathbf{Y}_1$, the intermediate 1/3 data as $\mathbf{Y}_2$, and the latter 1/3 data as $\mathbf{Y}_3$. In addition, the RBFNN is optimized by running 10 times and the parameters of IMOEA remain unchanged at each run.

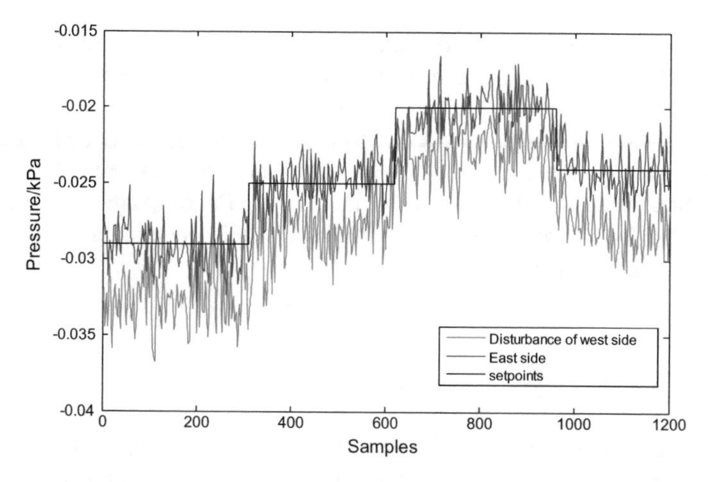

Fig. 6.26 Outputs of PRC8112A (main channel) and its disturbance on pressure PRC8112B

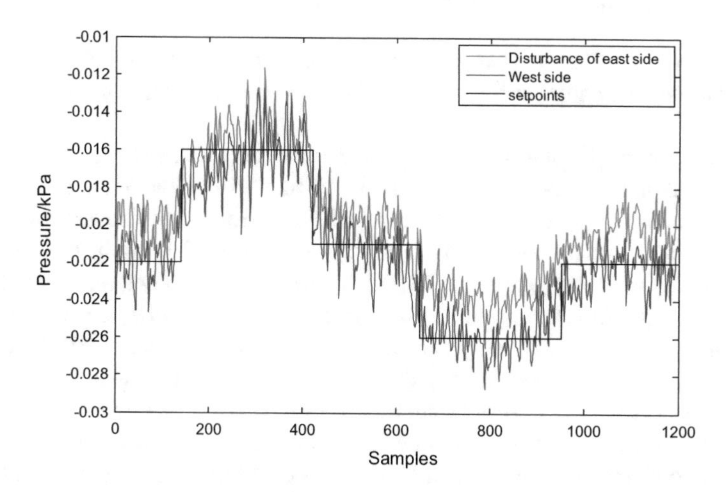

Fig. 6.27 Outputs of PRC8112B (main channel) and its disturbance on pressure PRC8112A

The IMOEA method is compared with NSGA-II, IMOEA with fixed input layer (IMOEA-Fix), IRNA-GA with WSO method [19], and artificial neural network with LM algorithm (LMANN). The details of the compared algorithms are given as follows:

(1) The encoding/decoding method, crossover, mutation, and zero $\boldsymbol{c}_i$ deletion of IMOEA are adopted in NSGA-II, while the individuals in the mating pool are selected in terms of the non-dominated sorting rank and its spread evenness information.

(2) For IMOEA-Fix, all the parameters and operators are the same as those of IMOEA except for the structure of the input layer, where m is set as 3, n as 2, and they are fixed during the whole optimization procedure.
(3) A weighted sum of objectives based on IRNA-GA that has obtained good results of temperature modeling in Sect. 6.4 is chosen to be compared. The objective function in Eq. (6.17) for IRNA-GA is similar to Eq. (6.7) and that in Ref. [19], however, the weight coefficient is decreased to 0.01 by trial and error focusing on the modeling accuracy.

$$J = f_1 + \lambda f_2 \tag{6.17}$$

(4) For LMANN, an enumeration method is applied to select the NN structure, i.e., the number of the neurons in the input layer $(m + n)$ is enumerated from 2 to 10 and the number of the neurons in the hidden layer (n_h) is enumerated from 1 to 60, and there are totally 25×60 combinations of ANN.

The selecting criterion in step 7 is used to choose the ultimate RBFNN among Pareto individuals and also for selecting the best one in 10 runs of IRNA-GA and LMANN.

To illustrate the population diversity and distribution evenness, the Pareto frontier with the maximal number of individuals in 10 runs using 3 MOEAs are shown in Figs. 6.28, 6.29, 6.30,and 6.31, where the modeling error f_1 is in the horizontal coordinate and the structure complexity f_2 is in y-coordinate. Obviously, the objectives are conflicting with each other; the RBFNN with simpler structure, that is smaller f_2, has weaker approximation capability, that means larger f_1, and vice versa. When f_2 is less than 10, f_1 grows quickly because of too simple structure of RBFNN. Since only zero c_i deletion is used to keep the rationality of the RBFNN in NSGA-II, the value of f_2 in NSGA-II is larger than that of IMOEA as shown in Figs. 6.28, 6.29, 6.30, and 6.31. Moreover, in Figs. 6.29 and 6.30, it is obvious that IMOEA is nearer to the Pareto frontier compared with NSGA-II because the local search operator and pruning operator are beneficial to produce more individuals and decrease the structure complexity. In Fig. 6.31, though IMOEA-fix has more individuals, only three

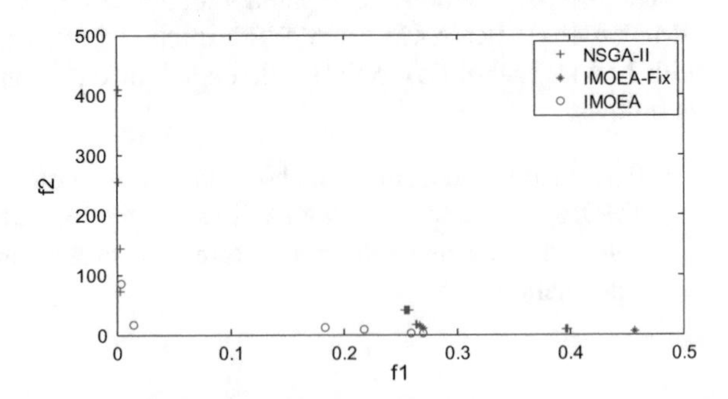

Fig. 6.28 Pareto frontier for PRC8112A

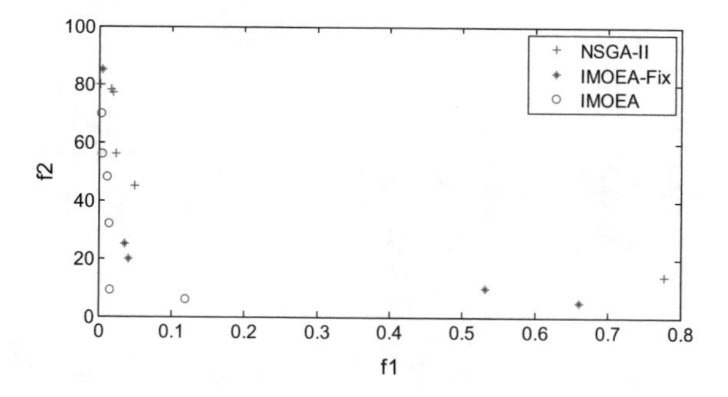

Fig. 6.29 Pareto frontier for the disturbance of PRC8112B

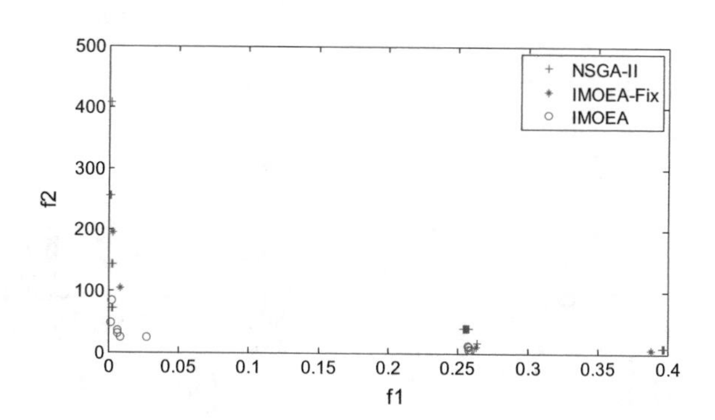

Fig. 6.30 Pareto frontier for PRC8112B

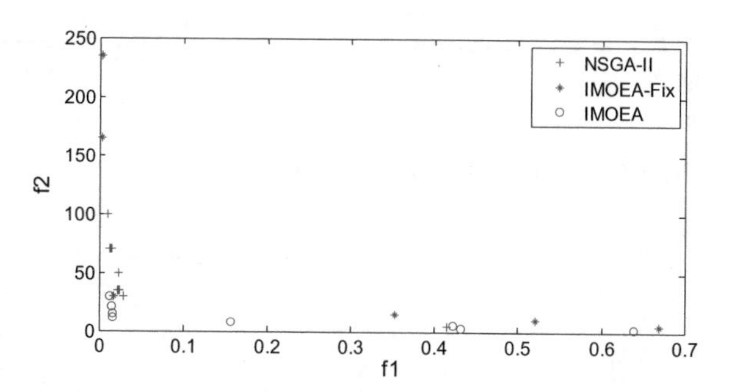

Fig. 6.31 Pareto frontier for the disturbance of PRC8112A

individuals have good modeling accuracy where f_1 is less than 0.1. Since the structure of the input layer is fixed, the performance of the RBFNN is restricted greatly. The maintaining scheme of IMOEA has kept the elitist archive size in a rational range during the optimization process. However, the number of individuals in the Pareto frontier in 3 MOEAs is relatively small, i.e., the size of the elitist archive is less than N_e and the evenness distribution problem is not serious, thus, the maintaining scheme will not be implemented in most evolution generations.

The best RBFNNs of five methods are obtained after selecting in terms of modeling accuracy and their modeling errors are plotted in Figs. 6.32, 6.33, 6.34, and 6.35. It is obvious that the errors of IRNA-GA are much larger than those of MOEA

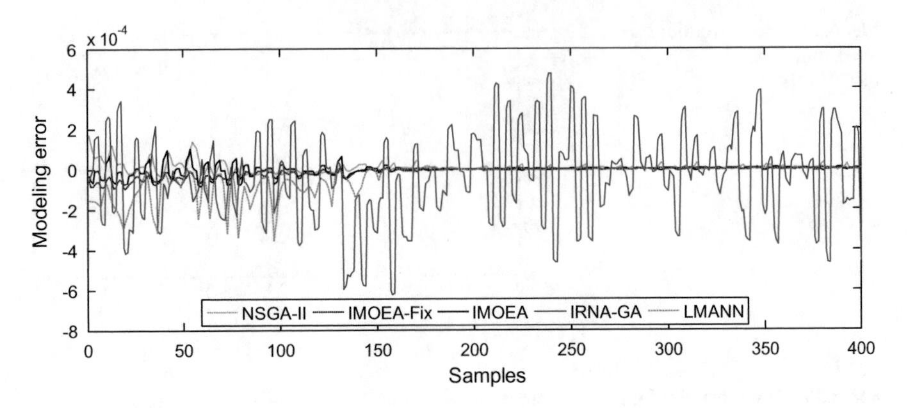

Fig. 6.32 Errors of best RBFNN for PRC8112A

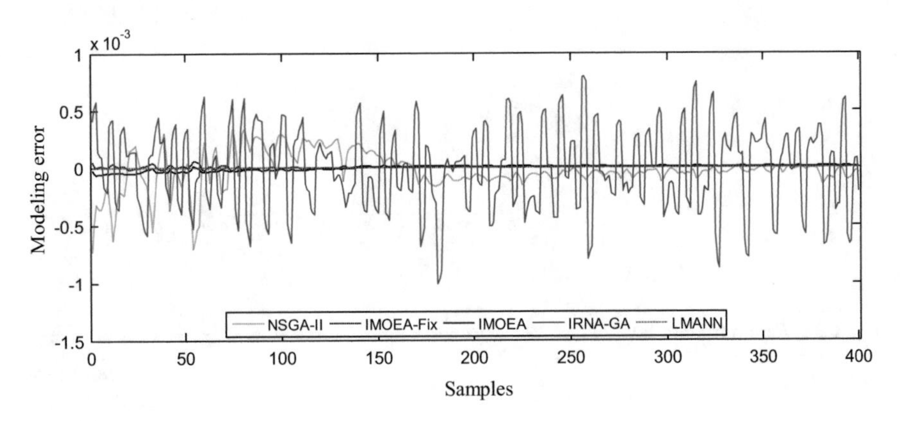

Fig. 6.33 Errors of best RBFNN for PRC8112B

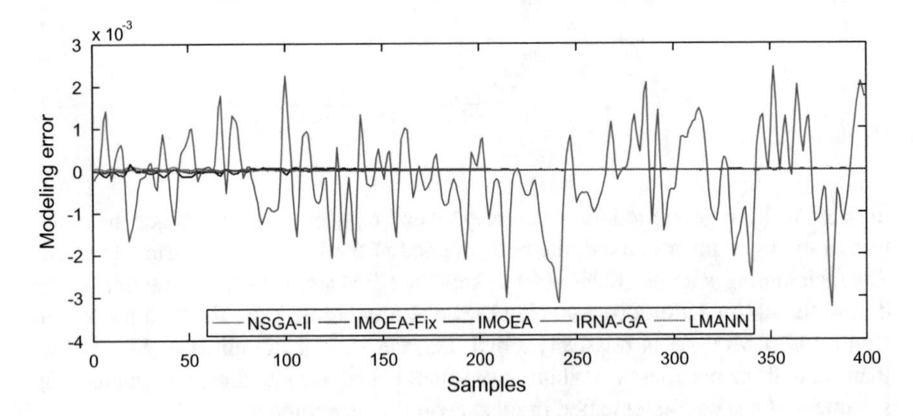

Fig. 6.34 Errors of best RBFNN for the disturbance of PRC8112B

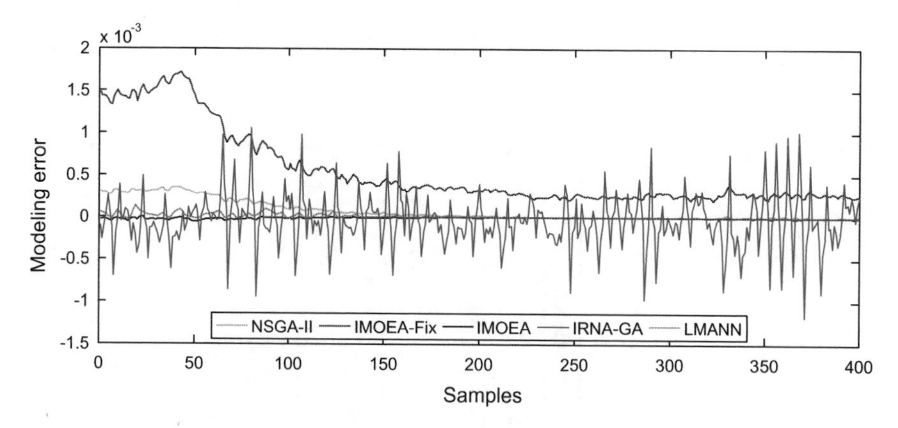

Fig. 6.35 Errors of best RBFNN for the disturbance of PRC8112A

partly because the structure simplification in Eq. (6.17) is considered during the optimization processes. As a typical and widely used MOEA, NSGA-II can obtain satisfactory modeling accuracy as shown in Fig. 6.34. However, by introducing local search, prolong and pruning operators, IMOEA has obtained smaller errors than other GA-based methods. As for IMOEA-fix, if the input layer is not set appropriately beforehand as shown in Fig. 6.35, its modeling error will be larger than those of IRNA-GA and NSGA-II. Otherwise, its modeling error of IMOEA-fix is similar to that of IMOEA.

The statistical results about the structure of the input layer, the number of hidden nodes of the best RBFNN, and their RMSEs of six methods among 10 runs are listed in Tables 6.2 and 6.3, respectively. In Table 6.2, LMANN has obtained the best results of PRC8112B and its disturbance channel using the enumeration method, however, the average running time is quite long, which depends on the range of m, n, and n_r, and the enumeration method may be impracticable with large variable range. LMANN-fix used the same structure of the optimal RBFNN's, however, its results are worse than those of the enumerated LMANN because the structure, weight learning algorithm, and radial basis function are quite different between ANN and RBFNN, and the optimal structure of the RBFNN is not suitable for ANN. Moreover, ANN will be greatly affected by the initialization of the weight vector that is implemented automatically by the toolbox, and its average RMSE ($\overline{\text{RMSE}}$) in Table 6.3 is much larger than their RMSEs in Table 6.2. The RMSE of IRNA-GA in Table 6.2 is several times larger than those in three MOEAs except for the disturbance model of PRC8112A by IMOEA-fix, which is consistent with Fig. 6.34. The variation range of n_r by IRNA-GA in Table 6.3 is relatively smaller than that of MOEA because the WSO method takes into account the structure simplification in the optimization process and the average running time ($\bar{T}$) of the single-objective optimization is relatively shorter. Compared with the implementation of NSGA-II, IMOEA can obtain better accuracy with simpler structure, because several new operators are carried out, however, its $\bar{T}$ is the longest in the MOEAs. As for IMOEA-Fix, the

Table 6.2 The comparison of best simulation results by 6 methods

Methods	PRC8112A				Disturbance of PRC8112B				PRC8112B				PRC8112B			
	m	n	n_h	RMSE	m	n	n_h	RMSE	m	n	n_h	RMSE	m	n	n_h	RMSE
LM ANN	1	1	9	3.5e − 5	2	2	6	5.73e − 6	3	2	10	5.96e − 6	1	2	6	2.25e − 5
LMANN-Fix	1	2	9	7.65e − 5	1	1	11	1.41e − 4	3	2	7	3.53e − 5	1	3	18	8.18e − 5
IRNA-GA	2	1	7	2.28e − 4	2	2	6	3.69e − 4	2	3	10	2.89e − 4	3	2	11	3.05e − 4
NSGA-II	2	1	12	7.42e − 5	1	2	10	1.94e − 5	4	1	23	2.41e − 5	2	4	19	1.32e − 4
IMOEA-Fix	3	2	13	2.53e − 5	3	2	13	8.77e − 5	3	2	24	2.22e − 5	3	2	12	3.76e − 4
IMOEA	1	2	9	2.47e − 5	1	1	11	6.01e − 6	3	2	7	3.62e − 5	1	3	18	1.51e − 5

Table 6.3 The performance comparison of 6 methods in 10 runs

Methods	PRC8112A					Disturbance of PRC8112B					PRC8112B					PRC8112B				
	m	n	n_h	$\bar{T}$	$\overline{RMSE}$	m	n	n_h	$\bar{T}$	$\overline{RMSE}$	m	n	n_h	$\bar{T}$	$\overline{RMSE}$	m	n	n_h	$\bar{T}$	$\overline{RMSE}$
LM ANN	2 ± 1	2 ± 1	6 ± 3	3150	8.15e − 5	2 ± 1	2 ± 1	4 ± 2	2850	1.18e − 4	2 ± 1	2 ± 1	6 ± 4	2550	3.47e − 4	2 ± 1	2 ± 1	6 ± 4	2820	2.51e − 4
LMANN-Fix	1 ± 0	2 ± 0	9 ± 0	21	3.96e − 4	1 ± 0	1 ± 0	11 ± 0	19	7.13e − 4	3 ± 0	2 ± 0	7 ± 0	17	2.5e − 4	1 ± 0	3 ± 0	18 ± 0	19	8.74e − 4
IRNA-GA	3 ± 1	3 ± 2	10 ± 8	651	9.81e − 4	3 ± 1	3 ± 1	16 ± 12	853	1.52e − 3	3 ± 1	3 ± 1	15 ± 9	607	1.12e − 3	3 ± 1	3 ± 1	13 ± 8	569	1.05e − 3
NSGA-II	3 ± 1	3 ± 1	28 ± 19	1029	4.46e − 4	3 ± 2	3 ± 2	34 ± 24	1152	5.19e − 4	3 ± 2	3 ± 2	26 ± 20	1027	2.19e − 4	3 ± 1	4 ± 1	37 ± 22	925	1.75e − 4
IMOEA-Fix	3 ± 0	2 ± 0	19 ± 10	1043	5.03e − 3	3 ± 0	2 ± 0	26 ± 13	1022	2.76e − 3	3 ± 0	2 ± 0	30 ± 24	1012	2.84e − 3	3 ± 0	2 ± 1	19 ± 16	982	1.41e − 3
IMOEA	3 ± 2	2 ± 1	15 ± 10	1829	7.32e − 5	3 ± 2	3 ± 2	26 ± 15	1774	4.25e − 5	3 ± 1	2 ± 1	15 ± 10	1783	1.23e − 4	2 ± 1	4 ± 1	29 ± 12	1624	4.91e − 4

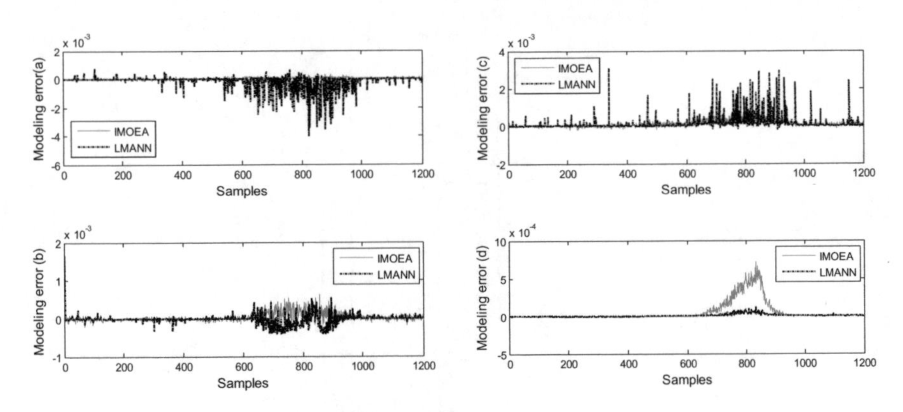

Fig. 6.36 Comparison IMOEA RBFNN with LMANN in the case of noisy data **a** PRC8112A **b** PRC8112B **c** Disturbance of PRC8112B **d** Disturbance of PRC8112A

complicated input layer structure seems to require more hidden nodes to obtain the similar modeling precision compared with IMOEA. Moreover, one order higher of $\overline{\text{RMSE}}$ in Table 6.3 illustrates that $m = 3, n = 2$ is not appropriate to construct the chamber pressure model. Hence, the input layer will affect both the model structure and modeling accuracy greatly.

In order to illustrate the generalization capability of the optimization methods, 10% of noise levels are added to the sampled data, and other parameters are kept unchanged. The modeling errors of IMOEA RBFNN are compared with the enumerated LMANN and plotted in Fig. 6.36. The RMSEs of enumerated LMANN are 5.10e − 4, 3.83e − 4, 1.26e − 4, 1.67e − 5, while the IMOEA RBFNN's are 6.70e − 5, 2.32e − 4, 1.178e − 4 and 1.39e − 4, respectively. Though small RMSEs of PRC8112B and its disturbance of PRC8112B have been derived by the enumerated LMANN in Table 6.2, much larger error has been observed from the noisy data as shown in Fig. 6.36 and reflected in their RMSEs, which may be caused by overfitting. IMOEA based RBFNN is more robust than LMANN in the aspect of generalization capability. It can be concluded that IMOEA is superior to the other four methods with respect to the smaller errors in model accuracy and simpler structure of RBFNN, however, the algorithm is more complicated and its running time is longest.

6.6 PCA and INSGA-II Based RBFNN Disturbance Modeling of Chamber Pressure

6.6.1 RV Criterion in PCA Variable Selection

The key variables are important to be found for multiple variables system modeling, the RV criterion in principal component analysis has been utilized to measure the similarity of the selected subset. If the value of RV criterion is the largest among all possible subsets, the optimal subset will be obtained. In order to calculate the RV criterion, the following augmentation of notation has been listed in Table 6.4.

The optimal solution for a given subset, P, is equivalent to maximize the RV criterion as follows [22]:

$$f_1 = \sqrt{\frac{tr((S_P^{-1}[S^2]_P)^2)}{tr(S^2)}} \tag{6.18}$$

If all variables are selected, the maximum value of f_1 reaches 1. Since f_1 is to be maximized, the objective is then changed into the minimization problem by reciprocal operation of Eq. (6.18), shown as follows:

$$J_1{=}1/f_1. \tag{6.19}$$

Once the selected variables are determined in terms of J_1, they are used as the inputs of the system model.

When the RBFNN model is constructed and its parameters are trained, its modeling accuracy can be evaluated according to the sum of RMSEs of the training and testing data

$$J_2 = \sqrt{\sum_{k=1}^{N_1} |y_1(k) - \hat{y}_1(k)|^2 \Big/ N_1} + \sqrt{\sum_{k=1}^{N_2} |y_2(k) - \hat{y}_2(k)|^2 \Big/ N_2} \tag{6.20}$$

The two objectives J_1, J_2 can be optimized simultaneously. The encoding method and various operators in an improved NSGA-II for variable selection, RBFNN structure, and parameter optimization are then designed to solve this multi-objective optimization problem.

Table 6.4 Notation for RV criterion

Parameters	Description
X	an $N \times M$ data matrix for N objects measured on M variables
$X(P)$	the $N \times p$ vector of X for the selected variables in P
S	the covariance matrix for the full data matrix, X
S^2	the product of the covariance matrix and itself, $S^2 = SS$
S_P	the $p \times p$ submatrix of S corresponding to the selected variables in P
$[S^2]_P$	the $p \times p$ submatrix of S^2 corresponding to the selected variables in P

6.6.2 *Encoding of RBFNN*

For simplicity, n in the input layer is set as 2, while m for one input variable is set as 1, here, there are at most six disturbance variables according to prior knowledge of coke furnace, m is then limited to [1, 6]. Once the key variables are selected, the input nodes are then determined. The number of the hidden neurons (n_r) and the parameters of Gaussian functions $\mathbf{c}_i, \sigma_i, i = 1, \ldots, n_r, 1 \le n_r \le D$, with D being the maximal number of the hidden nodes, are to be optimized. The encoding for different variables selection and RBFNN is then designed, and the ith chromosome is shown as follows:

$$\mathbf{C}_i = \begin{bmatrix} c_{11} & c_{21} & c_{31} & \cdots & c_{81} & \sigma_1 \\ \vdots & \vdots & \vdots & \vdots & \vdots & \vdots \\ c_{1n_r} & c_{2n_r} & c_{3n_r} & \cdots & c_{8n_r} & \sigma_{n_r} \\ 0 & 0 & 0 & \cdots & 0 & 0 \\ 0 & 0 & 1 & \cdots & 1 & 0 \end{bmatrix} \tag{6.21}$$

Here $1 \le i \le N$, $(m + n) \in [3, 8]$ and the elements in rows $[1, n_r]$ can be obtained as follows:

$$\begin{cases} c_{ij} = y_{\min} + r(y_{\max} - y_{\min}) & 1 \le i \le 2\ 1 \le j \le n_r \\ u_{\min} + r(u_{\max} - u_{\min}) & 3 \le i \le 8\ 1 < j \le n_r \\ \sigma_j = r w_{\max} & 1 \le j \le n_r \end{cases} \tag{6.22}$$

where r is randomly generated in [0.01,1], $u_{\min}$, $u_{\max}$ $y_{\min}$, $y_{\max}$, and $w_{\max}$ are the same as those in Eqs. (6.12)–(6.13).

The last row in $\mathbf{C}_i$ delegates which variable of columns 3–8 will be selected, thus it is encoded by binary encode, and the valid bits are located at [3 8], for example

$$c_{D+1} = [0\ 0\ 0\ 0\ 1\ 1\ 0\ 1\ 0\ 0] \tag{6.23}$$

It means that u_3, u_4, u_6 are selected, and columns $\mathbf{c}_5, \mathbf{c}_6, \mathbf{c}_8$ are valid centers of the Gaussian functions.

Once $\boldsymbol{C}_i$ is obtained, both the structure and the parameters of RBFNN can be determined, and the weight $\boldsymbol{w}$ can be further calculated by RLS based on the training data.

6.6.3 *Operators of INSGA-II*

(1) Selection operator

When the NSGA-II is implemented, the rank and crowding distance of individuals can be obtained. The individuals with rank 1 are regarded as the elitists and chosen as the parents. The individuals at each frontier from rank 1 are selected into the parent population one by one until exceeding the population size N. Then, the crowing distance in the current frontier is compared by sorting in descend and the individuals with larger crowing distance are selected into the parent population. If the size is still less than the population size, the Roulette wheel selection operator is implemented to select half of the rest of the population in terms of J_1 and the other half of the rest of the population in terms of J_2.

(2) Crossover and mutation operators

The crossover and mutation operators are carried out among the selected population to produce the offspring.

In Fig. 6.37, the crossover operator with probability p_c is executed between the individuals $\mathbf{C}_i$ and $\mathbf{C}_i'$, and the location is randomly generated between [1 9]. The parameters of the radial basis function are changed and the selected variables are also changed in the offspring. Note that the number of the hidden nodes cannot be changed by using this crossover operator.

The element in Eq. (6.21) is mutated with the probability p_m. When the mutation operator is implemented, the elements are produced in terms of Eq. (6.22) and the elements in Eq. (6.23) performs a logic NOT operation, that is, 1 to 0 and 0 to 1. The new structure of RBFNN and different key variables can be obtained.

In addition to crossover and mutation, the prolong and pruning are also designed for improving the searching capability of NSGA-II and keeping the rationality of RBFNN.

(3) Prolong and pruning operators

Due to the fact that the number of the hidden nodes is not changed and some irrational structure may be produced by random operators, the prolong and pruning operators are then designed. If the number of the hidden neuron is less than 2, the prolong

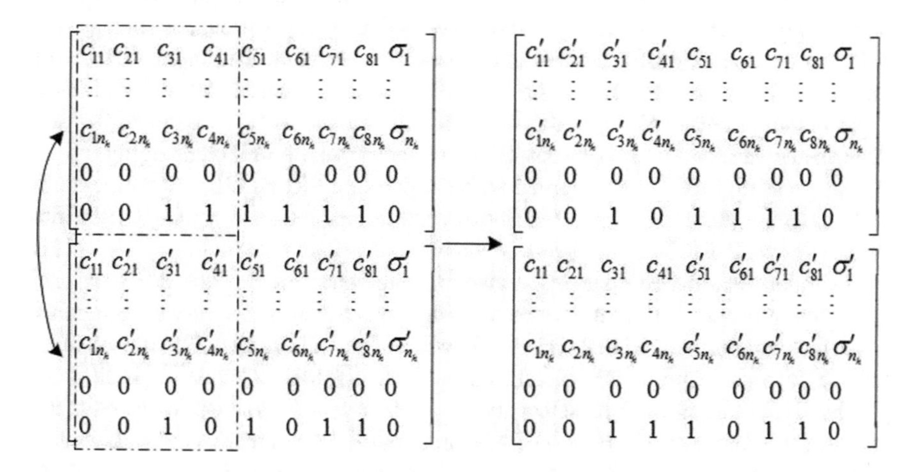

Fig. 6.37 Example of the crossover operator

operator is executed, i.e., a random number between [1, D-2] is produced randomly as the newly added neuron and the elements of the new neuron are calculated in terms of Eq. (6.22). If the neuron has only one nonzero element in $\boldsymbol{c}_i$, the pruning operator is implemented. Here, the neuron will be deleted and the number of the hidden neuron is reduced.

6.6.4 The Procedure of Improved NSGA-II

The whole procedure of the INSGA-II applied for RBFNN optimization is shown as follows:

Step 1: Initialize the population size N, the maximum generations G, the operators' probabilities p_c, p_m, the system parameters $u_{\min}$, $u_{\max}$, $y_{\min}$ and $y_{\max}$, then generate N chromosomes randomly.

Step 2: Select variables in terms of Eq. (6.18) and calculate J_1.

Step 3: Construct the RBFNN and calculate the value of J_2.

Step 4: Implement INSGA-II and select the parent population in terms of the front rank, crowing distance, and Roulette wheel selection operator.

Step 5: Implement the crossover and mutation operators with p_c and p_m, then the prolong and pruning operators are carried out for the offspring.

Step 6: Repeat steps 2–5 until G is met.

6.6.5 Main Disturbance Modeling of Chamber Pressure

The chamber pressure operation in the coke furnace is critical to guarantee burning security. However, its system model for advanced control is a highly complex task because nonlinear characteristics, time delay, and a lot of disturbances such as fuel volume, the coupling of pressures, etc., coexist in the unit. The input variable of the main channel is known, but the main disturbance model is especially difficult to be obtained because of the above various disturbances. How to select the key disturbance variables and construct its disturbance model is still challenging.

The pressure PRC8112A coupled with the pressure PRC8112B, the temperature in the chamber TR8109A, TR8109B, the oxygen content AR8102, ARC8101, and the external flow XLF103, are sampled. Meanwhile, the other side pressure PRC8112B with similar disturbance variables is also collected and stored in the PAI database. Since the values of different variables in Figs. 6.38a and 6.39a vary considerably large, they are normalized to [0, 1] and shown in Figs. 6.38b and 6.39b. It is obvious that the dynamic response is complex, and accurate disturbance modeling is difficult.

The INSGA-II is used to select the main disturbances and optimize both the structure and the parameters of RBFNN such that the nonlinear dynamic behavior of the disturbance of chamber pressure can be captured. The population size N is set to 60, the maximal evolution generation G is 1000, and the operator probabilities p_c

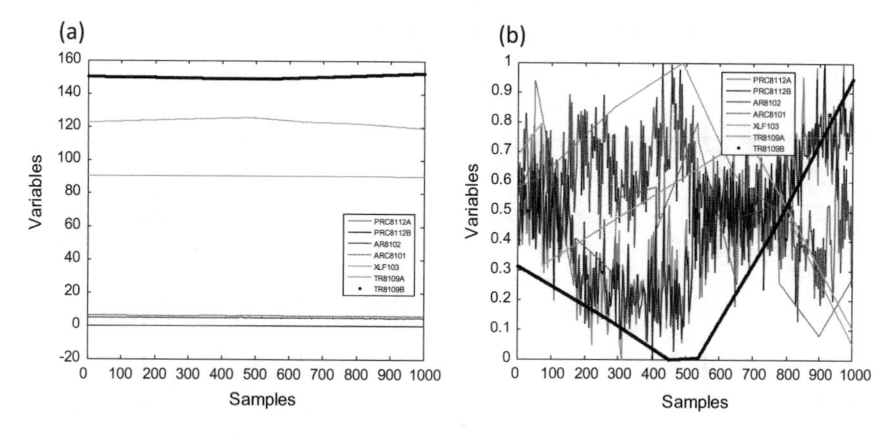

Fig. 6.38 **a** Original variables sampled in the PAI database for PRC8112A, **b** normalized variables

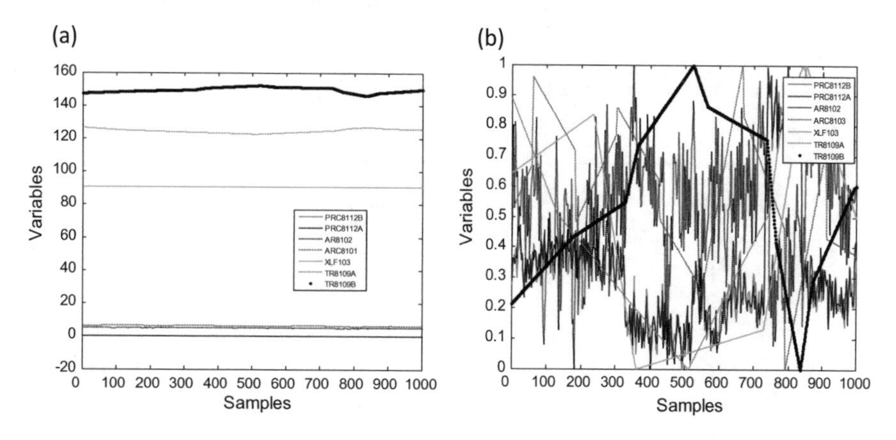

Fig. 6.39 **a** Original variables sampled in the PAI database for PRC8112B, **b** normalized variables

and p_m are set to 0.9 and 0.1, N_1, N_2 of the training data ($\mathbf{Y}_1$) and the testing data ($\mathbf{Y}_2$) are set as 400, 400, N, M in $\boldsymbol{X}$ are 6 and 800, respectively. The maximal number of the hidden nodes (D) is set to 30, $[u_{\min}, u_{\max}]$ is [0, 1] and $[y_{\min}, y_{\max}]$ is [0, 1]. The optimization result for PRC8112A and PRC8112B is a group of Pareto optimal solutions and shown in Figs. 6.40 and 6.41, respectively.

It can be seen in Figs. 6.40 and 6.41 that the RMSE of RBFNN model J_2 becomes larger with the value of RV criterion J_1 close to 1, where the number of the selected variables has been changed from 1 to 6. Though the RMSE becomes larger, the difference between the maximal and minimal values of J_1 in the Pareto frontier is not large, e.g., PRC8112A's is 0.18 and PRC8112B's is 0.35. J_2 is then used to select the final solution. The individual with the minimal value of J_2 is chosen as the final solution, *i.e.*, the individuals where (J_1, J_2) is (1.18, 0.071) in Fig. 6.40, and (J_1, J_2) is (1.18, 0.071) in Fig. 6.41, are chosen. In the selected individual, c_{D+1} is

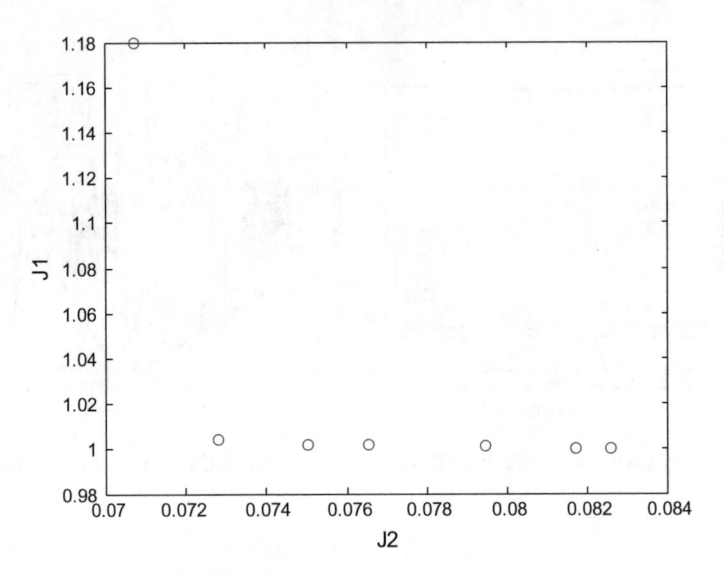

Fig. 6.40 Pareto front for PRC8112A

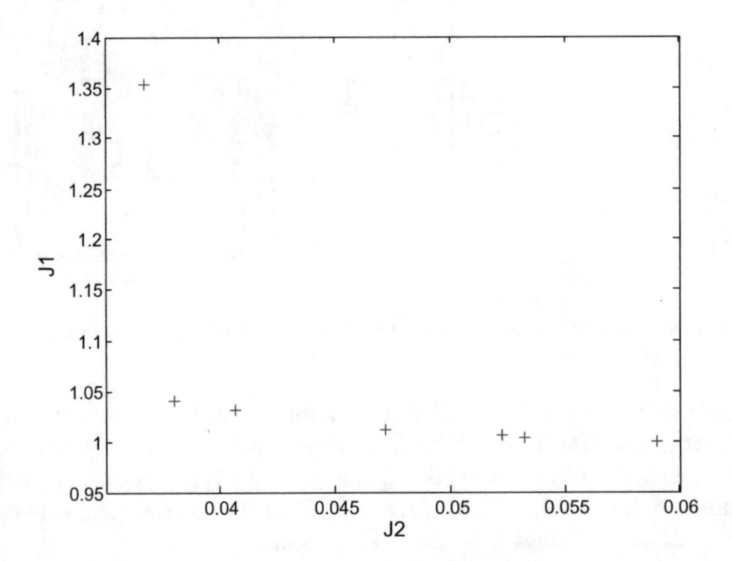

Fig. 6.41 Pareto front for PRC8112B

[00100000], which means PRC8112B is the main disturbance for PRC8112A and PRC8112A is the main disturbance for PRC8112B. Therefore, the input vector of RBFNN is $[y(k-1), y(k-2), u_1(k)]$. The model output and its modeling errors for PRC8112A and PRC8112B are plotted in Figs. 6.42 and 6.43.

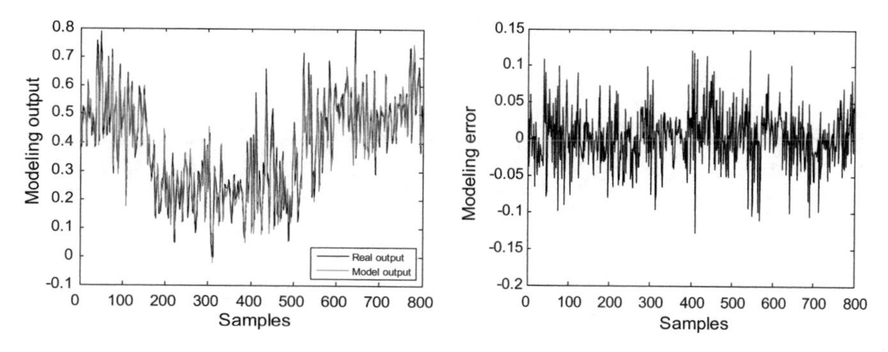

Fig. 6.42 Outputs and errors of RBF disturbance model for PRC8112A by proposed method

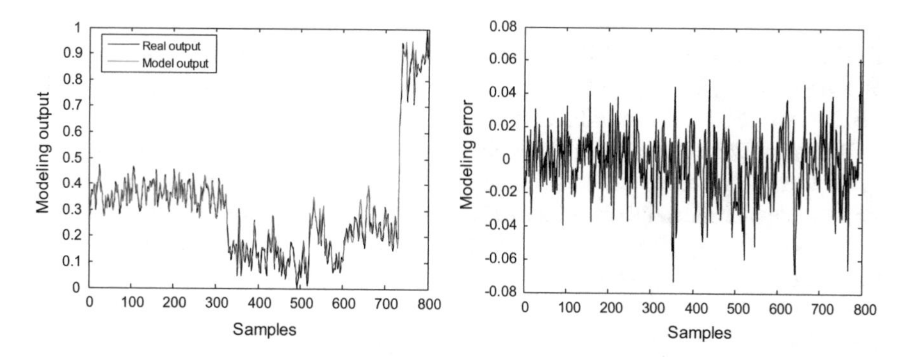

Fig. 6.43 Outputs and errors of RBF disturbance model for PRC8112B by proposed method

We also selected several methods to select the key variables and construct the disturbance model, which are PCA variable selection method [47] with RBFNN optimized by an improved GA [7] aimed at minimizing J_2 (PCAGA RBF), simulated annealing for variable selection in terms of RV criterion [23] and RBFNN modeling (SAPCA RBF), and the multilayer perceptron (MLP) neural network with least absolute shrinkage and selection operator to select the input variables (LASSO NN) [27].

In the PCAGA RBF method, the component with small eigenvalue, usually less than 0.7, is of less importance. Consequently, the variable that dominates it should be superfluous. Here, the eigenvector is [19.6760, 9.3106, 6.1678, 3.8346, 3.3994, 1.2042] for PRC8112A and [17.7668, 10.5847, 8.3911, 6.6600, 4.7317, 3.2693] for PRC8112B, obviously, all values in the eigenvector are larger than 0.7, thus, the disturbances need to be kept. The parameters of the radial basis function and the number of the hidden nodes are derived after optimization. The errors of the constructed RBFNN disturbance models for PRC8112A and PRC8112B are shown in Fig. 6.44.

When using SAPCA RBF method, the third disturbance and the fifth disturbance are selected as the main disturbance for PRC8112A and PRC8112B, respectively.

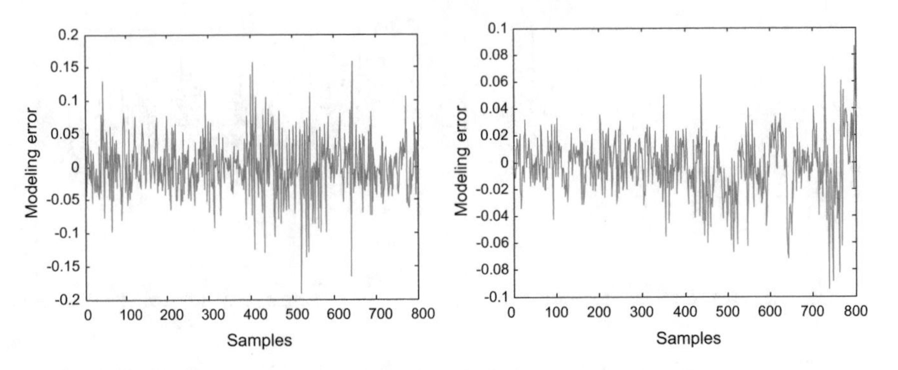

Fig. 6.44 The modeling errors of PRC8112A and PRC8112B by PCARBF method

The input vector of the RBFNN for the two-side disturbance model is $[y(k-1), y(k-2), u_3(k)]$ and $[y(k-1), y(k-2), u_5(k)]$, respectively. The modeling error of the RBFNN for PRC8112A and PRC8112B are plotted in Fig. 6.45.

When applying the LASSO NN method, the MLP neural network is first obtained by the Levenberg-Marquard learning algorithm, then the LASSO is applied to select the variables. Three main disturbances $[u_1(k), u_2(k), u_5(k)]$ for PRC8112A and four variables $[u_1(k), u_2(k), u_4(k), u_5(k)]$ for PRC8112B are obtained, and their modeling errors are given in Fig. 6.46.

It can be seen from Figs. 6.42, 6.43, 6.44, 6.45, and 6.46, that the INSGA-II method has obtained the best modeling accuracy. In order to be convenient to show the different performances of the above methods, RMSEs of the training data and test data, which are denoted as RMSE_1 and RMSE_2, the parameters of the RBFNN, the running time, and the RV criterion are listed in Table 6.5.

In Table 6.5, the number of disturbances is only one with the less similarity RV value for the INSGA-II method, while the PCA eigenvalue analysis method selected all disturbance. For the SAPCA RBF method, larger RV value is obtained than the proposed method because the RV criterion is optimized independently by SA.

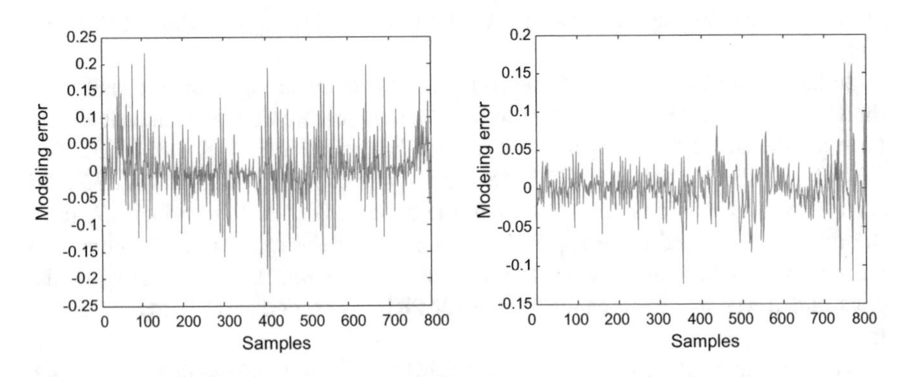

Fig. 6.45 The modeling errors of PRC8112A and PRC8112B by SAPCA RBF method

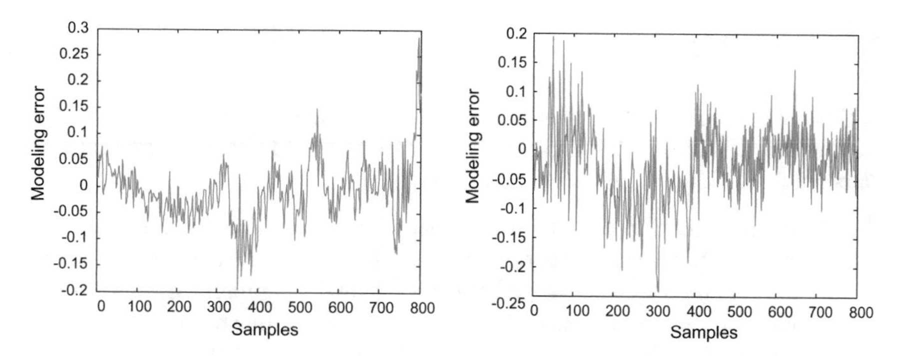

Fig. 6.46 The modeling errors of PRC8112A and PRC8112B by LASSO NN method

Table 6.5 The comparison results of 4 methods

Methods	Plant	No. of disturbances	No. of input nodes	No. of hidden nodes	$RMSE_1$	$RMSE_2$	Running time (s)	RV
Proposed	PRC8112A	1	3	7	0.019	0.020	5035	0.847
	PRC8112B	1	3	6	0.0154	0.0213	5106	0.739
PCAGA RBF	PRC8112A	6	8	20	0.032	0.043	2096	1
	PRC8112B	6	8	22	0.0145	0.0258	2198	1
SAPCA RBF	PRC8112A	1	3	6	0.0235	0.0356	606	0.953
	PRC8112B	1	3	4	0.0427	0.0477	609	0.767
LASSO NN	PRC8112A	3	5	15	0.0542	0.0599	3.92	0.991
	PRC8112B	4	6	15	0.0829	0.0422	3.89	0.996

However, the final $RMSE_1$ and $RMSE_2$ of the disturbance models are inferior to the INSGA-II's. Besides, the number of the hidden nodes of PCAGA RBF is the largest because of the complicated input structure by using eight inputs. Since the computation of INSGA-II is complicated, the running time of INSGA-II is the longest among the four methods, while the running time of the LASSO NN method without using the random search algorithm is the shortest. In summary, the multi-objective optimization method considering the RV criterion coordinated with modeling accuracy can gain a group of solutions to be selected for specific purposes. The values of $RMSE_1$ and $RMSE_2$ using the INAGA-II method are the smallest with a simple structure of using less input and hidden nodes. The modeling method is efficient in key variable selection, such as the disturbances selection, and complex system model construction.

6.7 Summary

In this chapter, temperature, pressure, and its disturbance models of coke furnace are constructed using RBF Neural Network and improved GA. All data sets were gathered from the same industrial equipment.

(1) To improve the approximation and generalization performance, an improved RNAGA is first developed to optimize the RBF neural network structure and its corresponding parameters of radial basis functions. A pruning operator is also designed to simplify the RBFNN structure. The simulation results show that the constructed RBFNN model optimized by IRNA-GA can obtain good prediction accuracy with a relatively simple network structure.
(2) An IMOEA is used for the RBFNN modeling of the chamber pressure. The encoding method is designed for the structure of the input layer, the hidden layer, and the parameters of the Gaussian basis function. The local search operators are helpful to improve the search capability, and the prolong and pruning operators are beneficial to change the hidden layer structure. The adaptive archive maintaining is applied to retain the elitists and maintain their evenness. It is efficient and easy to be implemented in industrial processes with a little prior knowledge.
(3) The disturbance selection using the RV criterion of principle analysis and RBFNN modeling for nonlinear processes are optimized by using an INSGA-II, where the RV criterion and modeling error are optimized simultaneously. In addition, the encoding, prolong, and pruning operators are adopted to make NSGA-II suitable for RBFNN optimization. Among a set of Pareto solutions, RMSE value is used to choose the final result in the Pareto frontier. The main disturbance can be selected successfully and the RBFNN has established the disturbance model with satisfactory accuracy.

References

1. Broomhead, D.S., and D. Lowe. 1988. Multivariable functional interpolation and adaptive networks. *Complex Systems* 2 (3): 321–355.
2. Wei, C., and J. Qiao. 2014. Passive robust fault detection using RBF neural modeling based on set membership identification. *Engineering Applications of Artificial Intelligence* 28 (1): 1–12.
3. Wilamowski, B.M., et al. 2015. A novel RBF training algorithm for short-term electric load forecasting and comparative studies. *IEEE Transactions on Industrial Electronics* 62 (10): 6519–6529.
4. Reiner, P., and B.M. Wilamowski. 2015. *Efficient incremental construction of RBF networks using quasi-gradient method.* 150: 349–356.
5. Ahmadizar, F., et al. 2015. Artificial neural network development by means of a novel combination of grammatical evolution and genetic algorithm. *Engineering Applications of Artificial Intelligence* 39: 1–13.
6. Wu, J., L. Jin, and M. Liu. 2015. Evolving RBF neural networks for rainfall prediction using hybrid particle swarm optimization and genetic algorithm. *Neurocomputing* 148 (2): 136–142.

7. Zhang, R., J. Tao, and F. Gao. 2014. Temperature modeling in a coke furnace with an improved RNA-GA based RBF network. *Industrial Engineering Chemistry Research* 53 (8): 3236–3245.
8. Wang, Y., and P. Yao. 2003. Simulation and optimization for thermally coupled distillation using artificial neural network and genetic algorithm. *Chinese Journal of Chemical Engineering* 11 (3): 307–311.
9. Yang, T., H.C. Lin, and M.L. Chen. 2006. Metamodeling approach in solving the machine parameters optimization problem using neural network and genetic algorithms: A case study. *Robotics Computer Integrated Manufacturing* 22 (4): 322–331.
10. Blanco, A., M. Delgado, and M.C. Pegalajar. 2001. A real-coded genetic algorithm for training recurrent neural networks. *Journal of the International Neural Network Society* 14 (1): 93–105.
11. Delgado, M., and M.C. Pegalajar. 2005. A multiobjective genetic algorithm for obtaining the optimal size of a recurrent neural network for grammatical inference. *Pattern Recognition* 38 (9): 1444–1456.
12. Yang, L. and J. Yen. 2010. An Adaptive Simplex Genetic algorithm. in *Genetic & Evolutionary Computation Conference.*
13. Esposito, A., et al. 2000. Approximation of continuous and discontinuous mappings by a growing neural RBF-based algorithm. *Neural Networks the Official Journal of the International Neural Network Society* 13 (6): 651–665.
14. Sarimveis, H., et al. 2004. A new algorithm for developing dynamic radial basis function neural network models based on genetic algorithms. *Computers Chemical Engineering Journal* 28 (1–2): 209–217.
15. Guang-Bin, H., P. Saratchandran, and S. Narasimhan. 2005. A generalized growing and pruning RBF (GGAP-RBF) neural network for function approximation. *IEEE Transactions on Neural Networks* 16 (1): 57–67.
16. Du, D., L. Kang, and M. Fei. 2010. A fast multi-output RBF neural network construction method. *Neurocomputing* 73 (10): 2196–2202.
17. Han, H.G., Q.L. Chen, and J.F. Qiao. 2011. An efficient self-organizing RBF neural network for water quality prediction. *Neural Networks the Official Journal of the International Neural Network Society* 24 (7): 717–725.
18. Chen, X., and N. Wang. 2009. A DNA based genetic algorithm for parameter estimation in the hydrogenation reaction. *Chemical Engineering Journal* 150 (2): 527–535.
19. Tao, J., and N. Wang. 2007. DNA computing based RNA genetic algorithm with applications in parameter estimation of chemical engineering processes. *Computers Chemical Engineering Journal* 31 (12): 1602–1618.
20. Dayal, B.S., and J.F. Macgregor. 1997. Improved PLS algorithms. *Journal of Chemometrics* 11 (1): 73–85.
21. Zhang, R., J. Tao, and F. Gao. 2016. A new approach of takagi-sugeno fuzzy modeling using an improved genetic algorithm optimization for oxygen content in a coke furnace. *Industrial Engineering Chemistry Research* 55 (22): 6465–6474.
22. Zhang, R., et al. 2016. New minmax linear quadratic fault-tolerant tracking control for batch processes. *IEEE Transactions on Automatic Control* 61 (10): 3045–3051.
23. Brusco, M.J. 2014. A comparison of simulated annealing algorithms for variable selection in principal component analysis and discriminant analysis. *Computational Statistics & Data Analysis* 77: 38–53.
24. Li, J., C. Duan, and Z. Fei. 2016. A Novel Variable Selection Approach for Redundant Information Elimination Purpose of Process Control. *IEEE Transactions on Industrial Electronics* 63 (3): 1737–1744.
25. Zhang, R. and J. Tao,. 2017. A nonlinear fuzzy neural network modeling approach using an improved genetic algorithm. *IEEE Transactions on Industrial Electronics 65*(7): 5882–5892.
26. Andersen, C.M., and R. Bro. 2010. Variable selection in regression—a tutorial. *Journal of Chemometrics* 24 (11–12): 728–737.
27. Sun, K., et al. 2016. Design and application of a variable selection method for multilayer perceptron neural network with LASSO. *IEEE Transactions on Neural Networks and Learning Systems 28*(6): 1386–1396.

28. Enrique, R. and S. Josep María, Romero. 2008. Performing feature selection with multilayer perceptrons. *IEEE Transactions on Neural Networks 19*(3): 431–441.
29. Souza, F.A.A., et al. 2013. A multilayer-perceptron based method for variable selection in soft sensor design. *Journal of Process Control* 23 (10): 1371–1378.
30. Estévez, P.A., et al. 2009. Normalized mutual information feature selection. *IEEE Transactions on Neural Networks* 20 (2): 189–201.
31. Zhang, R., et al. 2016. Decoupled ARX and RBF neural network modeling using PCA and GA optimization for nonlinear distributed parameter systems. *IEEE Transactions on Neural Networks Learning Systems* 29 (2): 457–469.
32. Pacheco, J., S. Casado, and S. Porras. 2013. Exact methods for variable selection in principal component analysis: Guide functions and pre-selection. *Computational Statistics Data Analysis* 57 (1): 95–111.
33. Puggini, L., and S. Mcloone. 2017. Forward selection component analysis: Algorithms and applications. *IEEE Transactions on Pattern Analysis Machine Intelligence* 39 (12): 2395–2408.
34. Chen, J. 2004. Computational aspects of algorithms for variable selection in the context of principal components. *Computational Statistics Data Analysis* 47 (2): 225–236.
35. Cadima, J.F.C.L., and I.T. Jolliffe. 2001. Variable selection and the interpretation of principal subspaces. *Journal of Agricultural Biological Environmental Statistics* 6 (1): 62–79.
36. Brusco, M.J. 2014. A comparison of simulated annealing algorithms for variable selection in principal component analysis and discriminant analysis. *Computational Statistics & Data Analysis* 77 (9): 38–53.
37. Li, C.J., et al. 2018. Near-optimal stochastic approximation for online principal component estimation. *Mathematical Programming* 167 (1): 75–97.
38. Wei-Shi, Z., L. Jian-Huang, and P.C. Yuen. 2005. GA-fisher: A new LDA-based face recognition algorithm with selection of principal components. *IEEE Transactions on Systems, Man and Cybernetics Part B* 35 (5): 1065–1078.
39. Brusco, Michael J. 2017. A comparison of simulated annealing algorithms for variable selection in principal component analysis and discriminant analysis. *Computational Statistics & Data Analysis* 77: 38–53.
40. Nchare, M., B. Shen, and S.G. Anagho. 2012. Co-processing vacuum residue with waste plastics in a delayed coking process: Kinetics and modeling. *China Petroleum Processing Petrochemical Technology* 14 (3): 44–49.
41. Bello, O.O., et al. 2006. Effects of operating conditions on compositional characteristics and reaction kinetics of liquid derived by delayed coking of Nigerian petroleum residue. *Brazilian Journal of Chemical Engineering* 23 (3): 331–339.
42. Zhang, R., and S. Wang. 2008. Support vector machine based predictive functional control design for output temperature of coking furnace. *Journal of Process Control* 18 (5): 439–448.
43. Zhang, R., and J. Tao. 2017. Data-driven modeling using improved multi-objective optimization based neural network for coke furnace system. *IEEE Transactions on Industrial Electronics* 64 (4): 3147–3155.
44. Zhang, R., Q. Lv, J. Tao, et al. 2018. Data driven modeling using an optimal principle component analysis based neural network and its application to a nonlinear coke furnace. *Industrial and Engineering Chemistry Research* 57 (18): 6344–6352.
45. Tao, J., X. Chen, and Z. Yong. 2012. Constraint multi-objective automated synthesis for CMOS operational amplifier. *Neurocomputing* 98 (18): 108–113.
46. Li, Y., C. Fang, and Q. Ouyang. 2004. Genetic algorithm in DNA computing: A solution to the maximal clique problem. *Chinese Science Bulletin* 49 (9): 967–971.
47. Martin, N. and H. Maes. 1979 *Multivariate analysis*. London: Academic Press.

Chapter 7
GA Based Fuzzy Neural Network Modeling for Nonlinear SISO System

Abstract Fuzzy neural networks are quite useful for nonlinear system identification with only input/output data information available. A fuzzy neural network and its improved framework are proposed and the improved genetic algorithms are designed for the structure and parameter optimization to catch the unknown plant dynamics. The hybrid encoding/decoding, neighborhood search operator and maintaining operator are presented to optimize the structure of the input layer, fuzzy rule layer and the parameters of the membership functions together. The liquid level and oxygen content modeling problems in the industrial coke furnace described in Ch. 6 are utilized to compare the performance of several methods. Simulation results show that GA optimized fuzzy neural network is superior in modeling precision and generalization capability. Fuzzy neural networks are quite useful for nonlinear system identification with only input/output data information available. A fuzzy neural network and its improved framework are proposed and the improved genetic algorithms are designed for the structure and parameter optimization to catch the unknown plant dynamics. The hybrid encoding/decoding, neighborhood search operator and maintaining operator are presented to optimize the structure of the input layer, fuzzy rule layer and the parameters of the membership functions together. The liquid level and oxygen content modeling problems in the industrial coke furnace described in Ch. 6 are utilized to compare the performance of several methods. Simulation results show that GA optimized fuzzy neural network is superior in modeling precision and generalization capability.

7.1 Introduction

Takagi-Sugeno (T-S) fuzzy neural network is a universal approximation tool, which is widely used in system modeling [1–5]. By using fuzzy modeling approach with satisfactory prediction accuracy [6–10], a large number of nonlinear systems have been approximated. When considering the performance of fuzzy neural networks, T-S fuzzy system and their improved forms are especially outstanding. Some newly presented fuzzy frameworks, such as interval type-2 radial basis function neural network [11], fuzzy neural network with correlated fuzzy rules [12], four layers

J. Tao et al., *DNA Computing Based Genetic Algorithm*,
https://doi.org/10.1007/978-981-15-5403-2_7

network featured Takagi-Sugeno-Kang fuzzy architecture with multivariate Gaussian kernels [13], wavelet fuzzy neural network [14], etc., were proposed and are still in the process of exploration to obtain a compact fuzzy neural network with the advantage of smaller number of fuzzy rules.

Once the framework of the fuzzy model is fixed, focusing on the structure optimization and parameter identification is the most important part in system modeling. As for structure evolving learning method, a starting point of 2^n fuzzy rules was initialized to deal with the exponential increase of fuzzy rules [15]; The initial rule base in the interval type-2 fuzzy neural network was empty, and the online clustering method was presented to generate fuzzy rules that flexibly partition the input space [16]. Only one fuzzy rule was started in a fuzzy system identification problem, if the simplified structure evolving method was applied [17]. In terms of spiking intensity and relative mutual information, a set of fuzzy rules were obtained via learning the structure and parameters simultaneously in a self-organizing fuzzy system [18]. In an adaptive neuro-fuzzy inference system, using new adding/pruning techniques could derive a high accuracy with compact structure [19]. Moreover, a fuzzy clustering method was used to generate fuzzy rules, which flexibly partitioned the input space and implemented the structure identification [20, 21]. An adaptive second order gradient learning algorithm was presented to decide the widths, centers and output weights [22]. However, the structure evolving algorithm are usually in more coarse partition for the fuzzy region among these methods; As for the clustering method, only the input space is considered when used in system modeling; while for the back propagation algorithm, it is prone to trap into local optimum and is limited to the given system structure. In addition, the inefficient input variables cannot be eliminated in the above methods.

Generally, the structure identification should be the difficult section in fuzzy modeling. As a global optimization method, GA and other evolutionary computing algorithms, such as Tabu search algorithm [23], particle swarm optimization (PSO) [24], artificial bee colony algorithm [25], and differential evolution [26] etc., are capable of obtaining a global optimal fuzzy neural network. Moreover, the structure operations can be easily involved in the gene operators [27–29]. However, most of them only focused on the partition of input space, i.e., partial structure identification. How to determine the whole structure which includes the inputs, the linguistic partitioning, the fuzzy rule set and the consequent part, is still quite challenging, especially for the nonlinear system with more complex dynamics, such as the oxygen content, liquid level of the coke furnace. It is also a difficult problem to develop a simple but efficient fuzzy model that can be used in system identification and controller design in the industrial field.

A fuzzy model based on input-output data is desired to have both precise modeling accuracy and simple structure [30–33]. It can be regarded as a bi-objective optimization problem, which can be solved by multi-objective optimization algorithms, such as weighted sums of objectives (WSO) [34], multi-objective evolution algorithms (MOEAs), etc. [35–37]. NSGA-II is superior to several representative algorithms among MOEAs [38]. However, its computing complexity is relevant to o(MN), where M is the number of objectives and N is the population size. Its Pareto optimal set includes a set of solutions, and need to be selected by decision makers. The bi-objective optimization problem can be transformed to single objective problem by the WSO method, and its weighting coefficient is critical to the final optimization solution. For T-S fuzzy model, its model information can also be obtained and the modeling accuracy can be predicted by expert *experience*. Thus, the weighting coefficient is easy to choose.

In this chapter, GA is used to optimize the structure and parameters of fuzzy model. Besides, autoregressive with exogenous input (ARX) plus Tanh function fuzzy model is presented to obtain satisfied nonlinear approximating capability. The specific encoding/decoding method, crossover, mutation, neighborhood search and maintain operators are designed together for the synchronous optimization of the structure and parameters on the premise of modeling accuracy. And the oxygen content and liquid level modeling in the industrial coke furnace are applied and compared with typical fuzzy modeling methods.

7.2 T-S Fuzzy Model

T-S fuzzy model possesses the powerful approximating capability with good predictive feature. A nonlinear mapping between the past input-output data and the predicted output is given below:

$$\hat{y}(k) = f(\mathbf{X}(k)) \tag{7.1}$$

where $\mathbf{X}(k) = [y(k-1), \cdots, y(k-n), u(k-d), u(k-d), \cdots, u(k-d-m)]$, m and n are the maximal lags considered for the input and output terms, d is the discrete time delay, and f represents the nonlinear relation of the fuzzy model.

7.2.1 T-S Fuzzy ARX Model

By using the ARX model structure, the T-S fuzzy ARX model often interpolates local linear time-invariant (LTI) ARX submodel, and the jth IF-THEN fuzzy rule is shown as follows:

Rule j: If $x_1(k)$ is A_{1j} and $x_2(k)$ is A_{2j} and … and $x_s(k)$ is A_{sj}, then.

$$f_j(k) = \mathbf{B}^{\mathrm{T}}\mathbf{X}(k),\ j = 1, 2, \cdots, M, M \le \prod_{i=1}^{s} m_i$$

where $B_j = [a_1^j, a_2^j, \cdots, a_n^j, b_1^j, b_2^j, \cdots, b_m^j]^{\mathrm{T}}$, the input vector $\boldsymbol{x}(k) = [x_1(k), \ldots, x_s(k)]$ is usually a subset of $\boldsymbol{X}(k)$, namely, $\boldsymbol{x}(k) \in \boldsymbol{X}(k)$, m_i is the number of membership functions of $x_i(k)$, and M is the number of fuzzy rules. The final output of the fuzzy model by a weighted mean defuzzification can be expressed as:

$$\hat{y}(k) = \frac{\sum_{j=1}^{M} \alpha_j[\mathbf{x}(k)] f_j(k)}{\sum_{j=1}^{M} \alpha_j[\mathbf{x}(k)]} \tag{7.2}$$

where $\alpha_j[\mathbf{x}(k)]$ delegates the overall value of the premise part of the jth implication for the input $\boldsymbol{x}(k)$ in the fuzzy inference system (FIS) A_j, $A_j = \prod_{i=1}^{s} A_{ij}$, and $\alpha_j[\mathbf{x}(k)]$ can be calculated as:

$$\alpha_j[\mathbf{x}(k)] = \mu_1^j \mu_2^j \cdots \mu_s^j \tag{7.3}$$

A Gaussian membership function is chosen and shown as follows:

$$\mu_i^j = \exp\left[-\frac{||x_i - c_{ij}||^2}{\sigma_{ij}^2}\right] \tag{7.4}$$

where c_{ij} and σ_{ij} are the center and width of the Gaussian function in A_{ij} respectively. Define fuzzy basis function (FBF) as:

$$\varphi_j[\mathbf{x}(k)] = \frac{\alpha_j[\mathbf{x}(k)]}{\sum_{i=1}^{M} \alpha_i[\mathbf{x}(k)]} \tag{7.5}$$

As a linear combination of FBFs for the fuzzy consequent ARX submodel, the output $\hat{y}(k)$ can be rewritten in the following form:

$$\hat{y}(k) = \sum_{j=1}^{M} \varphi_j[\mathbf{x}(k)] f_j(k) \tag{7.6}$$

The input vector, the number of fuzzy rules and its parameters of membership functions determine the fuzzy premise part. The ARX submodel structure and its parameters comprise the consequent part. Once the complete fuzzy premise part and ARX submodel structure are determined, RLS method in Ch. 6 can be utilized to determine the parameters of ARX submodel in terms of input-output data set.

Denote

$$\boldsymbol{\theta} = [\mathbf{B}_1^{\mathrm{T}} \mathbf{B}_2^{\mathrm{T}} \cdots \mathbf{B}_M^{\mathrm{T}}]^{\mathrm{T}} \tag{7.7}$$

$$\boldsymbol{\Phi}(k) = [\varphi_1[\mathbf{x}(k)]\mathbf{X}(k)^{\mathrm{T}},\ \varphi_2[\mathbf{x}(k)]\mathbf{X}(k)^{\mathrm{T}}, \cdots,\ \varphi_M[\mathbf{x}(k)]\mathbf{X}(k)^{\mathrm{T}}]^{\mathrm{T}} \tag{7.8}$$

Substitute Eqs. 7.7 and 7.8 into Eq. 7.6 yields:

$$\hat{y}(k) = \mathbf{\Phi}(k)^{\mathrm{T}}\boldsymbol{\theta} \tag{7.9}$$

If there are N_1 sampling outputs $\mathbf{Y} = [y(1), y(2), \ldots, y(N_1)]$, the value of $\boldsymbol{\theta}$ can be obtained by RLS method after N_1 iterations:

$$\begin{cases} \boldsymbol{\theta}(k) = \boldsymbol{\theta}(k-1) + \mathbf{K}(k)[y(k) - \mathbf{\Phi}^{\mathrm{T}}(k)\boldsymbol{\theta}(k-1)] \\ \mathbf{K}(k) = \mathbf{P}(k-1)\mathbf{\Phi}(k)[\mathbf{\Phi}^{\mathrm{T}}(k)\mathbf{P}(k-1)\mathbf{\Phi}(k) + 1]^{-1} \\ \mathbf{P}(k) = \mathbf{P}(k-1) - \mathbf{K}(k)\mathbf{K}^{\mathrm{T}}(k)[\mathbf{\Phi}^{\mathrm{T}}(k)\mathbf{P}(k-1)\mathbf{\Phi}(k) + 1] \end{cases} \tag{7.10}$$

where $k = 1, 2, \ldots, N_1$, $\mathbf{K}(0)$ is set as relatively small values of $(m+n)M$-by-1 vector and $\mathbf{P}(0)$ is set as big values of $(m+n)M$-by-$(m+n)M$ matrix.

7.2.2 T-S Fuzzy Plus Tah Function Model

The fuzzy model in Sec. 7.2.1 is essentially a combination of the local linear submodels. Taking the local nonlinear characteristics of system model into consideration, a discrete time form of the system dynamics can be constructed by a linear ARX model plus a nonlinear function f:

$$\begin{aligned} y(k) = &-a_1 y(k-1) - \cdots - a_n y(k-n) + b_1 u(k-d) + \cdots \\ &+ b_m u(k-d-m) + f(y(k), u(k)) \end{aligned} \tag{7.11}$$

where f is a smooth and bounded nonlinear function. According to the universal approximation theorem [39], a fuzzy neural network (FNN) $\phi(y(k), u(k), \boldsymbol{\theta})$ can approximate f with arbitrary precision, where $\boldsymbol{\theta}$ is the parameter vector of FNN. Suppose the neural network linearly parameterizable, i.e., $\phi(y(k), u(k), \boldsymbol{\theta})$ can be expressed as $\phi(y(k), u(k))\boldsymbol{\theta}$. If the approximation capability ε of the nonlinear function is given, the main objective of FNN modeling is to find a vector $\boldsymbol{\theta}$ such that:

$$\| f(y(k), u(k)) - \phi(y(k), u(k))\boldsymbol{\theta} \| \le \varepsilon \tag{7.12}$$

Accordingly, the objective now is to design a fuzzy neural network denoted as:

$$\begin{aligned} y(k) = &-a_1 y(k-1) - \cdots - a_n y(k-n) + b_1 u(k-d) \\ &+ \cdots + b_m u(k-d-m) + \phi(y(k), u(k))\boldsymbol{\theta} \end{aligned} \tag{7.13}$$

Remark 7.1: The advantage of the structure in Eq. 7.13 is that it is only to identify one simplified nonlinear function $\phi(y(k), u(k))$ that depends on (y, u), rather than identifying f in Eq. 7.11.

Given this model framework, it is necessary to obtain an algorithm and a fuzzy neural network, and the fuzzy neural network should use input/output samplings to identify the parameters of the ARX linear part, the nonlinear function ϕ and its corresponding parameters $\boldsymbol{\theta}$. Given the system to be identified described in Eq. 7.13, the rule of the fuzzy neural network is defined as follows:

Rule j: If $y(k)$ is A_{1j} and $u(k)$ is A_{2j} then $y^j(k) = \mathbf{B}_j^{\mathrm{T}}\mathbf{X}(k) + \rho(\mathbf{X}(k))\boldsymbol{\theta}_j$, $j = 1, 2, \cdots, M$

where $\mathbf{X}(k) = [y(k-1), \cdots, y(k-n), u(k-d), u(k-d), \cdots, u(k-d-m)]$, $\boldsymbol{\theta}_j = [\theta_{1j}, \theta_{2j}, \cdots, \theta_{m+n,j}]$, ρ is selected as a smooth Tanh function with the output domain $[-1, 1]$. A_{1j}, A_{2j} are Gaussian membership functions of fuzzy sets in Eq. 7.4. Note that the inputs of the fuzzy premise part are fixed, i.e., $s = 2$, $x_1 = y(k)$, $x_2 = u(k)$. Denote

$$\Theta = [B_1^{\mathrm{T}}, B_2^{\mathrm{T}}, LB_M^{\mathrm{T}}, \theta_1^{\mathrm{T}}, \theta_2^{\mathrm{T}}, L\theta_M^{\mathrm{T}}]^{\mathrm{T}}$$

$$\boldsymbol{\Phi}(k) = [\varphi_1(k)\mathbf{X}(k)^{\mathrm{T}} \cdots \varphi_R(k)\mathbf{X}(k)^{\mathrm{T}}, \varphi_1(k)\rho(\mathbf{X}(k))^{\mathrm{T}} \cdots \varphi_M(k)\rho(\mathbf{X}(k))^{\mathrm{T}}]^{\mathrm{T}}$$

The parameter calculation procedure for $\mathbf{B}_j$ and $\boldsymbol{\theta}_j$ is similar to Eqs. 7.9 and 7.10.

7.3 Improved GA based T-S Fuzzy ARX Model Optimization

If the system is nonlinear, we can select the T-S fuzzy ARX model at first. Then the model structure and parameters are optimized using an improved GA (IGA). The framework of the whole system optimization for the coke unit is shown in Fig. 7.1.

The structure incorporating parameter identification is the most difficult part in fuzzy modeling approaches. An IGA is proposed to simultaneously determine the structure and parameters of the fuzzy system using the hybrid encoding and several gene operators, also the structure complexity is taken into count for the modeling accuracy.

In Fig. 7.1, select the system output $y(k-1), \cdots, y(k-n)$ and the system input $u(k), \cdots, u(k-m)$ as the input vector of the T-S fuzzy ARX model, where the model prediction output $\hat{y}(k)$ and system output $y(k)$ is compared, and the modeling error $e(k)$ can be obtained. The modeling error and the T-S Fuzzy model structure complexity are two objectives of IGA, which can be changed into single objective by choosing the appropriate weighting coefficient in terms of WSO method.

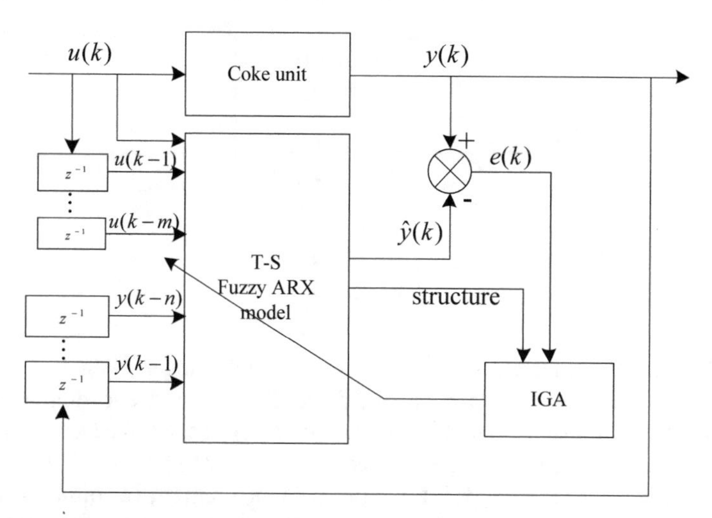

Fig. 7.1 Block diagram of IGA based T-S fuzzy model identification

It is clear that with the dimension increase of the input vector, the number of fuzzy rules, ARX submodel structure and the system complexity will increase exponentially. Moreover, the rules based on expert knowledge are usually complex, some even become impossible without necessary and enough knowledge especially for complex industrial applications. Determining the input variables, the linguistic partitioning, the rule set and ARX submodel structure together involves a complex search space, which is not an easy object to be optimized.

7.3.1 Hybrid Encoding Method

In the T-S fuzzy ARX model, considering the similarity of $u(k-1), \ldots, u(k-m)$ and $y(k-1), \ldots, y(k-n)$, the input vector $\boldsymbol{x}(k)$ is initially chosen as $[y(k-1), u(k-1)]$, which is denoted as m_1 and is to be optimized, d is set as 1, and m and n in $\mathbf{X}(k)$ are generally set in advance according to a priori knowledge. Since m and n will directly determine the approximation capability of the fuzzy model, they are also optimized in the limitation input vector of $[y(k-1), u(k-1)]$. Moreover, the number of fuzzy rules and their parameters in Eq. 7.6 are also included in the encoding method. The ith chromosome for encoding the whole fuzzy model is then designed:

$$C_i = \begin{bmatrix} c_{11} & c_{12} & \sigma_{11} & \sigma_{12} \\ c_{21} & c_{22} & \sigma_{21} & \sigma_{22} \\ \vdots & \vdots & \vdots & \vdots \\ c_{r1} & c_{r2} & \sigma_{r1} & \sigma_{r2} \\ \vdots & \vdots & \vdots & \vdots \\ 0 & 0 & 0 & 0 \\ n & 0 & m & 0 \end{bmatrix} \tag{7.14}$$

where $i = 1, 2, \cdots, N$, m_1, m and n are the positive integers with $1 \le m_1 \le 2$, $1 \le m \le 4$, $1 \le n \le 4$. m and n are adopted one-bit quaternary encoding (0, 1, 2, 3), where the decoding is just to add one to the quaternary encoding. The optimization of m_1 is essential to implement the selection of the input vector. If m_1 is 1, the input vector becomes $x(k) = [y(k\text{-}1)]$, c_{i2} in column 2 and σ_{i4} in column 4 are then set to zeros; else the input vector becomes $\boldsymbol{x}(k) = [y(k-1), u(k-1)]$. Here, m_1 is enumerated the above two cases. r is the number of fuzzy rules generated randomly between [1, 9], the rows in [r+1, 9] are also set to zeros. Though C_i is a 4-by-10 matrix, there are actually at most $r \times 4 + 2$ parameters to be optimized. The elements c_{ij}, σ_{ij} in Eq. 7.14 can be initialized as follows:

$$c_{ij} = \begin{cases} y_{\min} + \delta(y_{\max} - y_{\min}) & 1 \le i \le r,\ j = 1 \\ u_{\min} + \delta(u_{\max} - u_{\min}) & m_1 \ne 0,\ 1 \le i \le r,\ j = 2 \end{cases}$$

$$\sigma_i = \begin{cases} 0.1 + \delta(y_{\max} - 0.1) & 1 \le i \le r,\ j = 3 \\ 0.1 + \delta(u_{\max} - 0.1) & m_1 \ne 0,\ 1 \le i \le r,\ j = 4 \end{cases} \tag{7.15}$$

where δ is a random number produced between 0 and 1, $u_{\min}$ and $u_{\max}$ are the minimum and maximum of the system inputs, and $y_{\min}$ and $y_{\max}$ are the minimum and maximum of the system outputs. Therefore, the input vector, the fuzzy rules and ARX submodel structure are included in the given hybrid encoding method. The parameters $\boldsymbol{\theta}$ for ARX submodel can be obtained by RLS algorithm in terms of Eq. 7.10. Thus, N T-S fuzzy model can be obtained $(\mathbf{C}_1, \boldsymbol{\theta}_1), \cdots, (\mathbf{C}_N, \boldsymbol{\theta}_N)$.

7.3.2 Objectives of T-S Fuzzy Modeling

For the purpose of the improvement of the modeling accuracy and its generalization capacity, the samples are equally divided into two groups, where the former 1/2 data (Y_1) is selected to calculate the model parameter $\boldsymbol{\theta}$, and the latter 1/2 data (Y_2) is chosen to evaluate its generalization capacity at every generation. These two objectives considering the structure simplification and modeling precision are the same as those in Sec. 6.5.2.

The objective function is then derived using the weight-sum method, shown as follows:

$$\mathrm{Min}J(C_i) = \sqrt{\sum_{i=1}^{N_1}\left|Y_1(i) - \overset{\wedge}{Y_1}(i)\right|^2 \Big/ N_1} + \sqrt{\sum_{i=1}^{N_2}\left|Y_2(i) - \overset{\wedge}{Y_2}(i)\right|^2 \Big/ N_2} + \omega(m+n)r \quad (7.16)$$

The objective in Eq. 7.16 is composed of two performances of the fuzzy model. The first one is the sum of Root Mean Squared Error (RMSE) for Y_1 and Y_2, where $Y_1(i)$ $(i = 1, \ldots, N_1)$ are the samples in the training dataset Y_1, $\boldsymbol{\theta}$ can be gained according to Y_1, and $\hat{Y}_1(i)$ $(i = 1, \cdots, N_1)$ are then obtained as the predictions of T-S fuzzy model. Keep $\boldsymbol{\theta}$ invariant, $\hat{Y}_2(i)$ $(i = 1, \cdots, N_2)$ can be derived by the same fuzzy model. It shows that a testing procedure is incorporated in the objective function, which guarantees the generalization performance of the constructed fuzzy model. The second part expresses the structure complexity of the fuzzy system, and ω is a weighting coefficient between (0, 1], which reflects the importance degree of the structure complexity. Since the order of the magnitude of RMSE for the fuzzy model can be obtained easily, and the range of structure parameters (m, n, r) is known, ω is selected to propel the value of the second part ten times less than the order of magnitude of RMSE for the sake of the modeling accuracy. For example, the order of magnitude of RMSE is 10^{-2}, the order of magnitude of $(m + n)r$ is 10, then ω can be set as 10^{-4}.

7.3.3 *Operators of IGA for T-S Fuzzy Model*

In addition to traditional crossover and mutation operators, improved selection operator, the dynamical mutation probability and maintain operator have been designed for the optimization of fuzzy model, which may improve the weak local-search capability and avoid premature convergence of SGA.

(1) Selection operator

Roulette wheel method is mostly used as the selection operator, and the selection probability of an individual $\mathbf{C}_i$, can be calculated as follows.

$$p(\mathbf{C}_i) = \frac{f(\mathbf{C}_i)}{\sum_{i=1}^{N} f(\mathbf{C}_i)}, \quad f(C_i) = \frac{1}{J(C_i)} \quad (7.17)$$

In Eq. 7.17, the individual with better performance index, i.e., smaller value of the objective function, bigger probability to survive. A set of individuals can be selected as the parents for reproduction of offspring. To keep the population diversity, 3 N/4 parents are picked according to Roulette wheel method, while the remaining N/4 parents are chosen from the worst N/4 individuals to keep the population diversity.

The elitist, namely, the individual with the smallest value of objective function, is directly selected as the parent.

(2) Crossover and mutation operators

The crossover operation is executed with probability p_c which is between current individual C_i and the next individual C_{i+1}, and p_c is set as 0.9. The offspring $C_i^{'}$, $C_{i+1}^{'}$ are calculated after the crossover operator.

$$\begin{aligned} \mathbf{C}_i^{'} &= \alpha\mathbf{C}_i + (1-\alpha)\mathbf{C}_{i+1} \\ \mathbf{C}_{i+1}^{'} &= (1-\alpha)\mathbf{C}_i + \alpha\mathbf{C}_{i+1} \end{aligned} \tag{7.18}$$

where α is initiated randomly, specifically, $\alpha \in (0, 1)$, and n and m are rounded to the nearest integers. The input vector, ARX structure and the rules are changed dynamically during the crossover operation process. However, the operator will be prone to produce more fuzzy rules, and some irrational ones can also be yielded.

For a better exploration, the individual is mutated with different mutation probability p_{m_i}. To keep the better individual, the mutation probability is assigned in terms of the value of objective function, the better individual is set to a smaller mutation probability, shown as follows:

$$p_{m_i} = p_{m0} - \frac{i}{N}\Delta p_m \tag{7.19}$$

where p_{m0} is the initial mutation probability, Δp_m is the maximum change rate, $i = 1, \cdots, N$, and the individuals are sorted ascending according to the value of objective function. Once the mutation is operated, m, n are mutated within the range of quaternary encoding, r is kept invariant, and the elements of mutated individual are reproduced in terms of Eq. 7.15.

(3) Maintain operator

Essentially, GA is a random optimization algorithm, so there exists some irrational fuzzy systems during the optimization process. In the meantime, crossover operator cannot produce new structure of fuzzy rules, thus the maintain operator is designed.

1. Calculate $\Delta c_{ij} = \left|c_{ij} - c_{i,j+1}\right|$, if $\Delta c_{ij} < 0.03$, then c_{ij} is deleted, and the number of the fuzzy rules (r) is decreased.
2. If the number of the fuzzy rules is less than 2, a random Δr is produced satisfying $r + \Delta r \leqq 9$, and the elements in the new rules are produced according to Eq. 7.15.
3. If all the coefficients in $\boldsymbol{B}_j$ are less than 0.003, the submodel for rule j is regarded inactive, and the corresponding rule is deleted.

7.3.4 Optimization Procedure

The whole optimization process for IGA is described as the following steps:

Step 1: Setting m_1 as 1, the input vector becomes $x(k) = [y(k-1)]$, or Set m_1 as 2, the input vector becomes $x(k) = [y(k-1), u(\mathrm{k}-1)]$.

Step 2: Initialization of the maximal generation G and the population size N. Generation of the chromosomes randomly in the search space for the initial population.

Step 3: Decoding of the chromosome to generate N fuzzy model in terms of Eqs. 7.2–7.10 and computation of the performance J for each individual.

Step 4: Selection of the chromosomes to generate 3 $N/4$ parent chromosomes of the next generation, which is according to Roulette wheel selection. The worst $N/4$ individuals are directly inherited to keep population diversity, but trapping in the local optimization solution early should be avoided.

Step 5: Execution of the crossover operator. Repeat it for all the $p_c \times N/2$ pairs of parents, and implement the mutation operator with dynamical mutation probability.

Step 6: Carrying out maintain operator in the new individuals to improve the quality of fuzzy model generated by crossover and mutation operators.

Step 7: Repeat steps 3–6 until the set maximum evolution generation G is met. Moreover, elitism, the inclusion of the best individual in the next population, is used throughout the optimization process.

Step 8: Comparing the optimization result at cases of $m_1 = 1$ and $m_1 = 2$, choose the one with less value of fitness function as the final fuzzy model.

7.3.5 Computing Complexity Analysis

Consider the computation complexity for executing one iteration under the above optimization process. The time-consuming parts are mainly ascribed to the RLS and sort algorithms, and the worst-case complexity can be analyzed as follows:

1. RLS algorithm for one individual is $O(N_1(2M)(N_m + N_n))$ and N individuals have the complexity $O(NN_1(2M)(N_m + N_n))$.
2. Sorting algorithm of N individuals is $O(NM^2)$.

The overall computing complexity of the algorithm is $O(NN_1M(N_m + N_n))$, and the number of training data N_1 and the population size N will affect the running time. In fact, M and $(N_m + N_n)$ are more than r and $(m + n)$ in the evolutionary processes, then, the computing complexity can be much less than the worst-case complexity.

7.3.6 Oxygen Content Modeling by Fuzzy ARX Model

In order to validate the effectiveness of the optimal fuzzy ARX method, the IGA is run for 10 times. The parameters of IGA are set as follows: population size $N = 40$, maximal evolution generations $G = 1000$, the crossover probability $p_c = 0.9$, p_m is dynamic changing, p_{m0} is set to 0.2, and Δp_m is set to 0.1. The parameter range of the fuzzy ARX model is utilized with an expert knowledge, in other words, the maximal number of fuzzy rules is set as 9, the input vector is chosen from $[y(k-1), u(k-1)]$, and the maximal n and m are set as 4 and 4, respectively. At each running, the parameters of IGA, fuzzy ARX model and the training data set are kept invariant. In the case of oxygen content modeling, ω is set as 0.001, which is an order magnitude less than that of the expected RMSE of fuzzy model. The fuzzy model with the smallest RMSE of testing data is deemed as the best result. Comparing with a number of fuzzy models in the literature, mean square error (MSE) is calculated as the performance index of the IGA optimized fuzzy model, which is defined as the square of RMSE.

Oxygen content modeling is critical to realize the model-based advanced control. The main channel and its disturbance channel are constructed here. The experimental data is also gathered from the distributed control system (DCS) CS3000 in the same industrial coking unit of a refinery. 1200 input-output measurement samples are gained for oxygen content modeling, which is shown in Fig. 7.2a, b. At each sample, 2 inputs (the outlet oxygen content and the inlet blower valve opening for the main channel) and 2 outputs (the perturbation of oil flow and its corresponding oxygen content perturbation for the disturbance channel) are contained respectively. The input-output samples in Fig. 7.2 are normalized, and then divided equally into 2 groups, each one includes 600 input-output data.

Now that there are two objectives in nature, IGA is compared with a typical multi-objective genetic algorithm, namely, NSGA-II. The maintain operator is also applied in NSGA-II and the parameters of ARX submodel are derived using RLS method. The

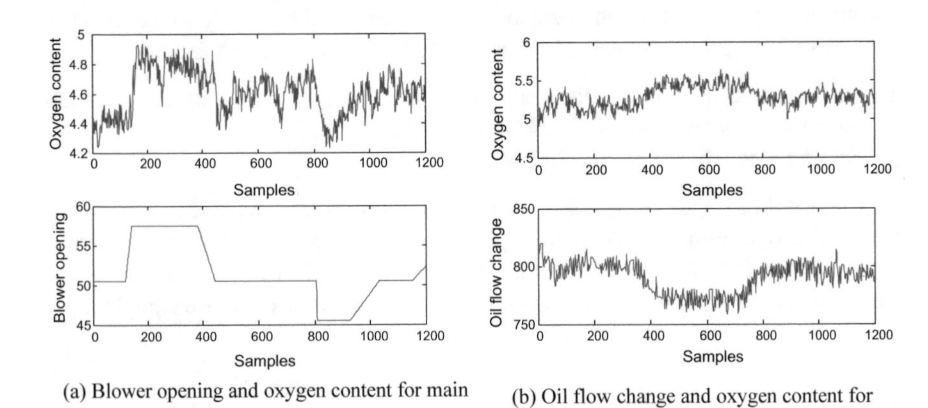

(a) Blower opening and oxygen content for main channel.

(b) Oil flow change and oxygen content for disturbance channel

Fig. 7.2 1200 input-output samples for system modeling

Pareto frontier of the main channel and disturbance channel are illustrated in Figs. 7.3 and 7.4. Since the range of structure parameter (m, n, r) is relatively small, there are only few solutions in the Pareto frontier. Furthermore, fuzzy c-means method is adopted to train the centers of Gaussian membership functions, while other fuzzy parameters, such as input vector, the number of fuzzy rules and ARX submodel structure, are set to be the same as the best results optimized by IGA.

The best fuzzy models obtained by the mentioned three methods are listed in Table 7.1, where RMSE is calculated for the testing data (Y_2). In Table 7.1, the fuzzy model of IGA can obtain better prediction precision than NSGA-II's in terms of RMSE. However, the structure of fuzzy model using NSGA-II is simplifier than that

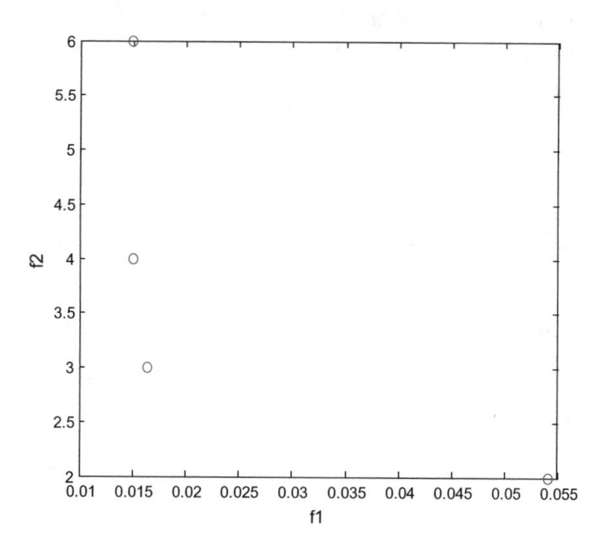

Fig. 7.3 The pareto front of main channel by NSGA-II

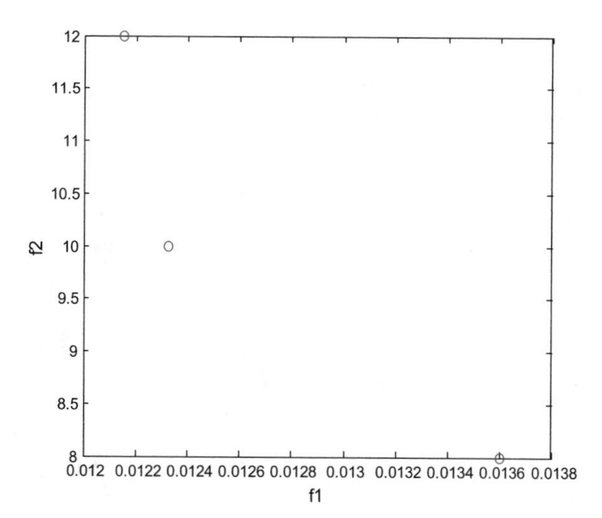

Fig. 7.4 The pareto front of disturbance channel by NSGA-II

Table 7.1 Comparison of the best simulation results with 3 methods

Methods	Main channel	Disturbance channel
	m_1 n m r RMSE	m_1 n m r RMSE
IGA	2 3 1 3 4.5e−3	1 3 3 3 4.4e−3
NSGA-II	1 1 1 3 6.1e−3	1 1 2 4 7.0e−3
c-means	1 3 1 4 4.5e−3	1 3 3 3 4.4e−3

of IGA. Because fuzzy c-means method utilizes the optimized structure parameters of IGA, the RMSEs of main channel and disturbance model by using IGA are similar to those of c-means method. The statistical results in 10 runs are listed in Table 7.2. It can be seen that the mean of RMSE of IGA is superior to that of NSGA-II, partly because the fuzzy model with complex ARX submodel gains better approximation capability. The simple fuzzy model structure and satisfied modeling precision are obtained by IGA. Specifically, the running time of IGA is much shorter than that of NSGA-II.

To reflect the prediction accuracy of the established model, the comparisons of the predicted oxygen content with the measured data of the testing set for the main channel and its disturbance channel are given in Figs. 7.5, 7.6, 7.7 and 7.8,

Table 7.2 The performance comparisons of 3 methods for 10 runs

Methods	Main channel	Disturbance channel
	m_1 n m r $\bar{T}(s)$ $\overline{RMSE}$	m_1 n m r $\bar{T}(s)$ $\overline{RMSE}$
IGA	1.5±0.5, 2.5±0.5, 2±1, 3.5±0.5, 217, 5.4e−3	1.5±0.5, 1.5±0.5, 2±1, 3±1, 214, 6.1e−3
NSGA-II	1.5±0.5, 1±0, 1.5±0.5, 3.5±0.5, 757, 6.8e−3	1.5±0.5, 2±1, 2.5±0.5, 4.5±0.5, 851, 7.3e−3
c-means	1±0, 1±0, 3±0, 1±0, 0.171, 4.5e−3	1±0, 3±0, 3±0, 3±0 0.177, 4.6e−3

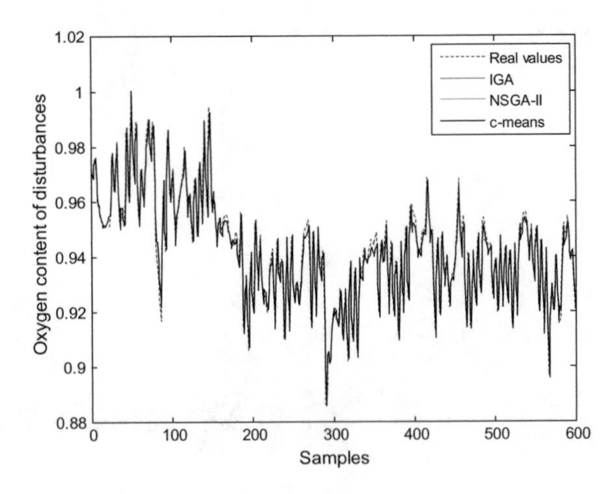

Fig. 7.5 Comparisons of disturbance channel outputs for 3 methods

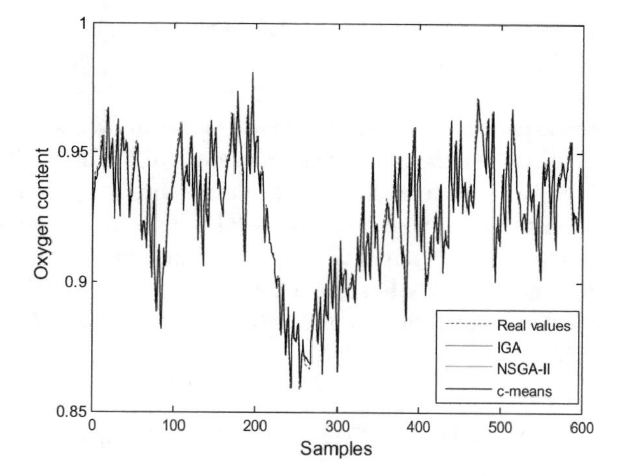

Fig. 7.6 Comparisons of main channel outputs for 3 methods

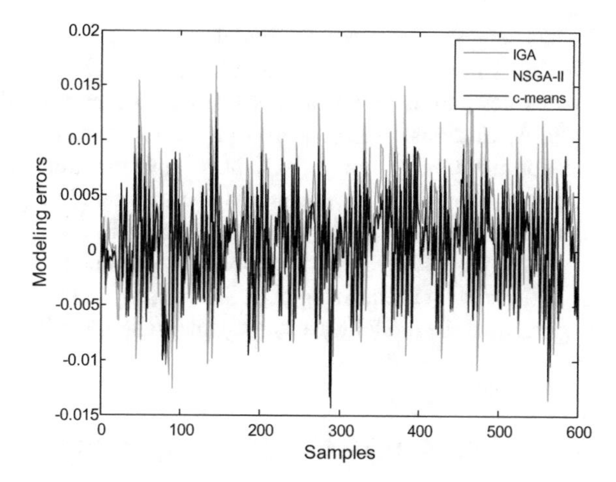

Fig. 7.7 Modelling errors of disturbance channel for 3 methods

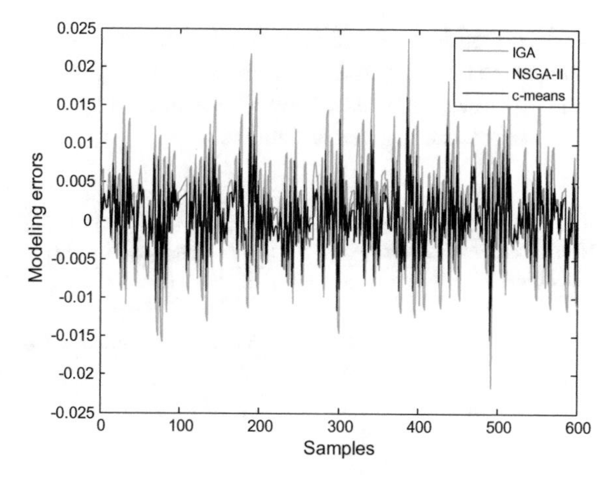

Fig. 7.8 Modelling errors of main channel for 3 methods

respectively. Figure 7.5 plotted the predicted yields comparing with measured outputs of the best model by using the 3 methods for the main channel, while the corresponding prediction errors are shown in Fig. 7.6. Figure 7.7 shows the fitting curve of the prediction values and the real values of the disturbance channel using the same methods, and the estimation errors are depicted in Fig. 7.8. Comparing Fig. 7.6 with Fig. 7.8, the maximal modeling errors obtained by c-means method and NSGA-II are larger than those obtained by IGA. The similar results can be observed in terms of the modeling errors of their disturbance models. It can be seen in Table 7.1 that IGA has managed to sustain the errors to considerably small values by optimizing the whole structure and parameters of the T-S fuzzy ARX model.

7.4 IGA Based Fuzzy ARX Plus Tanh Function Model

In this section, the fuzzy submodel is composed of the ARX plus Tanh function. The input layer, the structure and the parameters of the submodel, the number of fuzzy rules and its parameters of Gaussian functions will be solved simultaneously by IGA, which is quite difficult in nonlinear system modeling. The framework for the optimization of fuzzy ARX plus Tanh function model, i.e., a special type of fuzzy neural network (FNN) model, is illustrated in Fig. 7.9, where the input/output measurements $[u(k), y(k)]$ are utilized to design the fuzzy neural network using an improved GA (IGA). The structure and parameters of the proposed FNN can then be optimized by minimizing both the training and testing errors, which is beneficial to guarantee the modeling accuracy and generalization capability.

7.4.1 Encoding of IGA for Fuzzy Neural Network

As described in Sect. 7.3.1, m, n, r and the parameters of membership functions in FNN are to be optimized, and d is set according to a priori knowledge of the system. Since the fuzzy antecedent in rule j is limited to $y(k)$ and $u(k)$, the encoding of the ith chromosome is given as follows.

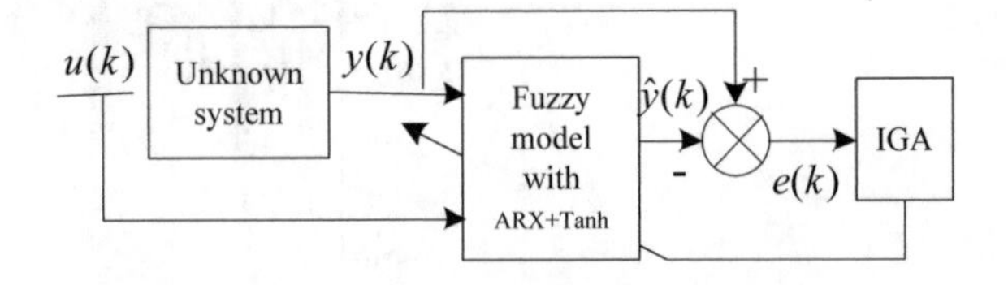

Fig. 7.9 Block diagram of IGA-based fuzzy system modelling

$$\mathbf{C}_i = \begin{bmatrix} c_{11} & c_{12} & \sigma_{11} & \sigma_{12} \\ c_{21} & c_{22} & \sigma_{21} & \sigma_{22} \\ \vdots & \vdots & \vdots & \vdots \\ c_{r1} & c_{r2} & \sigma_{r1} & \sigma_{r2} \\ \vdots & \vdots & \vdots & \vdots \\ c_{R+1,1} & c_{R+1,2} & c_{R+1,3} & c_{R+1,4} \\ c_{R+2,1} & c_{R+2,2} & c_{R+2,3} & c_{R+2,4} \end{bmatrix} \tag{7.20}$$

where $i = 1, 2, \cdots, N$. C_i is a 4-by-$R+2$ matrix, r is the number of the fuzzy rules with $1 \le r \le R$, and R is an empirical value denoted as the maximal number of fuzzy rules. The rows between $r+1$ and R are set to zeros. m, n are encoded by binary encoding format and located at the last two rows, that is, $c_{R+1,1}\, c_{R+1,2}\, c_{R+1,3}\, c_{R+1,4}$ and $c_{R+2,1}\, c_{R+2,2}\, c_{R+2,3}\, c_{R+2,4}$ are the binary encoding. Here $m \in [1, N_m], n \in [1, N_n]$ can be decoded as follows:

$$\begin{aligned} m &= < \frac{c_{12,1}2^3 + c_{12,2}2^2 + c_{12,3}2^1 + c_{12,4}2^0}{2^4 - 1} \times (N_m - 1) + 1 > \\ n &= < \frac{c_{13,1}2^3 + c_{13,2}2^2 + c_{13,3}2^1 + c_{13,4}2^0}{2^4 - 1} \times (N_n - 1) + 1 > \end{aligned} \tag{7.21}$$

where $<\cdot>$ rounds the element to the nearest integer.

The other $r \times 4$ elements in Eq. 7.20 are initialized as follows.

$$\begin{aligned} c_{ij} &= \begin{cases} y_{\min} + \delta(y_{\max} - y_{\min}) & 1 \le j \le r, i = 1 \\ u_{\min} + \delta(u_{\max} - u_{\min}) & 1 \le j \le r, \ i = 2 \end{cases} \\ \sigma_{ij} &= \begin{cases} 0.1 + \delta(y_{\max} - 0.1) & 1 \le j \le r, \ i = 1 \\ 0.1 + \delta(u_{\max} - 0.1) & 1 \le j \le r, \ i = 2 \end{cases} \end{aligned} \tag{7.22}$$

where δ is generated randomly between [0, 1], $u_{\min}$, $u_{\max}$, $y_{\min}$ and $y_{\max}$ are the same as those in Eq. 7.15.

Hence, the structure of input layer, fuzzy rules layer and the parameters of Gaussian functions are expressed by binary and decimal hybrid encoding method.

7.4.2 Operators of IGA for New Fuzzy Model

The selection, crossover and mutation operators are the same as those in Sec. 7.3.3. In order to improve the local search capability of GA, a neighborhood search operator is added to this IGA.

(1) Neighborhood search operator

An individual $\boldsymbol{C}_i$ is represented as a solution in the search space and a set of solutions surrounding $\boldsymbol{C}_i$ are defined as a neighborhood. A perturbation of $\boldsymbol{C}_i$ ($\Delta\mathbf{C}_i$) is obtained based on $\boldsymbol{C}_i$, which is produced with probability p_n.

$$\Delta C_i = \begin{bmatrix} \Delta c_{11} & \Delta c_{12} & \Delta\sigma_{11} & \Delta\sigma_{12} \\ \Delta c_{21} & \Delta c_{22} & \Delta\sigma_{21} & \Delta\sigma_{22} \\ \vdots & \vdots & \vdots & \vdots \\ \Delta c_{<r+\Delta r>,1} & \Delta c_{<r+\Delta r>,2} & \Delta\sigma_{<r+\Delta r>,1} & \Delta\sigma_{<r+\Delta r>,2} \\ \vdots & \vdots & \vdots & \vdots \\ \Delta c_{R+1,1} & \Delta c_{R+1,2} & \Delta c_{R+1,3} & \Delta c_{R+1,4} \\ \Delta c_{R+2,1} & \Delta c_{R+2,2} & \Delta c_{R+2,3} & \Delta c_{R+2,4} \end{bmatrix} \tag{7.23}$$

where $\Delta r = \delta r$, $< r + \Delta r > \in [1, R]$. If the number of fuzzy rules is increased, the elements of the new rules are produced in terms of Eqs. 7.21 and 7.22, otherwise, the elements of the decreased rules are set to zero. For the elements in the unchanged rules, $\Delta c_{ij} = \delta c_{ij}$ is added to $c_{ij}, i = 1, \cdots r$, where δ is randomly generated between $[-0.1, 0.1]$. $\Delta c_{R+1,i}$, $\Delta c_{R+2,i}$ $(i = 1, \cdots, 4)$ are set to 0, and $c_{R+1,i}$, $c_{R+2,i}$ are kept invariant.

(2) Maintain operator

In order to guarantee the rationality of FNN produced by IGA, the maintain operator is then designed to obtain the effective fuzzy rules and its parameters.

1. If the width σ_{ij} is zero, the rule will be deleted.
2. Sort the first column in ascending order, and if the distance of the adjacent centers dc_{ij} is too close, e.g., $dc_{ij} \leq \varepsilon$, c_{ij} will be deleted and the corresponding fuzzy rules are decreased.
3. If all the elements in Θ are smaller than 0.001, the submodel of the ith rule is considered useless and this rule is removed directly.

The procedure and complexity analysis are also similar to those in Sec. 7.3.4 and readers can refer to the relevant contents.

7.4.3 *Liquid Level Modeling by Nonlinear Fuzzy Neural Network*

The liquid level model in the industrial coke fractionation tower is also critical for advanced controller design. The detailed description of the coke unit can be found in [40], and the flow of the coke furnace is given in Ch. 6. The manipulated variable of the liquid level is the overall residual oil flow that goes into the convection rooms of the furnace. In addition, a lot of disturbances are influencing the process, such as the load changes, unsteady flames of the coke furnace, the amount of input fuel, the outer environment, and the coupling of the two pressure branches etc. The fluctuation of the liquid level is usually caused by these disturbances. Hence, there are also two types of model of liquid level to be constructed, namely, the main channel model and its disturbance channel model. 1600 input-output samples for constructing the model have been plotted in Figs. 7.10 and 7.11. The former 800 samples are selected as the training data and the latter 800 as the testing data. All the parameters of IGA are set the same as those in the case of oxygen content modeling, except w, which is set to 0.05. Some expert knowledge about the process is used to set the parameters of the fuzzy model, e.g., set the maximal number of the fuzzy rules (R) as 11, and N_m, N_n as 4 and 4, respectively. Similarly, 10 runs of the proposed approach are executed and the best model with the smallest J is shown in Fig. 7.12.

All data are divided into the training data ($\mathbf{Y}_1$) and testing data ($\mathbf{Y}_2$) to build the system model and test the performances of the fuzzy model. Herein, the training data

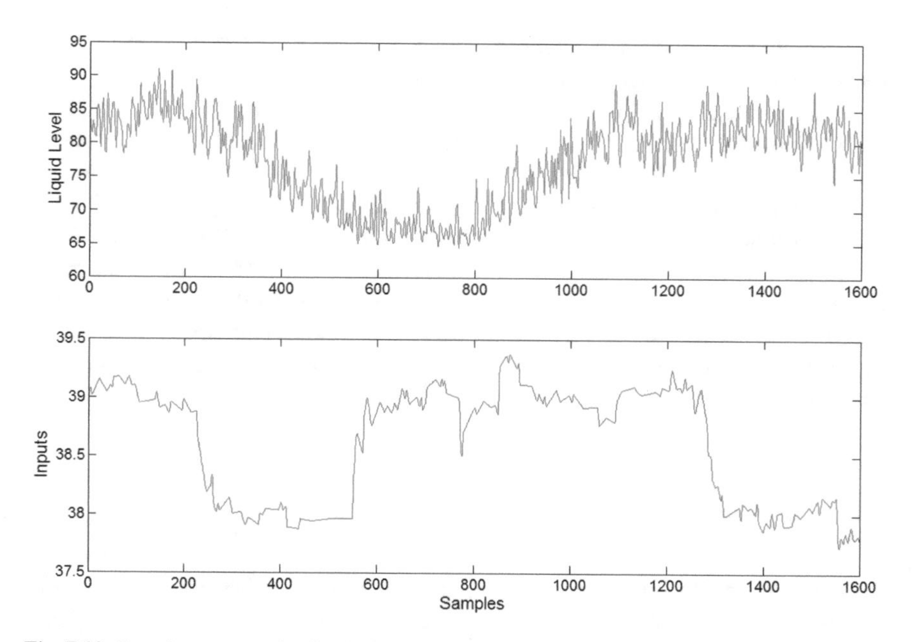

Fig. 7.10 Input/output samples for the liquid level main channel

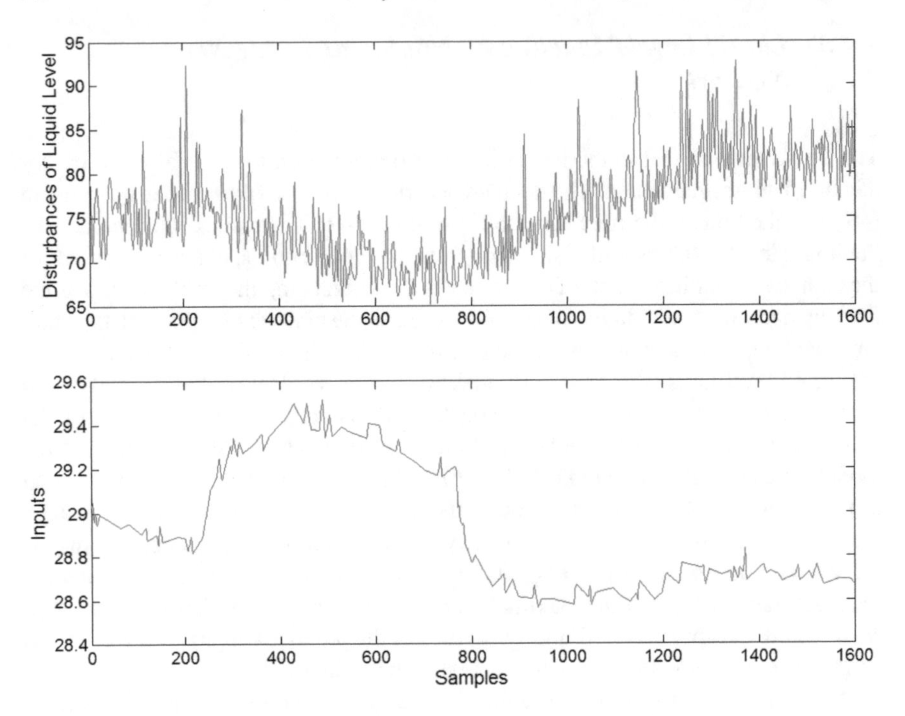

Fig. 7.11 Input/output samples for the liquid level disturbance channel

set is utilized to produce the coefficients $\mathbf{\Theta}$ of the FNN model, and the testing data set is adopted to verify its generalization capability.

When comparing the predicted outputs with the samples in Figs. 7.13 and 7.14, it is obvious that the constructed models can match the outputs of the two channels, and the training error is much less than the testing error. The best RMSE of the main channel for the training errors (0.3157) is less than that of testing errors (0.6080) in Table 7.3, where the RMSE of the training error is denoted as RMSE_1 and the RMSE of the testing error is RMSE_2. Similar results can be obtained for the disturbance channel model by using IGA method too. The statistical results in 10 runs are also shown in Table 7.3. It shows that the number of fuzzy rules is distributed between [4, 6], and the average values of RMSE from 0.3295 to 0.8193 are satisfactory comparing with the magnitude of system output.

7.5 Summary

An improved genetic algorithm was proposed to construct a T-S fuzzy ARX model and T-S ARX plus Tah function model, which consists of the input layer, the fuzzy rules layer and the output layer. The input variables and the fuzzy rules layer are considered to construct the premise part of the fuzzy model, then the output layer

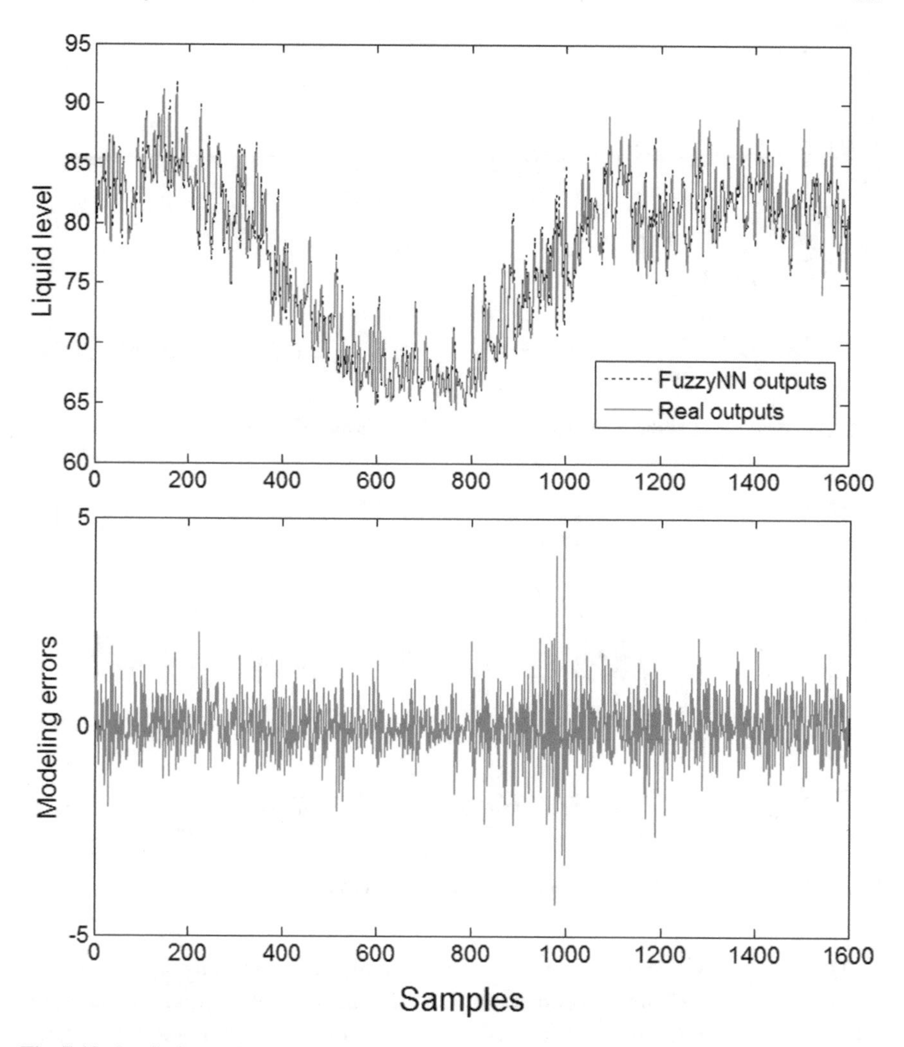

Fig. 7.12 Predictions of the built model and its modelling errors of main channel

is used to form the consequent part. In order to determine both the structure and parameters in the whole fuzzy system, the hybrid encoding is designed to encode the structure and parameters of fuzzy model in a chromosome. The selection, crossover, local search, mutation and maintain operators are used, especially a maintain operator including pruning and deleting are proposed to guarantee the validity of the fuzzy system. RLS is applied to obtain the coefficients of the submodel. Furthermore, to simplify the fuzzy model structure and improve the modeling accuracy, modeling accuracy, generalization performance and structure complexity of the fuzzy system are considered simultaneously in the objective function.

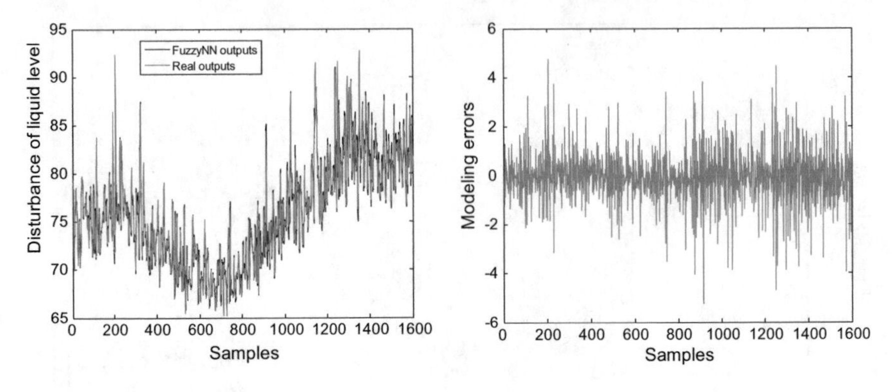

Fig. 7.13 Predictions of the built model and its modeling errors of main channel

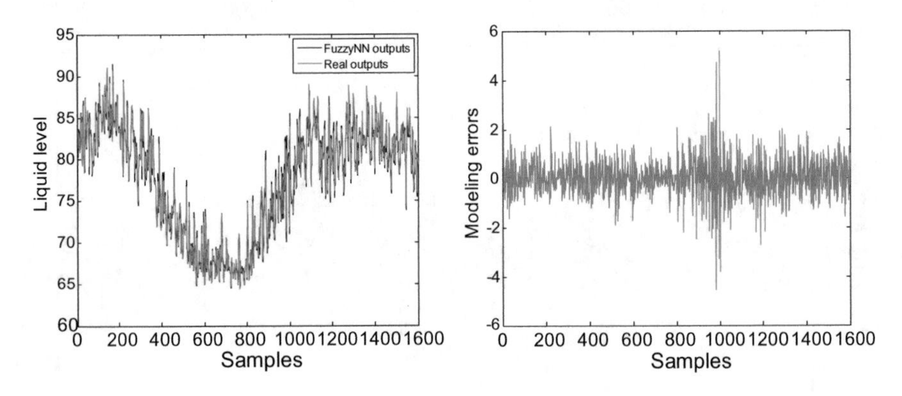

Fig. 7.14 Predictions of the built model and its modeling errors of disturbance channel

The fuzzy neural network optimization method is successfully applied to the oxygen content and liquid level modeling of an industrial coke furnace. Simulation results show that the optimization method has the ability to obtain a more compact structure with higher prediction accuracy compared with other works.

Table 7.3 The best fuzzy model and its statistics results in 10 runs

Results	Main channel					Disturbance channel				
	m	n	No. of rules	$RMSE_1$	$RMSE_2$	m	n	No. of rules	$RMSE_1$	$RMSE_2$
Best	3	3	4	0.3157	0.6080	3	1	4	0.4425	0.7148
Statistics	2.5±0.5	2.5±0.5	4.5±0.5	0.3295	0.6547	1.5±0.5	2.5±0.5	5.5±0.5	0.5429	0.8193

References

1. Kosko, B. 1994. Fuzzy systems as universal approximators. In *Proceedings of IEEE International Conference of Fuzzy Systems*.
2. Zeng, K., N.Y. Zhang, and W.L. Xu. 2000. A comparative study on sufficient conditions for Takagi-Sugeno fuzzy systems as universal approximators. *IEEE Transactions on Power Electronics* 8 (6): 773–780.
3. Park, C.W., and M. Park. 2004. Adaptive parameter estimator based on T-S fuzzy models and its applications to indirect adaptive fuzzy control design. *Information Sciences* 159 (1–2): 125–139.
4. Hyun, C.H., C.W. Park, and S. Kim. 2010. Takagi-Sugeno fuzzy model based indirect adaptive fuzzy observer and controller design. *Information Sciences* 180 (11): 2314–2327.
5. Elmetennani, S., and T.M. Laleg-Kirati. 2014. Fuzzy universal model approximator for distributed solar collector field control. In *2014 UKACC International Conference on Control*. IEEE.
6. Pedrycz, W. 2005. *Knowledge-Based Clustering: From Data to Information Granules*.
7. Riid, A., and E. Rüstern. 2011. Identification of transparent, compact, accurate and reliable linguistic fuzzy models. *Information Sciences* 181 (20): 4378–4393.
8. Nayak, P.C., K.P. Sudheer, and S.K. Jain. 2014. River flow forecasting through nonlinear local approximation in a fuzzy model. *Neural Computing Applications* 25 (7–8): 1951–1965.
9. Zhang, R., and J. Tao. 2017. A nonlinear fuzzy neural network modeling approach using improved genetic algorithm. *IEEE Transactions on Industrial Electronics* 65 (7): 5882–5892.
10. Zhang, R., J. Tao, and F. Gao. 2016. A new approach of Takagi-Sugeno fuzzy modeling using an improved genetic algorithm optimization for oxygen content in a coke furnace. *Industrial Engineering Chemistry Research* 55 (22): 6465–6474.
11. Rubio-Solis, A., and G. Panoutsos. 2015. Interval type-2 radial basis function neural network: a modeling framework. *Fuzzy Systems IEEE Transactions on* 23 (2): 457–473.
12. Ebadzadeh, M.M., and Armin Salimi-Badr. 2015. CFNN: correlated fuzzy neural network. *Neurocomputing* 148 (1): 430–444.
13. Pratama, M., et al. 2013. Data driven modeling based on dynamic parsimonious fuzzy neural network. *Neurocomputing* 110 (8): 18–28.
14. Hung, Y.C., et al. 2014. Wavelet fuzzy neural network with asymmetric membership function controller for electric power steering system via improved differential evolution. *IEEE Transactions on Power Electronics* 30 (4): 2350–2362.
15. Wang, D., X.J. Zeng, and J.A. Keane. 2010. A structure evolving learning method for fuzzy systems. *Evolving Systems* 1(2): 83–95.
16. Juang, C.F., and Y.W. Tsao. 2008. A self-evolving interval type-2 fuzzy neural network with online structure and parameter learning. *IEEE Transactions on Fuzzy Systems* 16 (6): 1411–1424.
17. Wang, D., et al. 2013. A simplified structure evolving method for Mamdani fuzzy system identification and its application to high-dimensional problems. *Information Sciences* 220 (1): 110–123.
18. Han, H., X.L. Wu, and J.F. Qiao. 2014. Nonlinear systems modeling based on self-organizing fuzzy-neural-network with adaptive computation algorithm. *IEEE Trans Cybern* 44 (4): 554–564.
19. Figueroa-García, J.C., C.M. Ochoa-Rey, and J.A. Avellaneda-González. 2013. Rule generation of fuzzy logic systems using a self-organized fuzzy neural network. *Neurocomputing* 151: 955–962.
20. Agrawal, S., B.K. Panigrahi, and M.K. Tiwari. 2008. Multiobjective particle swarm algorithm with fuzzy clustering for electrical power dispatch. *IEEE Transactions on Evolutionary Computation* 12 (5): 529–541.
21. Pedrycz, W., and H. Izakian. 2014. Cluster-centric fuzzy modeling. *IEEE Transactions on Fuzzy Systems* 22 (6): 1585–1597.

22. Han, H.G., L.M. Ge, and J.F. Qiao. 2016. An adaptive second order fuzzy neural network for nonlinear system modeling. *Neurocomputing* 214: 837–847.
23. Bagis, A. 2008. Fuzzy rule base design using Tabu search algorithm for nonlinear system modeling. *ISA Transactions* 47 (1): 32–44.
24. Liang, Z., et al. 2010. Automatically extracting T-S fuzzy models using cooperative random learning particle swarm optimization. *Applied Soft Computing* 10 (3): 938–944.
25. Lee, K.H., and K.W. Kim. 2015. Performance comparison of particle swarm optimization and genetic algorithm for inverse surface radiation problem. *International Journal of Heat Mass Transfer* 88: 330–337.
26. Hung, Y.C., et al. 2015. Wavelet fuzzy neural network with asymmetric membership function controller for electric power steering system via improved differential evolution. *IEEE Transactions on Power Electronics* 30 (4): 2350–2362.
27. Setnes, M., and H. Roubos. 2000. GA-fuzzy modeling and classification: complexity and performance. *IEEE Transactions on Fuzzy Systems* 8 (5): 509–522.
28. Chen, C.H., J.S. He, and T.P. Hong. 2013. MOGA-based fuzzy data mining with taxonomy. *Knowledge-Based Systems* 54: 53–65.
29. Ouarda, A., and M. Bouamar. 2014. A comparison of evolutionary algorithms: PSO, DE and GA for fuzzy C-partition. *International Journal of Computer Applications* 91 (10): 32–38.
30. Yang, T., et al. 2013. Fuzzy modeling approach to predictions of chemical oxygen demand in activated sludge processes. *Information Sciences* 235 (6): 55–64.
31. Barragán, A.J., et al. 2014. A general methodology for online TS fuzzy modeling by the extended Kalman filter. *Applied Soft Computing* 18 (4): 277–289.
32. David, R.C., et al. 2014. *An Approach to Fuzzy Modeling of Anti-lock Braking Systems.*
33. Lemos, A.P., W.M. Caminhas, and F.A.C. Gomide. 2011. Multivariable Gaussian evolving fuzzy modeling system. *IEEE Transactions on Fuzzy Systems* 19 (1): 91–104.
34. Das, I., and J.E. Dennis. 1997. A closer look at drawbacks of minimizing weighted sums of objectives for Pareto set generation in multicriteria optimization problems. *Structural Optimization* 14 (1): 63–69.
35. Deb, K. and H. Jain. 2012. Handling many-objective problems using an improved NSGA-II procedure. In *2012 IEEE Congress on Evolutionary Computation.* IEEE.
36. Santana-Quintero, L.V., and C.A.C. Coello. 2005. An algorithm based on differential evolution for multi-objective problems. *International Journal of Computational Intelligence Research* 1 (1): 151–169.
37. Sudeng, S., and N. Wattanapongsakorn. 2015. Post Pareto-optimal pruning algorithm for multiple objective optimization using specific extended angle dominance. *Engineering Applications of Artificial Intelligence* 38: 221–236.
38. Deb, K. 2012. *Optimization for Engineering Design: Algorithms and Examples.* PHI Learning Pvt. Ltd.
39. Tsekouras, G.E. 2005. On the use of the weighted fuzzy c-means in fuzzy modeling. *Advances in Engineering Software* 36 (5): 287–300.
40. Zhang, R., et al. 2014. Design and implementation of an improved linear quadratic regulation control for oxygen content in a coke furnace. *IET Control Theory and Applications* 8 (14): 1303–1311.

Chapter 8
PCA and GA Based ARX Plus RBF Modeling for Nonlinear DPS

Distributed parameter systems (DPSs) are difficult to model due to their nonlinearity and infinite-dimension characteristics. This chapter adopts principal component analysis (PCA) to derive a hybrid modeling strategy for modeling such systems. The strategy consists of a decoupled linear autoregressive exogenous (ARX) model and a nonlinear Radial Basis Function (RBF) neural network model. Using PCA, the spatial-temporal output is firstly divided into a few dominant spatial basis functions and finite-dimension temporal series. Then, the linear dynamic model of the dominant modes of the time series is constructed with a decoupled ARX model. The nonlinear residual is parameterized by RBF neural networks and GA is adopted to optimize the RBF neural network structure and parameters. A nonlinear spatial-temporal dynamic system will be finally obtained after the time/space reconstruction. In the simulation part, a catalytic rod and a heat conduction equation have been utilized to demonstrate the application process of the DPS modeling strategy.

8.1 Introduction

Distributed parameter systems (DPSs) are characterized by nonlinear partial differential equations (PDEs) with mixed or homogeneous boundary conditions, where the inputs, outputs and parameters vary spatially and temporally [1]. Modeling and control of such systems are difficult but also important. Due to the infinite-dimension characteristics of these systems, modeling and control methods for lumped parameter systems (LPSs) cannot be directly extended [2, 3]. Conventional spatial discretization methods, such as finite difference etc., will lead to approximation systems of high-order ordinary differential equations (ODEs). It is still a hot topic to develop methods to capture the dominant dynamics of DPSs [2–4].

J. Tao et al., *DNA Computing Based Genetic Algorithm*,
https://doi.org/10.1007/978-981-15-5403-2_8

Traditional model reduction methods rely on an accurate model, and this is not practical in industrial application. In view of this, data-driven methods can be considered [5–7]. Typically, PCA can be a choice and has been widely applied in DPSs [8–10]. However, it is also known that DPSs exist nonlinear behavior and traditional PCA cannot be directly used. In view of this, neural network can be chosen for nonlinear modeling [11, 12]. There have been some results with singular value decomposition (SVD) and proper orthogonal decomposition (POD) [8, 13]. Other nonlinear identification techniques using Wiener and Hammerstein were also proposed [14–16]. However, issues of simplifying the model and considering nonlinearity are still a hot topic. It is known that autoregressive models with exogenous inputs (ARX) are suitable for the subsequent controller design. For nonlinear systems, neural network a with exogenous input (NN-ARX) will show good nonlinear modeling in terms of root-mean-square error (RMSE) and mean absolute error (MAE) [17]. Moreover, searching algorithms, such as cuckoo search algorithm (CSA) and genetic algorithm (GA), can be used to optimize parameters [18, 19].

This chapter proposes an ARX model structure plus GA optimization based RBF neural networks for nonlinear DPSs. PCA is first used for spatial-temporal separation and a decoupled ARX model is obtained. RBF neural network is further used for nonlinear black box identification. GA is finally introduced to optimize neural network parameters to improve modeling accuracy and simple model structure.

8.2 DPS Modeling Issue

It is not possible to place infinite number of actuators and sensors to model and control DPSs for their infinite-dimension characteristics. In view of this, to map the infinite-dimension space into a new low-dimension eigenspace, PCA is first adopted using sampled spatio temporal data. For the linear part of the low-dimension time-series, a decoupled ARX model will be built. RBF neural networks will be used to approximate the nonlinear part, and its structure and parameters will be optimized with GA. The diagram is shown in Fig. 8.1.

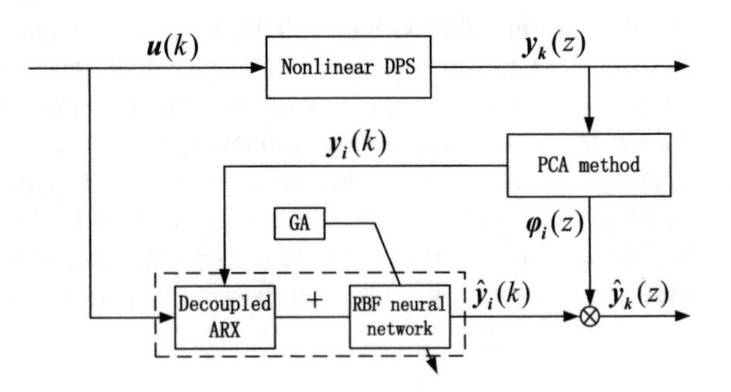

Fig. 8.1 Framework of PCA-base decoupled ARX plus GA optimized RBF neuralnetwork

Denote t and z as the continuous time and space variables, respectively, where $z \in \Omega \in \Re$ is an open connected and bounded region in the 1-dimension Euclidean space. The corresponding discrete time instants and locations in t and z are then defined as t_k and z_k, $k = 1, 2, \cdots$.

Here $\mathbf{u}(k) = [u_1(t_k), \ldots, u_m(t_k)]$ is the temporal inputs at m spatial locations $z_1, \ldots, z_m$ for time instant t_k. $\mathbf{y}_k(z) = [y(z_1, t_k), \ldots, y(z_N, t_k)]$ is the spatio-temporal output at N spatial locations $z_1, \ldots, z_N$. They are uniformly sampled at time instant t_k, $k \in [1, \ldots, L]$, and L is the time length.

In what follows, we will simplify the notation t_k as k. Here $\mathbf{y}_k(z)$ is an N-vector and referred as "snapshot". Using PCA, $\mathbf{y}_k(z)$ can be divided into n spatial basis functions $\varphi_i(z)$ and low-dimension temporal series $y_i(k)$, $i = 1, \ldots, n$, with $\varphi_i(z) = [\varphi_i(z_1), \varphi_i(z_2), \ldots, \varphi_i(z_N)]$, $\varphi_i(z) \in \varphi(z)$, $\varphi(z) = [\varphi_1(z), \varphi_2(z), \ldots, \varphi_L(z)]$. A decoupled ARX model plus RBF neural networks can then be used to predict $y_i(k)$ and the spatio-temporal dynamics $\hat{\mathbf{y}}_k(z)$ will be reconstructed.

8.2.1 Time/Space Separation via PCA

The dominant spatial patterns $\varphi(z)$ is derived using PCA for capturing almost all ensemble energy of the spatio-temporal outputs.

For simplicity, $y(z_i, k)$ $(i = 1, \ldots, N)$ is defined as a function of $y_k(z)$ at time instant k in space domain Ω. $\varphi(z)$ and $y_k(z)$ are simplified as φ and y_k, respectively. φ is maximize the averaged projection of y_k on it and expressed as a maximization problem [20, 21]:

$$\begin{aligned} &\max_{\varphi} \frac{< |(y_k, \varphi)|^2 >}{\|\varphi\|^2} \\ &\text{s.t. } \|\varphi\|^2 = 1 \end{aligned} \tag{8.1}$$

where $< \bullet >$ is an averaging operator, $|\bullet|$ denotes the modulus, $(\cdot, \cdot)$ is defined as the inner product and $\|\bullet\|$ is given as $\|\cdot\| = (\cdot, \cdot)^{\frac{1}{2}}$. $\|\varphi\|^2 = 1$ is a constraint to ensure that φ is unique. The Lagrange function is derived to solve Eq. 8.1:

$$J(\varphi) =< |(y_k, \varphi)|^2 > -\lambda(\|\varphi\|^2 - 1) \tag{8.2}$$

A necessary condition is that, for $\delta \in \Re$ and ψ as an arbitrary function, the functional derivative vanishes for all $\varphi + \delta\psi$:

$$\frac{\mathrm{d}}{\mathrm{d}\delta} J(\varphi + \delta\psi)_{|\delta=0} = 0 \tag{8.3}$$

Thus, we obtain:

$$
\begin{aligned}
&\frac{\mathrm{d}}{\mathrm{d}\delta} J(\varphi+\delta\psi)_{|\delta=0} \\
&= \frac{\mathrm{d}}{\mathrm{d}\delta}[< (y_k, \varphi+\delta\psi)(\varphi+\delta\psi, y) > -\lambda(\varphi+\delta\psi, \varphi+\delta\psi)]_{|\delta=0} \\
&= 2\mathrm{Re}[< (y_k, \psi)(\varphi, y_k) > -\lambda(\varphi, \psi)] \\
&=< \int_\Omega \psi(\varsigma) y_k(\varsigma)\mathrm{d}\varsigma \int_\Omega \varphi(z) y_k(z)\mathrm{d}z > -\lambda \int_\Omega \varphi(z)\psi(z)\mathrm{d}z \\
&= \int_\Omega \int_\Omega < y_k(\varsigma) y_k(z) > \varphi(\varsigma)\mathrm{d}\varsigma\psi(z)\mathrm{d}z - \int_\Omega \boldsymbol{\lambda}\varphi(z)\psi(z)\mathrm{d}z \\
&= \int_\Omega \left[\int_\Omega < y_k(\varsigma) y_k(z) > \varphi(\varsigma)\mathrm{d}\varsigma - \boldsymbol{\lambda}\varphi(z)\right]\psi(z)\mathrm{d}z = 0
\end{aligned} \tag{8.4}
$$

Note that $\psi(z)$ is an arbitrary function, the condition is reduced as:

$$
\int_\Omega < y_k(\varsigma) y_k(z) > \varphi(\varsigma)\mathrm{d}\varsigma = \lambda\varphi(z) \tag{8.5}
$$

By introducing the spatial two-point correlation function:

$$
R(z, \varsigma) =< y_k(z) y_k(\varsigma) >= \frac{1}{L}\sum_{k=1}^{L} y_k(z) y_k(\varsigma)
$$

Equation 8.5 can be rewritten as:

$$
\int_\Omega R(z, \varsigma)\varphi(\varsigma) d\varsigma = \lambda\varphi(z) \tag{8.6}
$$

The solution of Eq. 8.6 will require expensive computation, therefore, the method of snapshots is introduced [22]. Assuming that $\varphi(z)$ can be expressed as a linear combination of the snapshots:

$$
\varphi(z) = \sum_{k=1}^{L} \gamma_k y_k(z) \tag{8.7}
$$

Substituting Eq. 8.7 into Eq. 8.6, we obtain:

$$
\int_\Omega \frac{1}{L}\sum_{\kappa=1}^{L} y_\kappa(z) y_\kappa(\varsigma) \sum_{k=1}^{L} \gamma_k y_k(\varsigma)\mathrm{d}\varsigma = \lambda \sum_{k=1}^{L} \gamma_k y_k(z) \tag{8.8}
$$

A nice geometrical interpretation of this representation can be obtained particularly in the finite-dimension case [20], where the observations $y_k(z)$ are N-vectors.

Define $C_{\kappa k} = 1/L \int_\Omega y_\kappa(\varsigma) y_k(\varsigma)\mathrm{d}\varsigma$, $\gamma = [\gamma_1, \ \ldots \ , \ \gamma_L]^{\mathrm{T}}$, Eq. 8.8 will be rewritten as:

$$C\gamma = \lambda\gamma \tag{8.9}$$

Solving Eq. 8.9 with standard matrix theory, such as QR decomposition, the eigenvectors $\gamma = [\gamma_1, \cdots, \gamma_L]^T, \gamma_i = [\gamma_{i1} \cdots \gamma_{iL}], i = 1, \ldots, L$ will be obtained and used to construct the eigenfunctions $\varphi_1(z), \cdots, \varphi_L(z)$. Moreover, matrix C is symmetric and positive semi-definite with its eigenvalues $\lambda_\kappa, \kappa = 1, \cdots, L$ being real and non-negative.

Let the eigenvalues $\lambda_1 > \lambda_2 > \cdots > \lambda_L$, an 'energy' percentage for the associated eigenvalues, be assigned for every eigenfunction [8]:

$$E_i = \lambda_i \Big/ \sum_{j=1}^{L} \lambda_j \tag{8.10}$$

Here, the total 'energy' is calculated as the sum of the eigenvalues, and only the first few eigenfunctions that represent 99% of the 'energy' can capture the dominant dynamics of many spatio-temporal systems [14]. It can also be used to determine the value of n, $n < L$. That is, $\varphi_i(z), i = 1, \ldots, n$ are selected as the most representative characteristic eigenfunctions and $\varphi_i(z), i = n + 1, \ldots, L$ are neglected. $\varphi_i(z_j)$ can then be derived as:

$$\varphi_i(z_j) = \sum_{k=1}^{L} \gamma_{ik} y(z_j, k) \tag{8.11}$$

where $\hat{y}(z_j, k)$ can be expanded into n orthonormal spatial basis functions with temporal series $y_i(k)$, which is described as follows:

$$\hat{y}(z_j, k) = \sum_{i=1}^{n} \varphi_i(z_j) y_i(k), \; j = 1, 2, \ldots, N \tag{8.12}$$

Here, $\hat{y}(z_j, k)$ denotes the nth-order approximation of the spatio-temporal output, and $y_k(z)$ is approximated as $\hat{y}_k(z) = [\hat{y}(z_1, k), \cdots, \hat{y}(z_N, k)]$. Since $\varphi_i(z)$ is orthonormal, the temporal series $y_i(k)$ can be derived as:

$$y_i(k) = (\varphi_i(z), y_k(z)), i = 1, \ldots, n \tag{8.13}$$

The PCA time/space separation is summarized as follows:

Step 1: Obtain $L \times N$ data matrix of the snapshots $\mathbf{y}_k(z)$.
Step 2: Subtract the mean value for each dimension of $y_k(z)$.
Step 3: Calculate the covariance matrix C in Eq. 8.9.
Step 4: Find the eigenvectors and eigenvalues of Eq. 8.9 with QR decomposition.
Step 5: Extract the diagonal of R matrix as the eigenvalue vector and sort its elements in the decreasing order.

Step 6: Calculate the energy percentage with Eq. 8.10, and select n eigenvectors in the Q matrix that captures more than 99% of the system energy.
Step 7: Obtain n spatial basis functions with Eq. 8.11.
Step 8: Project the original time-space data with Eq. 8.13 to obtain $y_i(k)$.

8.2.2 Decoupled ARX Model Identification

We define n time series of the output and input as $\boldsymbol{y}(k) = [y_1(k), \cdots, y_n(k)]^{\mathrm{T}}$ and $\boldsymbol{u}(k) = [u_1(k), \cdots, u_n(k)]^{\mathrm{T}}$, respectively. The ith dynamic system within $\boldsymbol{u}(k)$ and $\boldsymbol{y}(k)$ can be simplified as a decoupled ARX model plus a nonlinear function [3]:

$$\begin{aligned} y_i(k) &= a_{1i}y_i(k-1) + \cdots + a_{n_y i}y_i(k-n_y) \\ &\quad + b_{1i}u_i(k-1) + \cdots b_{n_u i}u_i(k-n_u) + f_i(k) \\ &= (a_{1i}q^{-1} + \cdots a_{n_y i}q^{-n_y})y_i(k) + i = 1, \cdots, n \\ &(b_{1i}q^{-1} + \cdots b_{n_u i}q^{-n_u})u_i(k) + f_i(k) \end{aligned} \tag{8.14}$$

Here q is the time shift operator, n_y and n_u are the maximal output and input lags, respectively, $a_{1i}, \ldots, a_{n_y i}, b_{1i}, \ldots, b_{n_u i}$ are the model parameters, and $f_i(k)$ is the nonlinear part of the system.

The linear time invariant (LTI) system of Eq. 8.14 can be rewritten as:

$$\begin{bmatrix} y_1(k) \\ y_2(k) \\ \vdots \\ y_n(k) \end{bmatrix} = \begin{bmatrix} a_{11}q^{-1} + \cdots a_{n_y 1}q^{-n_y} \\ a_{12}q^{-1} + \cdots a_{n_y 2}q^{-n_y} \\ \vdots \\ a_{1n}q^{-1} + \cdots a_{n_y n}q^{-n_y} \end{bmatrix} \begin{bmatrix} y_1(k) \\ y_2(k) \\ \vdots \\ y_n(k) \end{bmatrix} + \begin{bmatrix} b_{11}q^{-1} + \cdots b_{n_u 1}q^{-n_u} \\ b_{12}q^{-1} + \cdots b_{n_u 2}q^{-n_u} \\ \vdots \\ b_{1n}q^{-1} + \cdots b_{n_u n}q^{-n_u} \end{bmatrix} \begin{bmatrix} u_1(k) \\ u_2(k) \\ \vdots \\ u_n(k) \end{bmatrix}$$

Denote $\mathcal{A}_i = diag(a_{1i}, \ldots, a_{ni})^{\mathrm{T}}, i = 1, \ldots, n_y$, $\mathcal{B}_i = diag(b_{1i}, \ldots, b_{ni})^{\mathrm{T}}, i = 1, \ldots, n_u$, we obtain:

$$\begin{cases} \mathbf{y}(k) = \mathcal{A}(q^{-1})\mathbf{y}(k) + \mathcal{B}(q^{-1})\mathbf{u}(k) \\ \mathcal{A}(q^{-1}) = \mathcal{A}_1 q^{-1} + \cdots + \mathcal{A}_{ny} q^{-n_y} \\ \mathcal{B}(q^{-1}) = \mathcal{B}_1 q^{-1} + \cdots + \mathcal{B}_{nu} q^{-n_u} \end{cases} \tag{8.15}$$

where $\mathcal{A}(q^{-1})$, $\mathcal{B}(q^{-1})$ are the $n \times n$ matrix polynomials.

In Eq. 8.15, the ith item $y_i(k)$ can also be rewritten as:

$$\begin{cases} y_i(k) = \theta_i \mathbf{H}_i(k) \\ \theta_i = [a_{1i} \ \cdots \ a_{n_y i} \ b_{1i} \cdots \ b_{n_u i}] \in \Re^{(n_y+n_u)} \\ \mathbf{H}_i(k) = [y_i(k-1) \cdots \ y_i(k-n_y) \\ \qquad u_i(k-1) \ \cdots \ u_i(k-n_u)]^T \end{cases}$$

The estimation of $\boldsymbol{\theta}_i$ can be derived through the recursive least squares (RLS) method [20]:

$$\begin{cases} \boldsymbol{\theta}_i(k) = \boldsymbol{\theta}_i(k-1) + \mathbf{K}(k)[y_i(k) - \mathbf{H}_\mathbf{i}^\mathrm{T}(k)\boldsymbol{\theta}_i(k-1)] \\ \mathbf{K}(k) = \mathbf{P}(k-1)\mathbf{H}_i(k)[\mathbf{H}_i^\mathrm{T}(k)\mathbf{P}(k-1)\mathbf{H}_i(k) + \mu]^{-1} \\ \mathbf{P}(k) = 1/\mu[\mathbf{I} - \mathbf{K}(k)\mathbf{H}_i^\mathrm{T}(k)]\mathbf{P}(k-1) \end{cases} \tag{8.16}$$

where $0 < \mu < 1$ is the forgetting factor, P(k) is a positive definite covariance matrix with $P(0) = \alpha^2 I$, I is an $(n_y + n_u) \times (n_y + n_u)$ identity matrix, α is a real number with sufficiently large value, $\boldsymbol{\theta}_i(0) = \boldsymbol{\varepsilon}$ with ε being a sufficiently small $n_y + n_u$ dimensional real vector and $\mathbf{K}(k)$ is a weight matrix, $\boldsymbol{\theta}_i, i = 1, \ldots, n$ can be obtained using n least squares estimations.

The procedure of the parameter identification of the decoupled ARX model is given as follows:

Step 1: Initialize n_y, n_u, $\boldsymbol{H}_i(1)$ and μ. Start from the first time series with $i = 1$.
Step 2: Initialize $\mathbf{P}(0)$ and $\boldsymbol{\theta}_i(0)$.
Step 3: Repeat to calculate $\boldsymbol{\theta}_i$ with Eq. 8.16 until k reaches L.
Step 4: Go to step 2 until i reaches n.
Step 5: Copy the values of $\boldsymbol{\theta}_i, i = 1, \ldots, n$ to $\mathcal{A}_1 \cdots \mathcal{A}_{ny}$, $\mathcal{B}_1 \cdots \mathcal{B}_{nu}$.

8.2.3 RBF Neural Network Modeling

When $\mathcal{A}_1 \cdots \mathcal{A}_{ny}$ and $\mathcal{B}_1 \cdots \mathcal{B}_{nu}$ are gained in Sec. 8.2.2, the outputs of the decoupled ARX model, $\hat{\mathbf{y}}(k) = [\hat{y}_1(k), \ldots \hat{y}_n(k)]$, can then be obtained. The deviation between $\mathbf{y}(k)$ and $\hat{\mathbf{y}}(k)$ is denoted as $\mathbf{f}(k)$:

$$\mathbf{f}(k) = \mathbf{y}(k) - \hat{\mathbf{y}}(k) \tag{8.17}$$

Here, $\mathbf{f}(k) = [f_1(k), \ldots, f_n(k)]$. See Fig. 8.2, the RBF neural network contains one input layer, one hidden layer and one output layer. Since Gaussian function [23] and thin plate spline function [24] are the typical radial basis functions, which has the fewest number of unknown parameters and shown in Ch. 6. There are a total of n RBF neural networks to model n nonlinear parts in Eq. 8.17.

The input vector of the ith RBF neural network is selected as its state variables [3]:

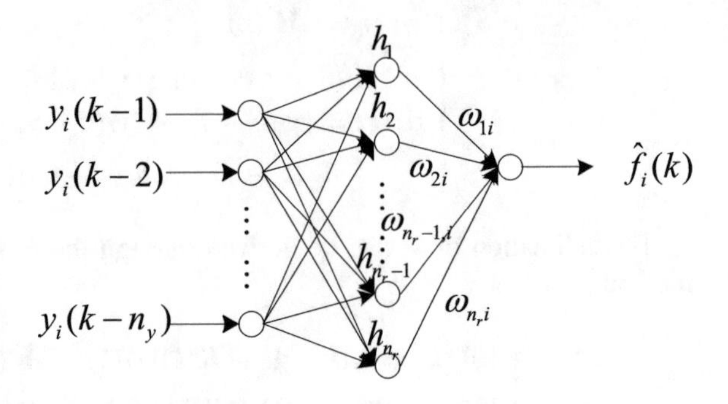

Fig. 8.2 RBF neural network modeling for prediction of $f_i(k)$

$$\mathbf{X}_i(k) = [y_i(k-1), \ldots y_i(k-n_y)],\ k = 1, \ldots, L.$$

The thin plate spline function as follow is used to obtain the output of the jth hidden node:

$$h_j(X_i(k)) = \left\|X_i(k) - \mathbf{c}_j\right\|^2 \lg(\left\|X_i(k) - \mathbf{c}_j\right\|),\ j = 1, 2, \ldots, n_r$$

Here $c_j = [c_{1j}, \ldots, c_{n_y j}]$ is the center of the jth RBF hidden node and $\left\|\mathbf{X}_i(k) - \mathbf{c}_j\right\| = \sqrt{\sum_{j=1}^{n_y}(y_i(k-j) - c_{ij})^2}$.

The output of the ith RBF neural network, denoted as $\hat{f}_i(k)$, is represented by a linear weighted sum of n_r hidden functions as:

$$\hat{f}_i(k) = \sum_{j=1}^{n_r} h_j(k)\omega_{ij} = \boldsymbol{\omega}_i\mathbf{h}(k) \tag{8.18}$$

where $\boldsymbol{\omega}_i = [\omega_{i1}, \ldots, \omega_{in_r}]$ is the weight vector connecting the hidden layer $\mathbf{h}(k) = [h_1(k), \ldots, h_{n_r}(k)]^{\mathrm{T}}$ to the output layer. RLS method can be used to calculate the estimation of $\boldsymbol{\omega}_i$:

$$\begin{cases} \boldsymbol{\omega}_i(k) = \boldsymbol{\omega}_i(k-1) + \mathbf{K}(k)[f_i(k) - \mathbf{h}^{\mathrm{T}}(k)\boldsymbol{\omega}_i(k-1)] \\ \mathbf{K}(k) = \mathbf{P}(k-1)\mathbf{h}(k)[\mathbf{h}^{\mathrm{T}}(k)\mathbf{P}(k-1)\mathbf{h}(k) + \mu]^{-1} \\ \mathbf{P}(k) = 1/\mu[\mathbf{I} - \mathbf{K}(k)\mathbf{h}^{\mathrm{T}}(k)]\mathbf{P}(k-1) \end{cases}$$

where $\mathbf{P}(k)$, $\mathbf{P}(0)$, μ and $\boldsymbol{\omega}_1(0) = \boldsymbol{\varepsilon}$ are the same as those in Eq. 8.16.

The relationship of the ith nonlinear part is shown as follows:

$$\hat{f}_i(k) = \sum_{j=1}^{n_r}(\left\|\mathbf{X}_i(k) - \mathbf{c}_j\right\|^2 \lg(\left\|\mathbf{X}_i(k) - \mathbf{c}_j\right\|))\omega_{ji} \tag{8.19}$$

The data set has been divided into a training data set and a testing data set with length of $L/2$. The sum of the absolute values of the prediction error for the ith RBF neural network is selected as:

$$e_s = \sum_{k=1}^{L/2} |f_i(k) - \hat{f}_i(k)| \tag{8.20}$$

The number of hidden nodes n_r and the centers c_j in the hidden functions are then optimized by GA.

The procedure of RBF neural network for modeling the ith nonlinear part is shown as follows:

Step 1: Initialize n_y, n_u, and calculate $\hat{y}_i(k)$ for the decoupled ARX model to obtain $f_i(k)$.
Step 2: Construct the input layer data $\mathbf{X}_i(k)$ and the output layer data $f_i(k)$, $k = 1, \cdots, L$.
Step 3: Divide the input/output data into a training set and a testing set with length of $L/2$. Optimize the number of hidden nodes (n_r) and the parameters of the hidden functions by using GA.
Step 4: Repeatedly calculate the output of the hidden layer $\mathbf{h}(k)$ until k reaches $L/2$.
Step 5: Initialize $\mathbf{P}(0)$, $\boldsymbol{\omega}_i(0)$, $\mathbf{h}(1)$, μ and compute $\boldsymbol{\omega}_i(k)$ with RLS algorithm until k reaches $L/2$.
Step 6: Compute the output of the RBF neural network with Eq. 8.19.

8.2.4 Structure and Parameter Optimization by GA

Note that the input vector $[y_i(k-1), \cdots y_i(k-n_y)]$ should be fixed, and the number of hidden nodes n_r and the centers of the hidden functions $\mathbf{c}_j$, $j = 1, \cdots, n_r$ will be optimized considering both modeling error and structure complexity:

$$f = e_s + \beta \cdot n_r,\ J = 1/f \tag{8.21}$$

where β is a weight coefficient between [0.0001, 0.1]. Generally, $\beta \cdot n_r$ should be proportional to the value of e_s. Larger β indicates that more importance is focused on the structure.

8.2.5 Encoding Method

The number of input nodes n_y is set in advance and there are $n_r \cdot n_y$ parameters to be optimized by GA. Here, the decimal encoding method is adopted and the decoding step is then not needed. The ith chromosome is given as follows:

$$\mathbf{C}_i = \begin{bmatrix} c_{11}^i & \dots & c_{1n_y}^i \\ c_{21}^i & \dots & c_{2n_y}^i \\ \vdots & \vdots & \vdots \\ c_{n_r1}^i & \dots & c_{n_rn_y}^i \\ 0 & \dots & 0 \end{bmatrix} \tag{8.22}$$

where $i = 1, \ldots, N_c$ with N_c as the population size, $1 \le n_r \le D$ with D as the maximal number of hidden nodes and set to 50, and $\mathbf{C}_i$ is a $n_y \times 50$ matrix. The rows below n_r are set to zeros and cannot be used as centers. The elements in Eq. 8.22 are:

$$c_{ij} = f_{\min} + r(f_{\max} - f_{\min}),\ 1 \le i \le n_r,\ 1 \le j \le n_y \tag{8.23}$$

where r is randomly generated between [0.1, 1], $f_{\min}$ and $f_{\max}$ are the minimum and maximum values of the outputs. RLS can be used to calculate the weights of the RBF neural network with the neural network structure and the parameters of the hidden functions derived in Eq. 8.22.

8.2.5.1 Operators

(1) Selection operator

The Roulette wheel method is adopted, and the selection probability depends on the reciprocal of the objective function in Eq. 8.21. Individuals with larger values of J will have more chances to be survived and a single spin of the Roulette wheel will select N_c individuals at the same time.

(2) Crossover operator

The crossover operator is executed with probability p_c between the current individual $\mathbf{C}_i$ and the next individual $\mathbf{C}_{i+1}$ at the former $N_c/2$ individuals. $\mathbf{C}_{i+1}$ is selected randomly in the latter $N_c/2$ individuals for keeping population diversity. The offspring chromosomes $\mathbf{C}_i^{'}$, $\mathbf{C}_{i+1}^{'}$ are produced through a multi-point crossover operator in Fig. 8.3. Suppose the number of the hidden nodes $n_r^{'}$ in $\mathbf{C}_{i+1}$ is larger than n_r in $\mathbf{C}_i$ and the crossover location r is generated randomly between 1 and n_r, then, $[\mathbf{c}_1^{\mathrm{T}}, \cdots, \mathbf{c}_r^{\mathrm{T}}]$ in $\mathbf{C}_i$ is exchanged with that in $\mathbf{C}_{i+1}$. After the crossover operation, the offspring $\mathbf{C}_i^{'}$, $\mathbf{C}_{i+1}^{'}$ are produced:

$$C_i^{'} = \begin{bmatrix} c_{11}^{i+1} & c_{12}^{i+1} & \cdots & c_{1n_y}^{i+1} \\ \vdots & \vdots & \cdots & \vdots \\ c_{r1}^{i+1} & c_{r2}^{i+1} & \cdots & c_{rn_y}^{i+1} \\ c^{i}_{r+1,1} & c^{i}_{r+1,2} & \cdots & c^{i}_{r+1,n_y} \\ \vdots & \vdots & \cdots & \vdots \\ c^{i}_{n_r1} & c^{i}_{n_r2} & \cdots & c^{i}_{n_rn_y} \\ \mathbf{0} & \mathbf{0} & \cdots & \mathbf{0} \end{bmatrix} \quad C_{i+1}^{'} = \begin{bmatrix} c_{11}^{i} & c_{12}^{i} & \cdots & c_{1n_y}^{i} \\ \vdots & \vdots & \cdots & \vdots \\ c_{r1}^{i} & c_{r2}^{i} & \cdots & c_{rn_y}^{i} \\ c^{i+1}_{r+1,1} & c^{i+1}_{r+1,2} & \cdots & c^{i+1}_{r+1,n_y} \\ \vdots & \vdots & \cdots & \vdots \\ c^{i+1}_{n_r1} & c^{i+1}_{n_r2} & \cdots & c^{i+1}_{n_rn_y} \\ \mathbf{0} & \mathbf{0} & \cdots & \mathbf{0} \end{bmatrix}$$

Fig. 8.3 Example of the crossover operator: The elements from line 1 to r in $\boldsymbol{C}_i$ are exchanged with those in $\boldsymbol{C}_{i+1}$

From Fig. 8.3, the crossover operator is likely to generate new parameters in the hidden layer. However, the structure of the hidden layer is kept unchanged.

(3) Regulation operator

Note that the crossover operator cannot produce new structures of the hidden layer, the regulation operator is then designed to be carried out with probability p_r to change the structure of the hidden layer. The new number of the hidden nodes is thus obtained as follows:

$$n_r^{'} = \lfloor n_r + r \cdot n_r \rfloor, \ 1 \leq n_r^{'} \leq D \tag{8.24}$$

where r is produced randomly between $[-0.5, 0.5]$, and $\lfloor \cdot \rfloor$ is a round down operation. The elements of the new added nodes from rows $n_r + 1$ to $n_r^{'}$ in Fig. 8.4 are created with Eq. 8.23, and the elements of the decreased nodes are set to zeros. The structure of the hidden layer is changed with n_r changing.

$$\begin{bmatrix} c_{11} \cdots c_{1n_y} \\ c_{21} \cdots c_{2n_y} \\ \vdots \quad \vdots \quad \vdots \\ c_{n_r,1} \cdots c_{n_r,n_y} \\ \mathbf{0} \ \cdots \ \mathbf{0} \end{bmatrix} \rightarrow \begin{bmatrix} c_{11} \cdots c_{1n_y} \\ c_{21} \cdots c_{2n_y} \\ \vdots \quad \vdots \quad \vdots \\ c_{n_r^{'}1} \cdots c_{n_r^{'}n_y} \\ \mathbf{0} \ \cdots \ \mathbf{0} \end{bmatrix}$$

Fig. 8.4 Example of regulation operator with the number of hidden nodes changed from n_r to $n_r^{'}$

(4) Mutation operator

A mutation operator with probability p_m is carried out. The new center c'_{ij} of the mutated element is produced as follows:

$$c'_{ij} = c_{ij} + r \cdot c_{ij} \tag{8.25}$$

where r is the same as that in Eq. 8.24. The new parameters of the hidden functions can then be derived by the mutation operator.

To further avoid getting stuck in the local optima, the dynamic probabilities of p_m and p_r are adopted as follows:

$$p_m = p_r = 0.1 + 0.1g/G \tag{8.26}$$

where g is the current evolution generation. What is noteworthy is that the mutation and regulation probability will gradually become larger. This will be helpful with keeping population diversity during the later evolution processes.

8.2.5.2 Optimization Process

The whole procedure of GA optimization for RBF neural network is given in the following steps:

Step 1: Set the population size N_c to 60, maximal generation G to 1000, β to 0.0001, and p_c to 0.8 and generate the initial population randomly.
Step 2: Calculate the outputs of the RBF neural network according to Sec. 8.2.3 and compute the value of the fitness function with Eq. 8.21.
Step 3: Perform a tournament selection to select the better individual with greater probability, calculate p_r and p_m, carry out the crossover, mutation and regulation operators to generate a new population.
Step 4: Keep the elitist individual in the new population.
Step 5: Go to Step 2 until G is reached.

8.3 Simulation Results

In this section, two DPSs are studied. One is a nonlinear catalytic rod in the chemical industry and the other is a heat conduction equation. The PDEs are solved using the finite difference method with $\{y(z_i, t_k)\}_{i=1k=1}^{120,100}$ spatial-temporal outputs produced to construct the low-order models. PCA is used to obtain the dominant components and temporal outputs. The decoupled ARX model will be formed using the temporal outputs and stimulus inputs. The deviation can then be derived for GA to optimize RBF neural networks.

The PCA-ARX method [16] and PCA-Hammerstein method [14] are adopted here for comparison. For the low-dimension temporal data set $\{\mathbf{u}(k), \mathbf{y}(k)\}$, the PCA-ARX model employed the following ARX form as $\mathbf{y}(k) = \mathcal{A}(q^{-1})\mathbf{y}(k) + \mathcal{B}(q^{-1})\mathbf{u}(k)$; while the PCA-Hammerstein algorithm utilized $\mathbf{y}(k) = \mathcal{A}(q^{-1})\mathbf{y}(k) + \mathcal{B}(q^{-1})\mathbf{v}(k)$, where $v(k)$ is the nonlinear static function of $u(k)$. Here $\mathcal{A}(q^{-1})$ and $\mathcal{B}(q^{-1})$ are $n \times n$ and $n \times m$ matrix polynomials, respectively.

The sums of the absolute error (SAE) and the RMSE are defined as follows.

$$\text{SAE} = \sum_{i=1}^{N} \sum_{k=1}^{L} |e(z_i, t_k)| \tag{8.27}$$

$$\text{RMSE} = \sqrt{\frac{1}{NL} \sum_{i=1}^{N} \sum_{k=1}^{L} |e(z_i, t_k)|^2} \tag{8.28}$$

where $e(z_i, t_k) = y(z_i, t_k) - \hat{y}(z_i, t_k)$.

8.3.1 Catalytic Rod

The reaction process is composed of a long, thin rod (catalytic rod) in a reactor shown in Fig. 8.5.

The reactor is fed with pure species A, and a zeroth order exothermic catalytic reaction of the form $A \to B$ takes place in the rod. The reaction is exothermic and a cooling medium is equipped for cooling. Here we firstly make assumptions of constant density, heat capacity, conductivity, temperature at both ends of the rod and an excess of species A, and then describe the spatio-temporal distribution of the dimensionless temperature as [14]:

$$\begin{aligned} \frac{\partial y(z,t)}{\partial t} &= \frac{\partial^2 y(z,t)}{\partial z^2} + \beta_T (e^{-\frac{\Upsilon}{1+y}} - e^{-\Upsilon}) \\ &+ \beta_u (\mathbf{b}^T(z)\mathbf{u}(t) - y(z,t)) \end{aligned} \tag{8.29}$$

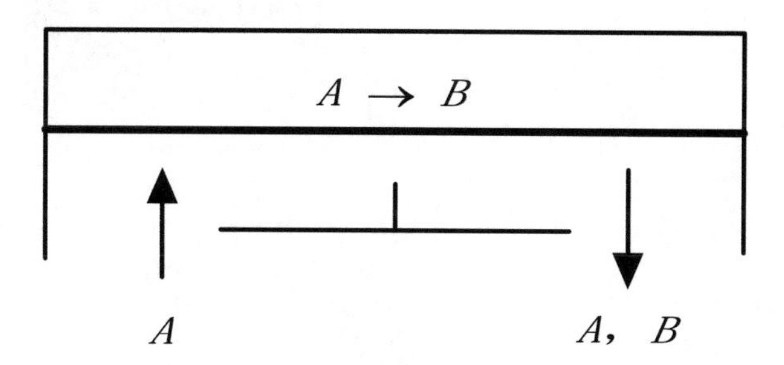

Fig. 8.5 A cataytic rod in a reactor with catalytic reaction form $A \to B$

subject to the Dirichlet boundary conditions:

$$y(0,t) = 0, \ y(Z,t) = 0$$

and the initial condition:

$$y(z,0) = y_0(z)$$

where $y(z, t)$, β_T, Υ, β_u, $\mathbf{b}(z)$, $\mathbf{u}(t)$, Z and T denote the dimensionless temperature in the reactor, the dimensionless reaction heat, the dimensionless activation energy, the dimensionless heat transfer coefficient, the actuator distribution, the manipulated input, i.e., the temperature of the cooling medium, the length of the reactor and the total chemical reaction time, respectively. The values of system parameters for the catalytic rod are shown in Table 8.1.

There are totally four actuators available $u(t) = [u_i(t), \ldots, u_4(t)]^{\mathrm{T}}$ with the space distribution function as $b_i(z) = H(z-(i-1)\pi/4) - H(z-i\pi)/4)$, $i = 1, \ldots, 4$, $\mathbf{b}(z) = [b_1(z), \ \cdots, \ b_4(z)]^{\mathrm{T}}$. Here, $u_i(t) = 1.1 + 5\sin(\frac{t}{10} + \frac{i}{10})$, $(i = 1, \ 2, \ 3, \ 4)$ is the standard Heaviside function. The input stimulus signals are $u_i(t) = 1.1 + 5\sin(\frac{t}{10} + \frac{i}{10})$, $i = 1, \ \cdots, \ 4$. The model shown in Fig. 8.6 is constructed by the production of the outputs of the catalytic rod.

Table 8.1 Typical values of the catalytic rod process

Parameters	β_T	β_u	Υ	$y_0(z)$	T	Z
The values	16	2	2	0.5	2	3

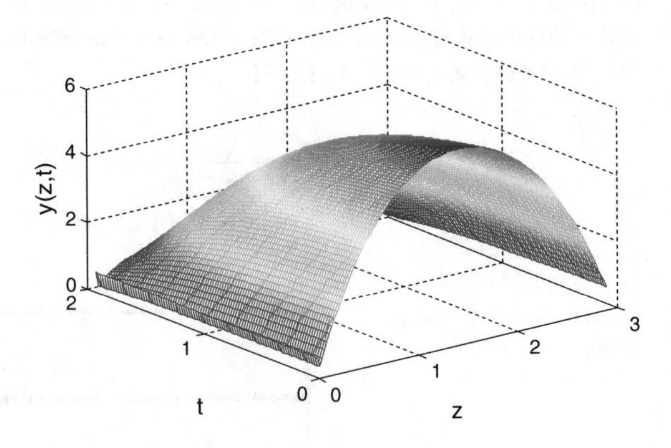

Fig. 8.6 The spatio temporal outputs of the catalytic rod with four stimulus input signals

8.3.1.1 PCA Time/Space Separation and ARX Modeling

The eigenvalues can be calculated based on PCA and the system energy for different numbers of dominant modes n, which is listed in Table 8.2. The system energy is more than 99% under $n = 3$ and $n = 4$ and a system model can be built with the information provided by this. The space basis functions $\varphi_i(z)$ and their temporal series $y_i(k)$ $(i = 1, \cdots, n)$ can also be obtained.

For $n = 4$, four actuators available in the reactor $\mathbf{u}(k) = [u_1(k), u_2(k), u_3(k), u_4(k)]^{\mathrm{T}}$ are used as system inputs and four temporal series are used as system outputs $\mathbf{y}(k) = [y_1(k), y_2(k), y_3(k), y_4(k)]^{\mathrm{T}}$. Set n_y to 2 and n_u to 2; the decoupled ARX model in Eq. 8.15 can be obtained as follows:

$$\mathbf{y}(k) = (\mathcal{A}_1 q^{-1} + \mathcal{A}_2 q^{-2})\mathbf{y}(k) + (\mathcal{B}_1 q^{-1} + \mathcal{B}_2 q^{-2})\mathbf{u}(k) \tag{8.30}$$

$\mathcal{A}_i = \text{diag}(a_{1i}, \cdots, a_{4i})$, $\mathcal{B}_i = \text{diag}(b_{1i}, \cdots, b_{4i})$, $i = 1, 2$.

The parameters in $\mathcal{A}_i$ and $\mathcal{B}_i$ are calculated with RLS algorithm in Eq. 8.16 as:

$$\begin{aligned}
\mathcal{A}_1 &= \text{diag}(1.9567, 1.8070, 1.8286, 1.3460)\\
\mathcal{A}_2 &= \text{diag}(-0.9574, -0.8110, -0.8443, -0.4277)\\
\mathcal{B}_1 &= \text{diag}(0.0387, -0.0016, 0.1682, -0.0661)\\
\mathcal{B}_2 &= \text{diag}(-0.0336, 0.0007, -0.1689, 0.0664)
\end{aligned}$$

For $n = 3$, to construct the three-input three-output decoupled ARX model, three actuators $\mathbf{u}(k) = [u_1(k), u_2(k), u_3(k)]^{\mathrm{T}}$ with the space distribution functions $b_i(z) = H(z - (i-1)\pi/3) - H(z - i\pi)3)$, $i = 1, 2, 3$ are chosen. The locations and the number of the actuators are different, so compared with that using four actuators, there exists a systematic deviation as shown in Fig. 8.7.

Here n_y and n_u are also set to 2, and three dominant time series are denoted as $\mathbf{y}(k) = [y_1(k), y_2(k), y_3(k)]^{\mathrm{T}}$, then the decoupled ARX model is similar to Eq. 8.30.

The parameters of $\mathcal{A}_1, \mathcal{A}_2, \mathcal{B}_1, \mathcal{B}_2$ are then derived and given as follows:

$$\begin{aligned}
\mathcal{A}_1 &= \text{diag}(a_{11}, a_{21}, a_{31}) = \text{diag}(1.9569, 1.7326, 1.8317)\\
\mathcal{A}_2 &= \text{diag}(a_{12}, a_{22}, a_{32}) = \text{diag}(-0.9575, -0.7376, -0.8459)\\
\mathcal{B}_1 &= \text{diag}(b_{11}, b_{21}, b_{31}) = \text{diag}(0.0453, -0.0122, 0.1549)\\
\mathcal{B}_2 &= \text{diag}(b_{12}, b_{22}, b_{32}) = \text{diag}(-0.0408, 0.0112, -0.1556)
\end{aligned}$$

Table 8.2 The energy of PCA with different n

The value of n	Energy(%)
1	95.3262
2	98.6537
3	99.7153
4	99.8996

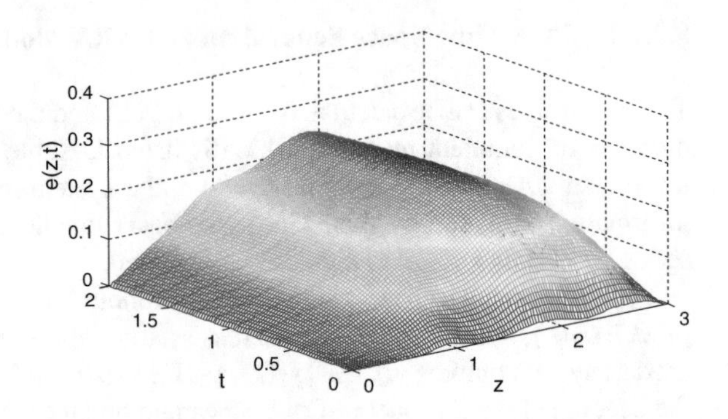

Fig. 8.7 The deviation of spatio temporal outputs using four actuators and three actuators

8.3.1.2 RBF Neural Network Optimized by GA

After constructing the decoupled ARX model, the nonlinear terms $\mathbf{f}(k) = [f_1(k), \cdots, f_n(k)]$ will be obtained with Eq. 8.17. Note that the weight coefficient β will affect the optimization result greatly, the first dominant time series $y_1(k)$ and its nonlinear part $f_1(k)$ for $n = 4$ are selected to show the effect of β. Here, the input layer is $[y_1(k-1), y_1(k-2)]$ and the output is $f_1(k)$. The input and output data are divided into two groups, where one group is used to train the hidden layer and its parameters of RBF neural network by using GA, and the other group is used to test the RBF neural network. GA is run 10 times for each value of β. The RBF neural network with the minimum value of SAE is regarded as the best one. The best results of RBF neural network with different values of β are shown in Table 8.3, where e_{s1} is SAE for the training data and e_{s2} is for the testing data.

It is obvious that when β is decreasing, the number of the hidden nodes will increase, and the modeling accuracy in Eq. 8.20 is also decreased. Herein, β is selected as 0.0001 to obtain good modeling accuracy.

After GA optimization, the seven best RBF neural networks with different n_r hidden nodes are obtained in the cases of $n = 4$ and $n = 3$. The results are shown in Table 8.4. It can be seen that although the testing error is larger than the training error, the values are quite small and satisfactory.

The outputs of the RBF neural network with its modeling errors in the cases of $n = 4$ and $n = 3$ are shown in Figs. 8.8, 8.9, 8.10, 8.11, 8.12, 8.13 and 8.14, respectively. In all of figures, the former fifty samples are selected as the training data and the latter fifty samples are the testing data. It can be seen that the order of the magnitudes has reached to 10^{-5} for the training and testing data.

Table 8.3 Optimization results of RBF for f_1 with different β

β	n_r	e_{s1}	e_{s2}
0.0001	22	0.009	0.011
0.01	17	0.0117	0.0217
0.1	8	0.0537	0.1218

Table 8.4 Results of RBF models for nonlinear functions

Index	$n = 4$				$n = 3$		
F	f_1	f_2	f_3	f_4	f_1	f_2	f_3
n_r	22	19	21	18	18	17	20
e_{s1}	0.009	0.008	0.010	0.021	0.028	0.024	0.013
e_{s2}	0.011	0.010	0.015	0.036	0.034	0.051	0.018

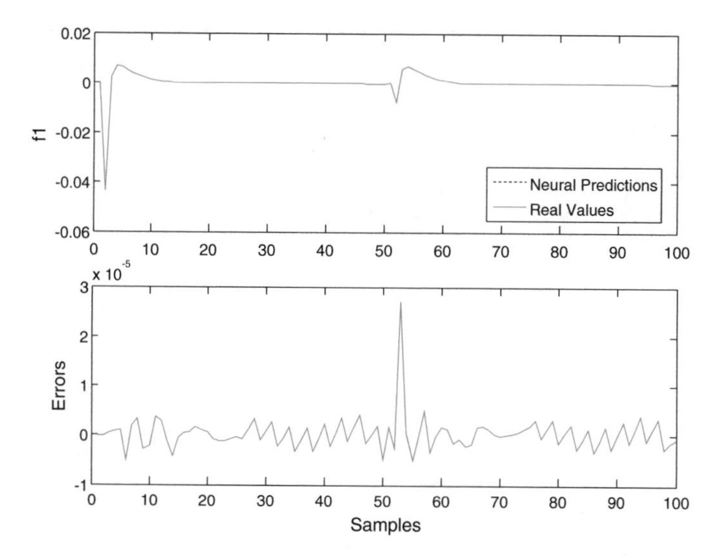

Fig. 8.8 Prediction of RBF network for f_1 and modeling errors of case $n = 4$

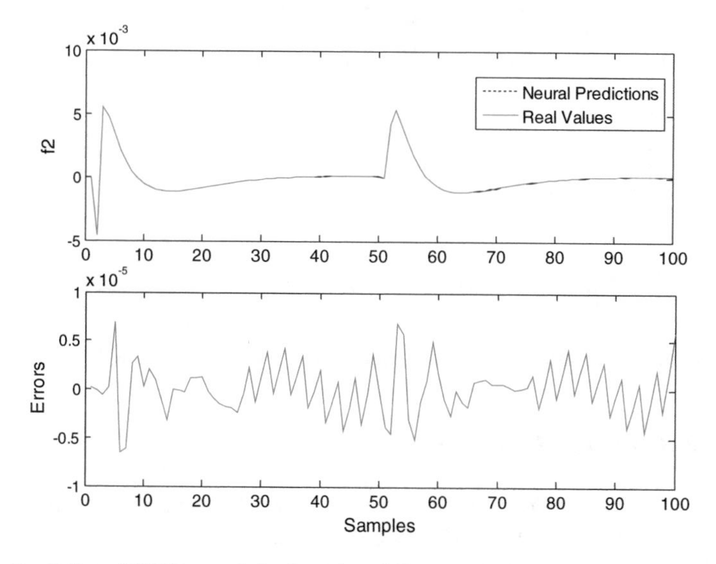

Fig. 8.9 Prediction of RBF network for f_2 and modeling errors of case $n = 4$

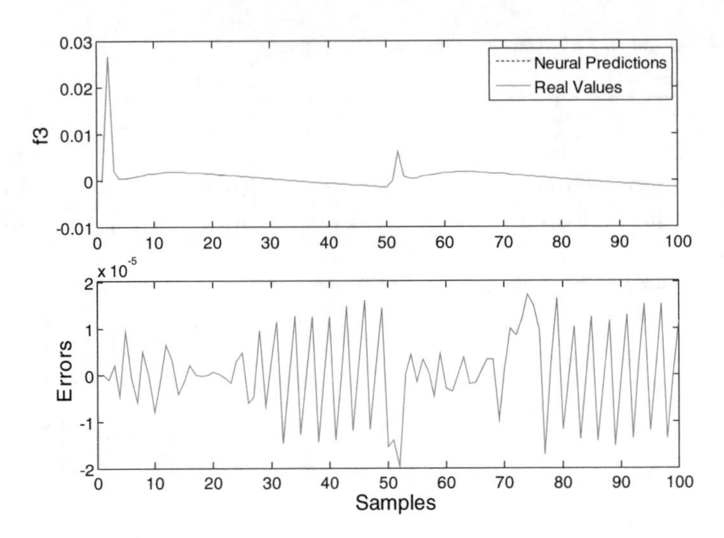

Fig. 8.10 Prediction of RBF network for f_3 and modeling errors of case $n = 4$

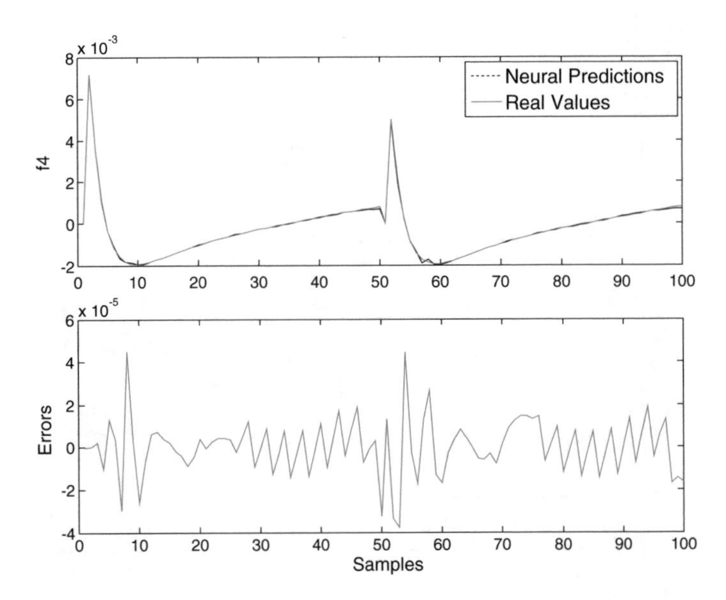

Fig. 8.11 Prediction of RBF network for f_4 and modeling errors of case $n = 4$

8.3.1.3 Performance Comparison

The decoupled ARX model plus RBF neural network has been obtained and $\hat{\mathbf{y}}_k(z)$ can be reconstructed with Eq. 8.12. For convenience, the proposed method is named as PCA-DARX-RBF. The reconstructed outputs in the cases of $n = 4$ are shown in Fig. 8.15 and the prediction errors $e(z, t)$ are presented in Fig. 8.16a. For the case of

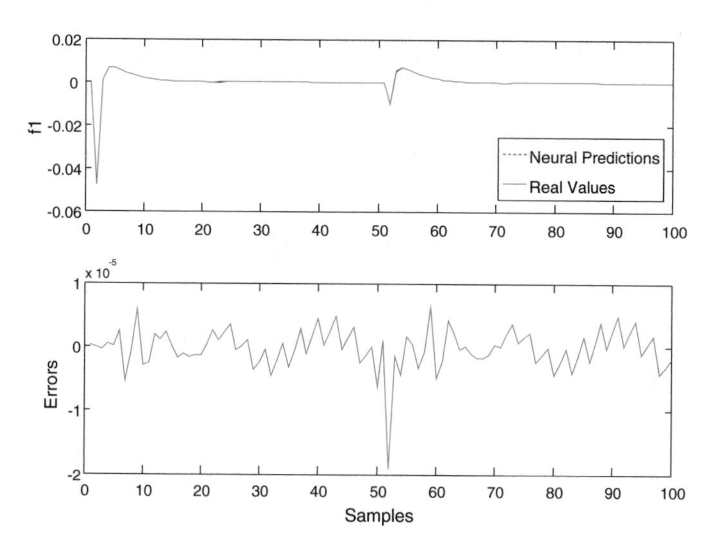

Fig. 8.12 Prediction of RBF network for f_1 and modeling errors of case $n = 3$

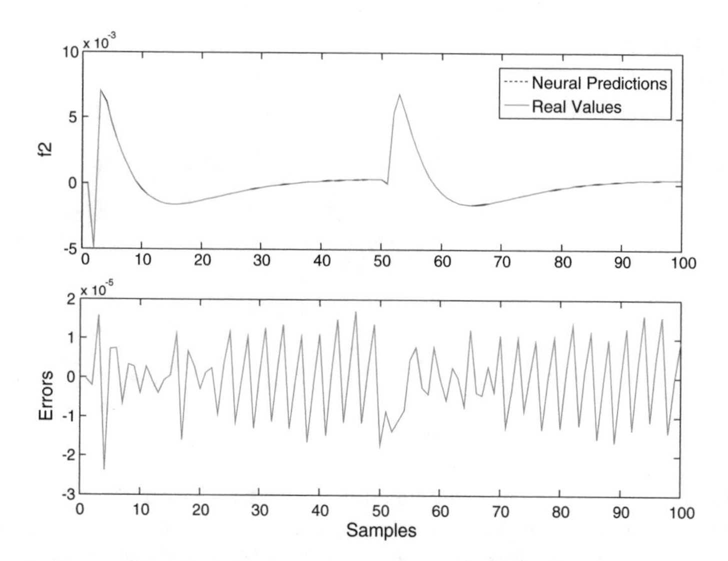

Fig. 8.13 Prediction of RBF network for f_2 and modeling errors of case $n = 3$

$n = 3$, the prediction error values are plotted in Fig. 8.16b. It can be seen that the PCA-DARX-RBF model can approximate the spatio-temporal dynamics of the original system satisfactorily. However, the error values increase with fewer dominant modes.

In Fig. 8.17, the spatial-temporal reconstruction error values for $n = 3$ and $n = 4$ show the efficiency of the PCA-ARX model. For the PCA-Hammerstein model, the literature [14] gave the expression of $v(k)$ for $n = 3$ with four input actuators and a

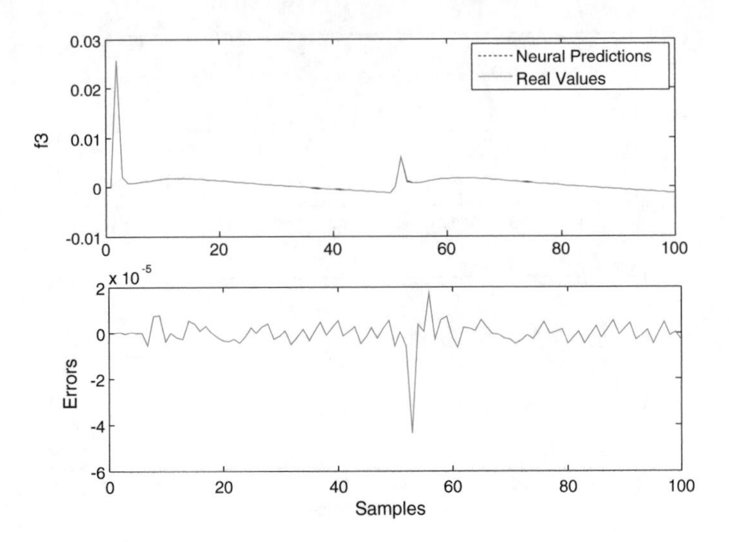

Fig. 8.14 Prediction of RBF network for f_3 and modeling errors of case $n = 3$

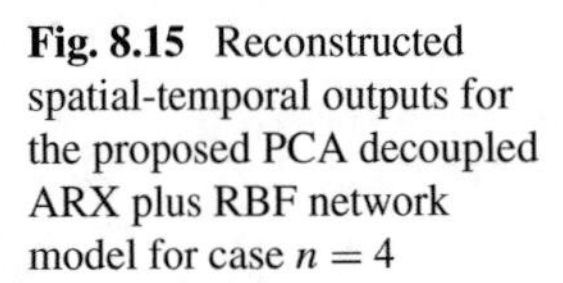
Fig. 8.15 Reconstructed spatial-temporal outputs for the proposed PCA decoupled ARX plus RBF network model for case $n = 4$

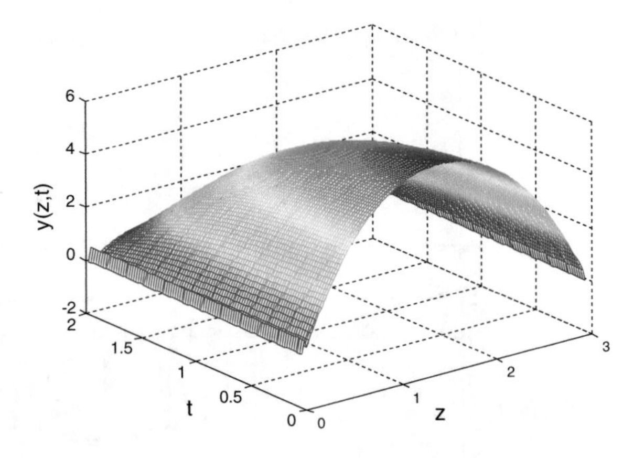

Gaussian function used as the static function. Its centers were uniformly distributed among the range of the input data, and the width was set to 1. Here, n_y and n_u are also set as 2. The reconstruction errors are smaller than those of [14] as shown in Fig. 8.18a, and the simulation result for $n = 4$ with the same $v(k)$ is shown in Fig. 8.18b. It can be seen that the PCA-Hammerstein model can approximate the spatio-temporal dynamics well.

The indexes of RMSE and SAE for the three methods are listed in Table 8.5. Since the PCA with $n \geq 3$ obtains more than 99% system energy, reconstruction accuracy is guaranteed in the whole spatio-temporal distribution. All of the methods in Figs. 8.16, 8.17 and 8.18 show satisfactory accuracy. However, the modeling errors with $n = 3$ are larger than those with $n = 4$. In Table 8.4, the values of RMSE and ASE with $n = 4$ are also less than those with $n = 3$.

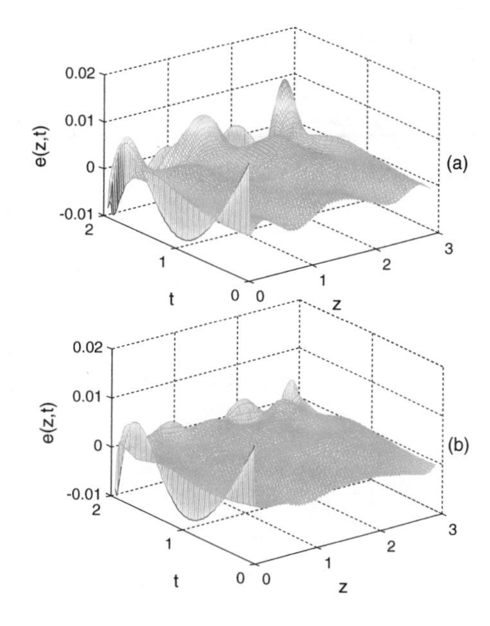

Fig. 8.16 Spatial-temporal modeling errors by PCA-DARX-RBF for the cases of $n = 3$ **a** and $n = 4$ **b**

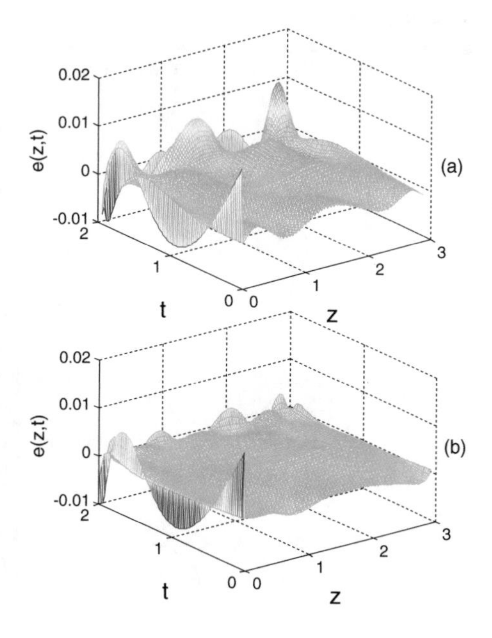

Fig. 8.17 Spatial-temporal modeling errors by PCA-ARX for the cases of $n = 3$ **a** and $n = 4$ **b**

It can be seen that PCA-Hammerstein has larger errors than those of PCA-ARX and PCA-DARX-RBF partly because of the structure of $v(k)$. The limitation of PCA-ARX model is that this model is confined to linear systems and fully coupled, while PCA-Hammerstein model is for the nonlinear and inter-coupling systems. Obviously, PCA-DARX-RBF can obtain good reconstruction accuracy for nonlinear DPSs.

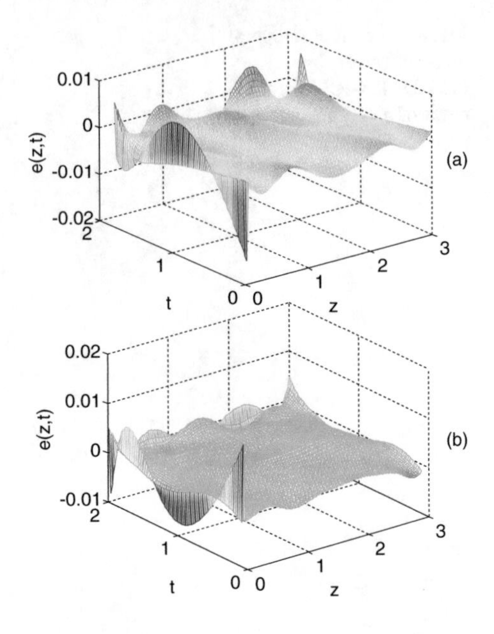

Fig. 8.18 Spatial-temporal modeling errors by PCA Hammerstein under the cases of $n = 3$ **a** and $n = 4$ **b**

Table 8.5 Index comparisons for two cases

Cases	Methods	RMSE	ASE
Catalytic rod	PCA-ARX $n = m$=4	7.77e − 4	4.8947
	PCA-ARX $n = m$=3	0.0017	12.1216
	PCA-Hammerstein $n = 3$,$m = 4$	0.0017	12.1815
	PCA-Hammerstein $n = m$=4	8.256e − 4	5.5348
	PCA-DARX-RBF $n = m$=4	9.25e − 4	5.2878
	PCA-DARX-RBF $n = m$=3	0.0017	11.3534
Heat conduction	PCA-ARX $n = m$=4	6.69e − 4	4.1655
	PCA-ARX $n = m$=3	0.0019	12.0925
	PCA-DARX $n = m$=4	7.51e − 4	4.2001
	PCA-DARX $n = m$=3	0.0019	12.0643
	PCA-DARX-RBF $n = m$=4	7.72e − 4	4.0954
	PCA-DARX-RBF $n = m$=3	0.0020	12.1424

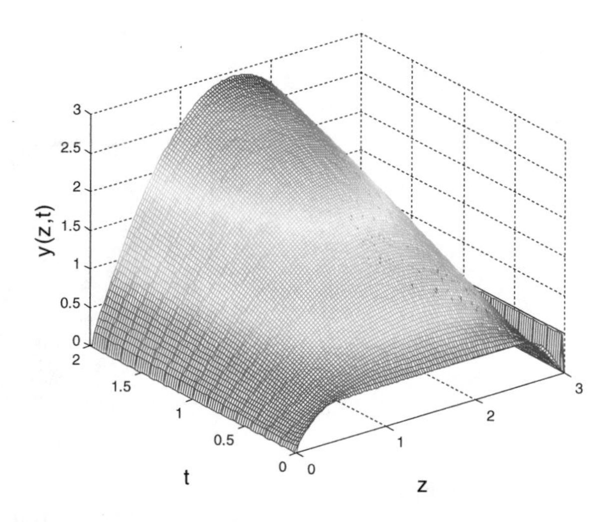

Fig. 8.19 The spatial-temporal outputs of the heat conduction equation with four stimulus input signal

8.3.2 *Heat Conduction Equation*

The dynamic parabolic PDE of a linear heat conduction equation is described as follows [25]:

$$\frac{\partial y(z,t)}{\partial t} = \frac{\partial^2 y(z,t)}{\partial z^2} + \mathbf{b}^{\mathrm{T}}(z)\mathbf{u}(t) \tag{8.31}$$

$$\text{s.t. } y(0,t) = 0,\ y(Z,t) = 0,\ t \in [0,T],\ y(z,0) = y_0(z),\ z \in [0,Z].$$

The parameters Z, T and their stimulus input signals $\mathbf{b}^T(z)\mathbf{u}(t)$ are set as the same as those in Sec. 8.3.1. The DPS outputs $\{y(z_i,t_k)\}_{i=1}^{120,}{}_{k=1}^{100}$ with four input stimulus signals are also produced and shown in Fig. 8.19. The systematic deviation between four inputs and three inputs is illustrated in Fig. 8.20.

By using PCA method, the system energy in the dominant modes $n = 2, 3, 4$ is gained 97.65%, 99.44% and 99.83%, respectively. Here the dominant modes $n = 3$ and $n = 4$ are selected, and the temporal series $y_i(k)$ $i = 1, \ldots, n$ can be obtained. In order to build the decoupled temporal model, the number of the stimulus inputs is equal to n. n_y and n_u in the ARX structure as the same as those in modeling the catalytic rod. The model coefficients for $n = 4$ are obtained as:

$$\begin{aligned}
A_1 &= \mathrm{diag}(1.4912, 0.8145, 1.8176, 1.3476)\\
A_2 &= \mathrm{diag}(-0.5063, 0.1784, -0.8290, -0.4186)\\
B_1 &= \mathrm{diag}(0.0296, -0.0051, -0.1339, 0.0613)\\
B_2 &= \mathrm{diag}(0.0369,\ 0.0013, 0.1345, -0.0615)
\end{aligned}$$

For $n = 3$, the model coefficients are:

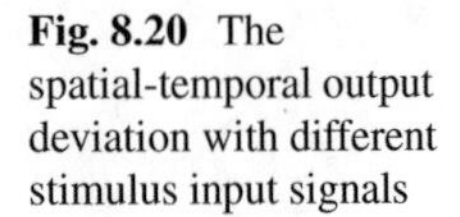
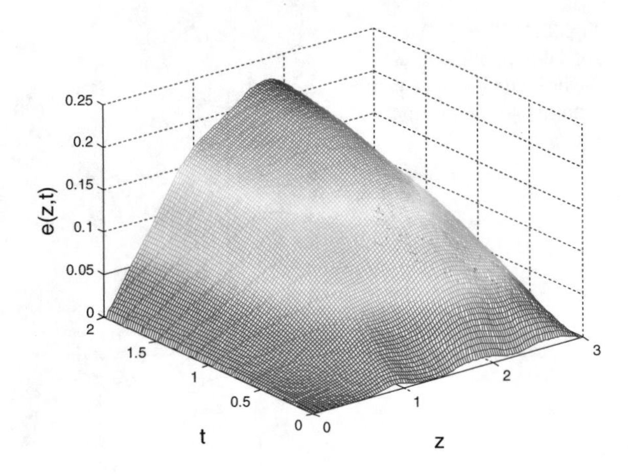

Fig. 8.20 The spatial-temporal output deviation with different stimulus input signals

$$
\begin{aligned}
A_1 &= \text{diag}(1.5112, 0.9337, 1.8093) \\
A_2 &= \text{diag}(-0.5248, 0.0502, -0.8184) \\
B_1 &= \text{diag}(0.0078, 0.0257, -0.1188) \\
B_2 &= \text{diag}(0.0485, -0.0288, 0.1194)
\end{aligned}
$$

The nonlinear terms for $n = 4$ are modeled and the predicted outputs are shown in Fig. 8.21. The output error values using PCA-DARX-RBF for $n = 4$ and $n = 3$ are shown in Fig. 8.22a, b. Note that the nonlinear parts are relatively small, the RBF neural networks are omitted and the algorithm is named as PCA-DARX. The spatial-temporal reconstruction error values are shown in Fig. 8.23a, b. Here the modeling error values for PCA-ARX are plotted in Fig. 8.24a, b.

It can be seen that the error of the four dominant modes is smaller than that of the three dominant modes. The RMSE and ASE for PCA-DARX-RBF, PCA-DARX and PCA-ARX are listed in Table 8.5. The nonlinearity is not serious and RBF neural network does not give any significant contribution, so RMSE for PCA-DARX-RBF is the largest. RMSE and ASE for PCA-DARX are somewhat larger than those of PCA-ARX because the simple decoupled ARX misses some system information. Therefore, for linear DPSs, RBF neural networks may not be helpful to improve modeling accuracy, and the PCA-DARX-RBF method can be simplified to PCA-DARX with satisfactory modeling accuracy.

8.4 Summary

This chapter proposes a hybrid modeling strategy for parabolic PDE systems. PCA is firstly utilized to derive the set of dominant spatial patterns and the temporal series. Then, a decoupled ARX form is constructed and the nonlinear unmodeled dynamics are further modelled with a GA optimization-based RBF neural network.

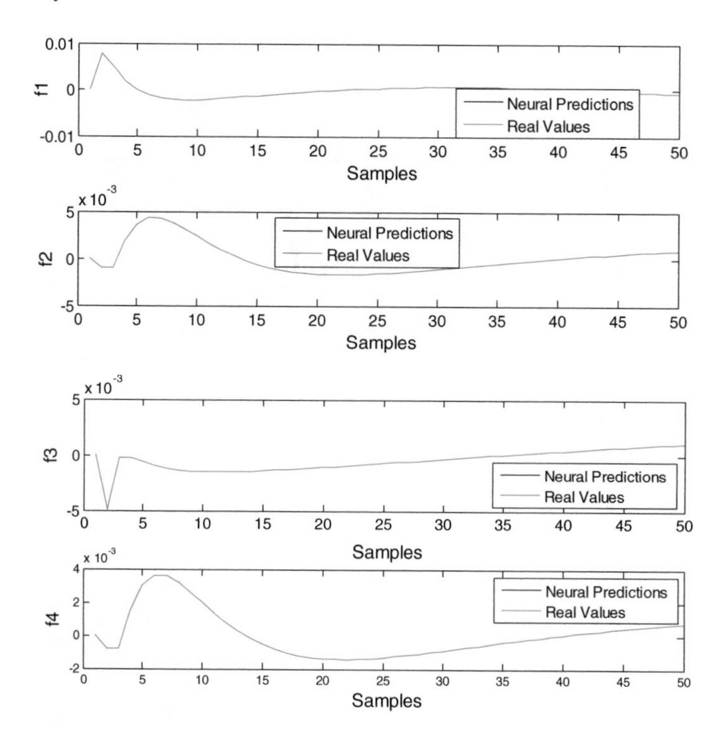

Fig. 8.21 The nonlinear terms predicted RBF network for $n = 4$

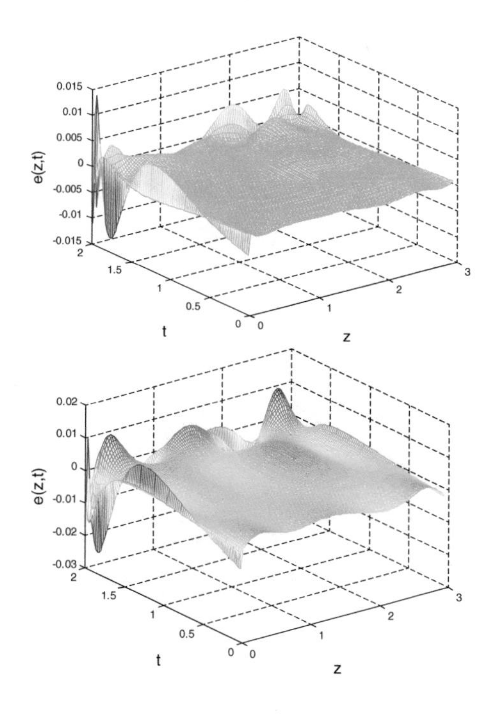

Fig. 8.22 Spatial-temporal modeling errors by PCA-DARX-RBF for $n = 4$ **a** and $n = 3$ **b**

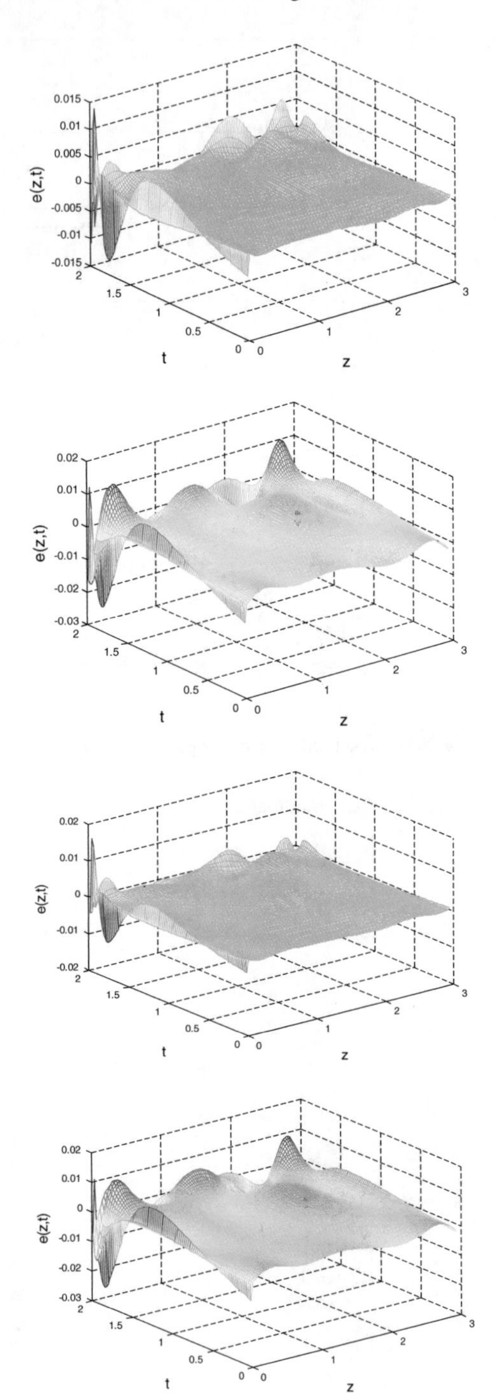

Fig. 8.23 Spatial-temporal modeling errors by PCA-DARX for $n = 4$ **a** and $n = 3$ **b**

Fig. 8.24 Spatial-temporal modeling errors by PCA-AR X for $n = 4$ **a** and $n = 3$ **b**

The satisfying modeling results of two DPS systems have been derived by using PCA-DARX-RBF method.

References

1. Zhang, R., et al. 2016. Decoupled ARX and RBF neural network modeling using PCA and GA optimization for nonlinear distributed parameter systems. *IEEE Transactions on Neural Networks Learning Systems* 29 (2): 457–469.
2. Baker, J., and P. Christofides. 2000. Finite-dimensional approximation and control of non-linear parabolic PDE systems. *International Journal of Control* 73(5): 439–456.
3. Luo, B., H.N. Wu, and H.X. Li. 2015. Adaptive optimal control of highly dissipative nonlinear spatially distributed processes with neuro-dynamic programming. *IEEE Transactions on Neural Networks Learning Systems* 26 (4): 684.
4. Christofides, P.D., and J. Chow. 2001. Nonlinear and robust control of PDE systems. *Applied Mechanics Reviews* 55 (2): B29–B30.
5. Zhang, R., and J.L. Tao. 2017. Data driven modeling using improved multi-objective optimization based neural network for coke furnace system. *IEEE Transactions on Industrial Electronics* 64 (4): 3147–3155.
6. Luo, B., H.N. Wu, and H.X. Li. 2014. Data-based suboptimal neuro-control design with reinforcement learning for dissipative spatially distributed processes. *Industrial Engineering Chemistry Research* 53 (19): 8106–8119.
7. Zhang, R., A. Xue, and S. Wang. 2011. Dynamic modeling and nonlinear predictive control based on partitioned model and nonlinear optimization. *Industrial Engineering Chemistry Research* 50 (13): 8110–8121.
8. Aggelogiannaki, E., et al. 2008. Nonlinear model predictive control for distributed parameter systems using data driven artificial neural network models. *Computers Chemical Engineering* 32(6): 1225–1237.
9. Yin, S., et al. 2013. Data-driven monitoring for stochastic systems and its application on batch process. *International Journal of Systems Science* 44(7): 1366–1376.
10. Wang, M., X. Yan, and H. Shi. 2013. Spatiotemporal prediction for nonlinear parabolic distributed parameter system using an artificial neural network trained by group search optimization. *Neurocomputing* 113 (7): 234–240.
11. Chairez, I., I. García-Peña, and A. Cabrera. 2009. Dynamic numerical reconstruction of a fungal biofiltration system using differential neural network. *Journal of Process Control* 19 (7): 1103–1110.
12. Zhang, R., et al. 2009. Neural network based iterative learning predictive control design for mechatronic systems with isolated nonlinearity. *Journal of Process Control* 19 (1): 68–74.
13. Shvartsman, S.Y., et al. 2000. Order reduction for nonlinear dynamic models of distributed reacting systems. *Journal of Process Control* 10 (2–3): 177–184.
14. Qi, C., and H.X. Li. 2009. A time/space separation-based Hammerstein modeling approach for nonlinear distributed parameter processes. *Computers Chemical Engineering* 33 (7): 1247–1260.
15. Qi, C., H.T. Zhang, and H.X. Li. 2009. A multi-channel spatio-temporal Hammerstein modeling approach for nonlinear distributed parameter processes. *Journal of Process Control* 19 (1): 85–99.
16. Hua, C., L.I. Ning, and L.I. Shao-Yuan. 2011. Time-space ARX modeling and predictive control for distributed parameter system. *Control Theory Applications* 28 (12): 1711–1716.
17. Kariminia, S., et al. 2016. Modelling thermal comfort of visitors at urban squares in hot and arid climate using NN-ARX soft computing method. *Theoretical Applied Climatology* 124 (3–4): 991–1004.

18. Zhang, R., J. Tao, and F. Gao. 2014. Temperature modeling in a coke furnace with an improved RNA-GA based RBF network. *Industrial Engineering Chemistry Research* 53 (8): 3236–3245.
19. Shamshirband, S., et al. 2016. Estimation of reference evapotranspiration using neural networks and cuckoo search algorithm. *Journal of Irrigation Drainage Engineering* 142 (2): 04015044.
20. Glauser, M. 1996. *Turbulence.* Dynamical Systems and Symmetry: Coherent Structures.
21. Armaou, A., and P.D. Christofides. 1999. Nonlinear feedback control of parabolic partial differential equation systems with time-dependent spatial domains ☆. *Journal of Mathematical Analysis Applications* 73 (17): 124–157.
22. Chorin, A. 1998. New perspectives in turbulence. *Quarterly of Applied Mathematics* 56 (4): 767–785.
23. Liu, J. 2013. *Radial Basis Function (RBF) Neural Network Control for Mechanical Systems.* Springer Science & Business Media.
24. Chen, S., C.N. Cowan, and P.M. Grant. 1991. Orthogonal least squares learning algorithm for radial basis function networks. *IEEE Transactions on Neural Network* 64 (5): 829–837.
25. Goldstein, J.A. 1985. *Semigroups of Operators and Applications.*

Chapter 9
GA-Based Controller Optimization Design

In this chapter, GA is used to optimize the controller design. First, a new PID controller is designed by using a non-minimal state-space model through predictive function control. The weighting matrix in the predictive function controller is optimized through GA so as to achieve a relatively desired closed-loop control performance. Secondly, a fuzzy neuron non-model controller is designed for a continuous casting process with strong nonlinearity and severe uncertainty, and its parameters are optimized through RNA-GA. Finally, a MOGA based on parameter stabilization space of the PID controller is used to control the first-order lag unstable process. The simulation results confirm the effectiveness of GA and its improved format in the optimization of the control system design problem.

9.1 Introduction

A robust and reliable control system is critical to obtain high-quality production for satisfying the requirement of industrial applications [1]. With the more and more complexity of the modern industrial processes, higher control performances are required for the control system design. All of the engineers are facing increasing challenges of how to choose the appropriate controller and optimize its parameters to meet different application requirements [2, 3]. The performances of the actual controller often have strict constraints, moreover, the whole control system may not have continuous and differentiable characteristics suitable for the traditional numerical optimization methods [4–5].

Because of PID controllers' simplicity, satisfactory performances, and high cost/benefit ratio, they are widely used in the industrial plants. For the purpose of simplifying the engineers' work, many PID parameter tuning methods have been devised during the past 70 years and applied widely in the industrial processes [6, 7]. PID parameter tuning is typically categorized by the type of process models, such as first-order plus dead time (FOPDT) [7–8] and integrator plus dead time

J. Tao et al., *DNA Computing Based Genetic Algorithm*,
https://doi.org/10.1007/978-981-15-5403-2_9

(IPDT) [9, 10], etc. Also, some PID tuning methods are applicable for both FOPDT and IPDT models [11−12]. For example, internal model control (IMC) based PID tuning method showed good robustness and set-point tracking, but poor response under disturbance for processes with dominant lags [10]; The improved disturbance rejection for IPDT processes could be obtained by PID parameter tuning but poor performance for processes with large time delay [11]. A compromise between robustness and stability control performances were discussed in [13]. Some model predictive control (MPC) based PID controllers have also been proposed in [14] and [15], which have made significant progress in both theory and practice among various advanced control strategies [16, 17]. Sometimes, there is great uncertainty in the controlled plant or its structure varies greatly. So that, the mathematical model of the practical process is difficult to be established. Thus, the application of model-based control may be limited to a restricted extent and its control performance will be deteriorated.

To solve the system modeling problem, the research has undergone from traditional input-output models to state-space models. To overcome the limitation of traditional state observers, non-minimal state-space models, fuzzy logic models, and other advanced models have been developed [18−19]. However, MPCs still need to overcome the troublesome nonlinearity and uncertainty of the system. Furthermore, the implementation of MPC is not as easy as that of PID controller, which possessed the advantages of finding effective control with a simple structure. The new PID controller combined with MPC and predictive function control (PFC) is also designed in [20] and [21].

Recently, many artificial intelligence (AI) techniques, such as fuzzy systems [22, 23], neural networks [24, 25], and neuro-fuzzy logic [26, 27] have been applied to improve the performances of PID controller. Neuron controllers are particularly suitable for the objects with the time-variance and uncertainty because they are independent on the model and have self-learning capability. The simulations studied on industrial process control such as turbine and electroslag remelting showed the excellent tracking performance and robustness of the controllers [28, 29].

Many evolutionary search methods, such as particle swarm optimization [30], differential evolution [31], simulated annealing (SA) [32], and genetic algorithms (GAs) [33] have received much interest due to their high potential to be applied to global optimization of controller optimization design. As an optimization technique, GAs neither need to know the property of the problem, nor need to have the strict mathematical characteristic requirement. Moreover, GA offers an effective and efficient optimization method based on selection, crossover, and mutation operators. Therefore, it can be easily applied to the problems involving non-differentiable functions and discrete search spaces.

In this chapter, GA is applied in optimizing three types of controllers to improve the control performance.

(1) Motivated by the extended non-minimal state-space models (ENMSS), a PID controller is designed based on PFC using these models. The resulting controller

is having the MPC framework in fact. In view of that, the performance is associated with the weighting matrix of the cost function, then GA is adopted to optimize the elements in these matrices.

(2) Considering the difficulty of system modeling, the model-free neuron controller combined with the fuzzy PI controller is constructed. The gain of the neuron controller is adjusted online by a fuzzy algorithm and the parameters of the proposed controller are optimized by RNA-GA. The results illustrate high precision mold-level control is reached and the proposed control method can efficiently control mold level for the plant with great uncertainty and grave nonlinearities.

(3) A MOGA based on stabilization subspaces that optimizes PID controller parameters for unstable FOPDT processes is developed. Two-level PID controller structure is utilized. The inner loop is used to stabilize the unstable system, and the outer loop of the PID controller is optimized by MOGA based on stabilization subspaces to improve the control performance. The simulation results of several unstable FOPDT plants show the efficiency of the proposed methods.

9.2 Non-minimal State-Space Predictive Function PID Controller

9.2.1 Process Model Formulation

For simplicity, the single-input single-output (SISO) model is adopted here, which is described as follows:

$$\Delta y(k+1) + L_1\Delta y(k) + L_2\Delta y(k-1) + \cdots + L_p\Delta y(k-p+1)$$

$$= S_1\Delta u(k) + S_2\Delta u(k-1) + \cdots + S_q\Delta u(k-q+1) \tag{9.1}$$

where $u(k)$, $y(k)$ are the input signals and process output at time instant K, p and q are the output and input orders, Δ is the difference operator, respectively.

Based on the strategy in [21], a non-minimal state vector is selected as

$$\begin{aligned}\Delta x_m(k)^{\mathrm{T}} = [\Delta y(k)\ \ \Delta y(k-1)\ \ \ldots\ \ \Delta y(k-p+1)\ \ \Delta u(k-1)\\ \Delta u(k-2)\ \ \ldots\ \ \Delta u(k-q+1)]\end{aligned} \tag{9.2}$$

Then a state-space model can be derived as follows:

$$\begin{aligned}\Delta x_m(k+1) &= A_m\Delta x_m(k) + B_m\Delta u(k)\\ \Delta y(k+1) &= C_m\Delta x_m(k+1)\end{aligned} \tag{9.3}$$

where

$$A_m = \begin{bmatrix} -L_1 & -L_2 & \cdots & -L_{p-1} & -L_p & S_2 & \cdots & S_{q-1} & S_q \\ 1 & 0 & \cdots & 0 & 0 & 0 & \cdots & 0 & 0 \\ 0 & 1 & \cdots & 0 & 0 & 0 & \cdots & 0 & 0 \\ \vdots & \vdots & \cdots & \vdots & \vdots & \vdots & \cdots & \vdots & \vdots \\ 0 & 0 & \cdots & 1 & 0 & 0 & \cdots & 0 & 0 \\ 0 & 0 & \cdots & 0 & 0 & 0 & \cdots & 0 & 0 \\ 0 & 0 & \cdots & 0 & 0 & 1 & \cdots & 0 & 0 \\ \vdots & \vdots & \cdots & \vdots & \vdots & \vdots & \cdots & \vdots & \vdots \\ 0 & 0 & \cdots & 0 & 0 & 0 & \cdots & 1 & 0 \end{bmatrix}$$

$$B_m = \begin{bmatrix} S_1 \; 0 \cdots 0 \; 1 \; 0 \cdots 0 \end{bmatrix}^{\mathrm{T}}$$

$$C_m = \begin{bmatrix} 1 \; 0 \; 0 \cdots 0 \; 0 \; 0 \; 0 \end{bmatrix}$$

Define the expected set point $r(k)$, the output error is further expressed as

$$e(k) = y(k) - r(k) \tag{9.4}$$

Considering Eqs. (9.3) and (9.4), the dynamics of output error are

$$e(k+1) = e(k) + C_m A_m \Delta x_m(k) + C_m B_m \Delta u(k) - \Delta r(k+1) \tag{9.5}$$

A new state variable $z(k)$ is further defined as

$$z(k) = \begin{bmatrix} \Delta x_m(k) \\ e(k) \end{bmatrix} \tag{9.6}$$

Thus, the final ENMSS process model is derived:

$$z(k+1) = Az(k) + B\Delta u(k) + C\Delta r(k+1) \tag{9.7}$$

where

$$A = \begin{bmatrix} A_m & \mathbf{0} \\ C_m A_m & 1 \end{bmatrix}; B = \begin{bmatrix} B_m \\ C_m B_m \end{bmatrix}; C = \begin{bmatrix} \mathbf{0} \\ -1 \end{bmatrix} \tag{9.8}$$

In Eq. (9.8), $\mathbf{0}$ is a zero vector with an appropriate dimension.

Remark 9.1 The process dead time can be incorporated when $S_1 = S_2 = \cdots = S_d$ $(d \leq q)$.

9.2.2 PID Controller Design

In this section, the PFC strategy is adopted for the design of PID controller, and the future state variable from sampling instant k is

$$z(k+P) = A^P z(k) + \psi \Delta u(k) + \theta \Delta R \tag{9.9}$$

where

$$\theta = \begin{bmatrix} A^{P-1}C & A^{P-2}C & \cdots & C \end{bmatrix};\ \psi = A^{P-1}B$$

$$\Delta R = \begin{bmatrix} \Delta r(k+1) & \Delta r(k+2) & \cdots & \Delta r(k+P) \end{bmatrix}^{\mathrm{T}}$$

$$r(k+i) = \alpha^i y(k) + (1-\alpha^i)c(k)$$

Here P is the prediction horizon, α is the smoothing factor, $c(k)$ is the set point. The cost function is selected as

$$\min J(k) = z(k+P)^{\mathrm{T}} Q z(k+P) \tag{9.10}$$

where Q is the weighting matrix.

The discrete PID controller is adopted as follows:

$$u(k) = u(k-1) + K_p(k)(e_1(k) - e_1(k-1)) + K_i(k)e_1(k)$$

$$+ K_d(k)(e_1(k) - 2e_1(k-1) + e_1(k-2))$$

$$e_1(k) = c(k) - y(k) \tag{9.11}$$

where $K_p(k)$, $K_i(k)$, $K_d(k)$ denote the proportional coefficient, the integral coefficient, and the derivative coefficient, respectively, $e_1(k)$ is the error between the process set point and the real output.

Equation (9.11) is further formulated as follows:

$$\begin{aligned} u(k) &= u(k-1) + w(k)^{\mathrm{T}} E(k) \\ w(k) &= [w_1(k), w_2(k), w_3(k)]^{\mathrm{T}} \\ w_1(k) &= K_p(k) + K_i(k) + K_d(k) \\ w_2(k) &= -K_p(k) - 2K_d(k) \\ w_3(k) &= K_d(k) \\ E(k) &= [e_1(k), e_1(k-1), e_1(k-2)]^{\mathrm{T}} \end{aligned} \tag{9.12}$$

From Eqs. (9.3)–(9.12), the optimal control action can be derived as follows:

$$w(k) = -\frac{\psi^{\mathrm{T}} Q(A^P z(k) + \theta \Delta R)E(k)}{\psi^{\mathrm{T}} Q \psi E(k)^{\mathrm{T}} E(k)} \tag{9.13}$$

where

$$\begin{aligned} K_p(k) &= -w_2(k) - 2K_d(k) \\ K_i(k) &= w_1(k) - K_P(k) - K_d(k) \\ K_d(k) &= w_3(k) \end{aligned} \tag{9.14}$$

However, w is to be infinite if $e_1(k)$ is reaching zero, but it is obviously unrealistic. Here a small output error limitation δ is chosen, which leads to the following realistic control formulation:

$$\begin{cases} K_p(k) = K_p(k-1) \\ K_i(k) = K_i(k-1) \ldots |e_1(k)| \le \delta \\ K_d(k) = K_d(k-1) \end{cases}$$
$$\begin{cases} K_p(k) = -w_2(k) - 2K_d(k) \\ K_i(k) = w_1(k) - K_P(k) - K_d(k) \ldots |e_1(k)| > \delta \\ K_d(k) = w_3(k) \end{cases} \tag{9.15}$$

Then

$$\begin{aligned} u(k) = u(k-1) &+ K_p(k)(e_1(k) - e_1(k-1)) + K_i(k)e_1(k) \\ &+ K_d(k)(e_1(k) - 2e_1(k-1) + e_1(k-2)) \end{aligned} \tag{9.16}$$

9.2.3 GA-Based Weighting Matrix Tuning

A basic GA is introduced to optimize the elements in Q in order to achieve the desired process responses. Note that Q is represented as

$$Q = \mathrm{diag}\{q_{j\,1}, q_{j\,2}, \ldots, q_{j\,p}, q_{j\,p+1}, q_{j\,p+2}, \ldots, q_{j\,p+q-1}, q_{j\,e}\}$$

Remark 9.2 It can be seen that $q_{j1}, q_{j2}, \ldots, q_{jp}$ are related to the responses of the process output, $q_{j\,p+1}, q_{j\,p+2}, \ldots, q_{j\,p+q-1}$ are related to the process input responses and $q_{i\,e}$ are the weights on the output error. It shows that these parameters can be tuned to achieve desired process performance. Note that the first element in Q has the largest impact on the closed-loop responses, thus the first one is optimized, and the others are set to zeros.

9.2.3.1 Encoding Method

The first element of Q is denoted by the binary encoding format, a 15-bit binary code just to be as follows:

$$[b_{14}, \ldots, b_0] = |1|0|0|1|1|0|1|0|1|1|1|1|0|1|1| \tag{9.17}$$

Then the first element in Q can be decoded and recalculated by

$$q_{11} = \sum_{i=0}^{14} b_i 2^i / 2^{15} (q_{\max} - q_{\min}) + q_{\min} \tag{9.18}$$

where $q_{\min} = 1$, $q_{\max} = 100$.

9.2.3.2 Fitness Function

Here the minimization of overshoot and rise time responses are considered as follows:

$$\min \quad \text{over shoot}(k) + \text{rise time}(k) \tag{9.19}$$

Thus, the following fitness function is expressed as

$$\text{Max} \quad 1/[\text{over shoot}(k) + \text{rise time}(k)] \tag{9.20}$$

9.2.3.3 Operators

(1) Selection operator

The selection probability of an individual is given by Roulette wheel method as follows:

$$P(c_l) = \frac{f(c_l)}{\sum_{l=1}^{N} f(c_l)} \tag{9.21}$$

where c_l is the individual of Q, $f(c_l)$ is its fitness value in Eq. (9.20),N is the population size.

(2) Crossover and mutation operators

The crossover and mutation probability are denoted as p_c and p_m. The single-point crossover operation is carried out between individuals c_u and c_v at probability p_c, then the offspring $c_u^{'}$, $c_v^{'}$ will be produced. The value of gene is negated when the mutation operation is implemented at probability p_m.

The whole optimization procedure of PFC-PID is described in the following steps.

Step 1 Initialization of the maximal generation G and the population size N. Generation of the chromosomes randomly in the search space for the initial population.

Step 2 Decoding of the chromosome to generate N Q weighting matrix and computation of the performance in Eq. (9.20) for each individual.

Step 3 Execution of the selection, crossover, and mutation operators to improve the quality of the PFC-PID controller.

Step 4 Repetition of steps 2 and 3 until the set maximum evolution generation G is met

9.2.4 The Chamber Pressure Control by PFC-PID

In this section, the chamber pressure control in the coke furnace is introduced. The chamber pressure is built as a FOPDT model derived by a step response test. The advantages are that FOPDT models are convenient for PID designs, and they can also be easily transformed into the difference equation models using sampling time $T_s = 20$ s. The process model under the operation conditions in Table 9.1 is modeled as

Table 9.1 Steady-state operating conditions

Coke furnaces	
Radiation output temperature	495 °C
Convection output temperature	330 °C
Chamber temperature	800 °C
Oxygen content	5%
Circulating oil flow	35 t/h
Coke fractionating tower	
Tower bottom temperature	350 °C
Tower liquid level	70%
Coke towers	
Tower top temperature	415 °C
Tower bottom temperature	300 °C
Temperature after cooling	85 °C
Tower top pressure	0.25 Mpa

$$G(s) = \frac{-0.02}{150s + 1} e^{-40\,s} \tag{9.22}$$

The simulations environments are set as follows. The step set point is commanded at $k = 0$, and an output disturbance with amplitude -0.1 is added to the process at $k = 100$ for output disturbance test. An input disturbance with amplitude -30 is also added to the process at $k = 100$ for input disturbance test. The permissible error δ is chosen as 10^{-4}. The recent ENMSSPFC-based PID design [21] proposed by Zhang et al. has shown its superiority over many methods and is chosen here for comparison. The common control parameters of two methods are the same, specially $P = 8$ and the weighting factor on the output error goes to be 1. However, the difference between the proposed PID and that of Zhang's is that the rest elements in Q can be optimized through GA. The population size of GA is set to 40, the maximal generation G is set to 1000, the mutation probability and the crossover probability are 0.05 and 0.8, respectively.

Since uncertainty exists in such processes and causes model/process mismatches, thus it is very important to evaluate the control performance under mentioned conditions. Here three cases of model/process mismatches are generated through Monte Carlo simulations, i.e., parameters in Eq. (9.22) are changed randomly each time. Three cases with real process parameter uncertainty are

Case 1 $K = -0.024, T = 109, \tau = 32.$
Case 2 $K = -0.025, T = 115, \tau = 48.$
Case 3 $K = -0.014, T = 162, \tau = 32.$

However, two controllers are still designed using the nominal process model described by Eq. (9.22).

Figures 9.1 9.2, 9.3, 9.4, 9.5, 9.6, 9.7, 9.8, 9.9, 9.10, 9.11, and 9.12 show the responses of the two controllers. It can be seen that the performances of the set-point tracking, output disturbance rejection, and input disturbance rejection of the PFC-PID are better than those of Zhang's. In the first case, the first element in Q is optimized as 26.35. Under such a weighting factor, the response of the PFC-PID is smoother than Zhang's controller. The output shows small overshoot and short rise time simultaneously. While for Zhang's controller, the responses are oscillating strongly with larger overshoot.

The first element in Q for case 2 is optimized as 20.24. As expected, the output responses show improved performances for both output tracking and disturbance rejections compared with Zhang's controller. This is due to the function of elements in Q, which are closely associated with the closed-loop control performance.

For case 3, the first element in Q is optimized as 18.63. It can be seen that Zhang's method shows continuous oscillation responses and they can never settle to their set point, which can also be revealed through the corresponding control signals. As to the proposed controller, the responses are smoother and the oscillations are nearly invisible. These results are more acceptable for industrial applications.

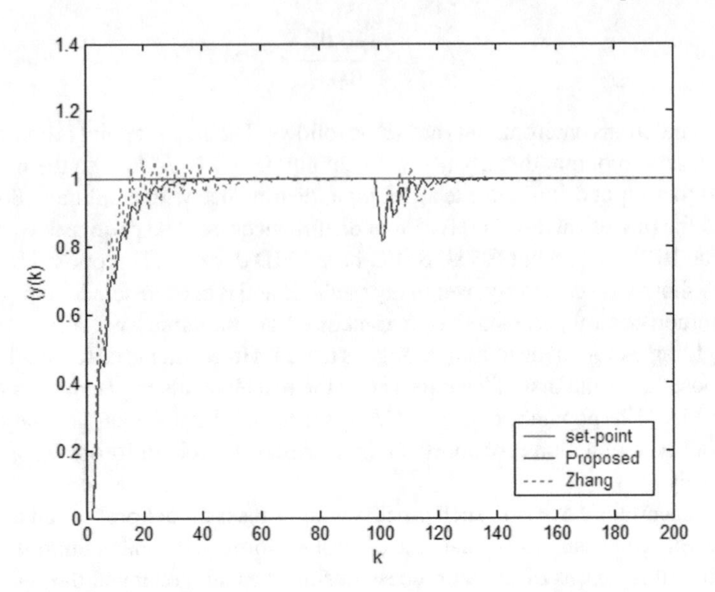

Fig. 9.1 Closed-loop output responses under output disturbance for case 1

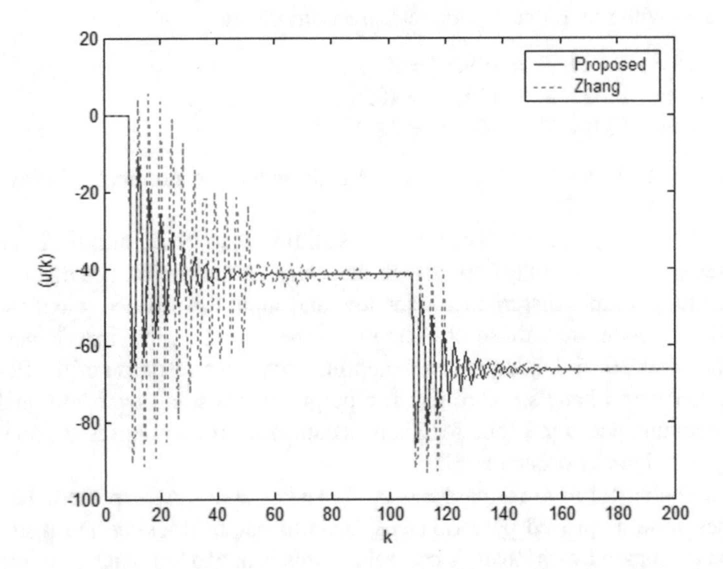

Fig. 9.2 Closed-loop control signals under output disturbance for case 1

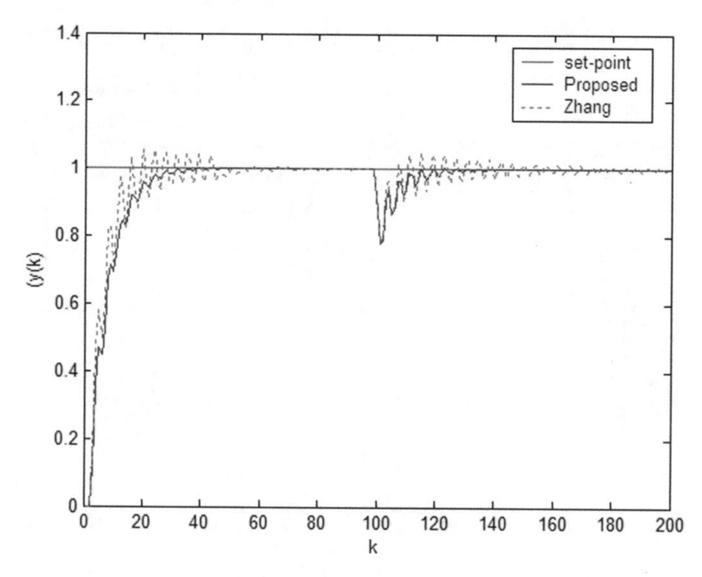

Fig. 9.3 Closed-loop output responses under input disturbance for case 1

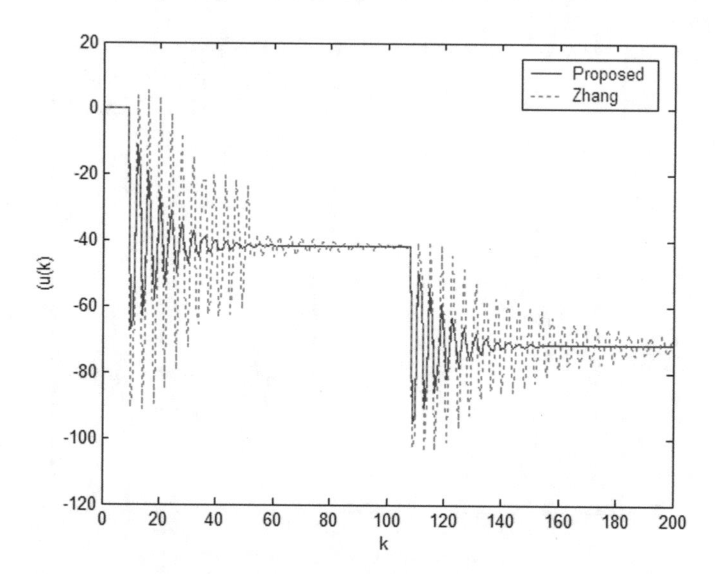

Fig. 9.4 Closed-loop control signals under input disturbance for case 1

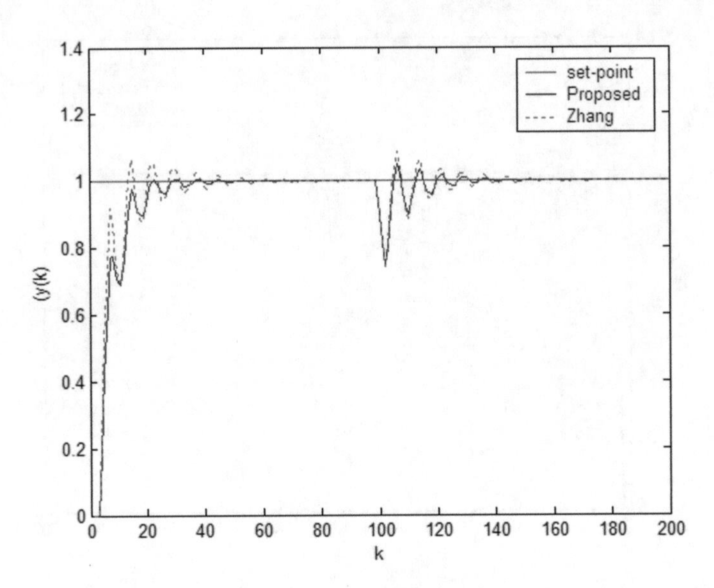

Fig. 9.5 Closed-loop output responses under output disturbance for case 2

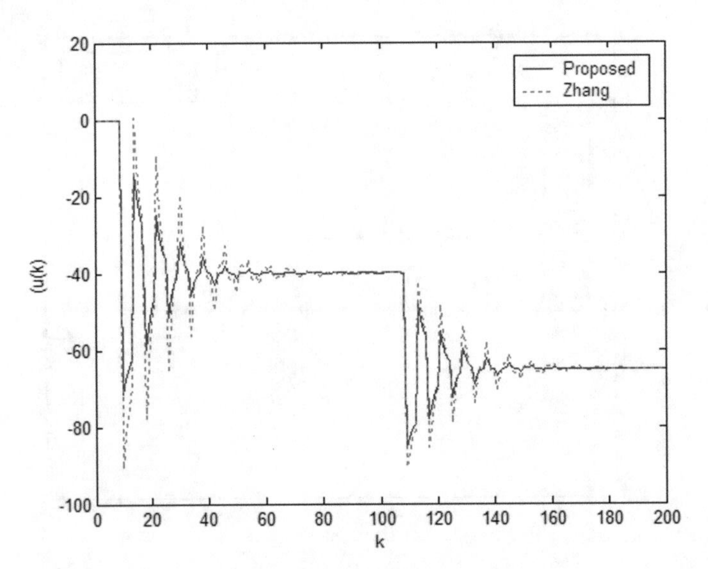

Fig. 9.6 Closed-loop control signals under output disturbance for case 2

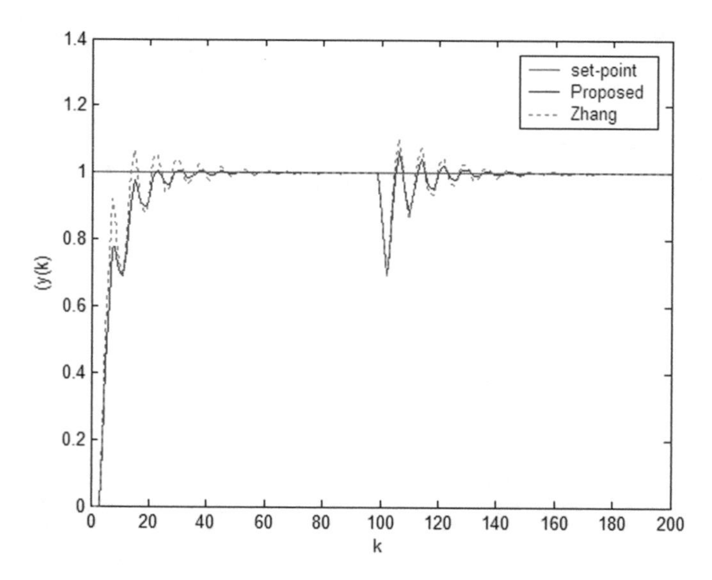

Fig. 9.7 Closed-loop output responses under input disturbance for case 2

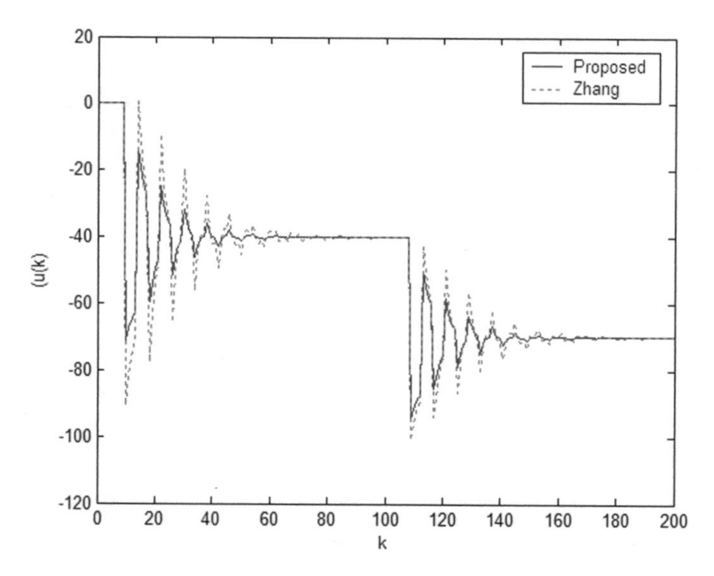

Fig. 9.8 Closed-loop control signals under input disturbance for case 2

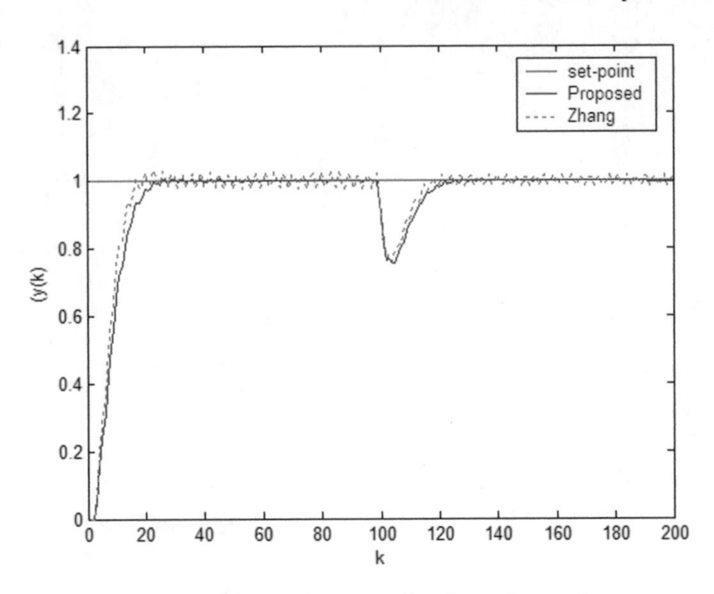

Fig. 9.9 Closed-loop output responses under output disturbance for case 3

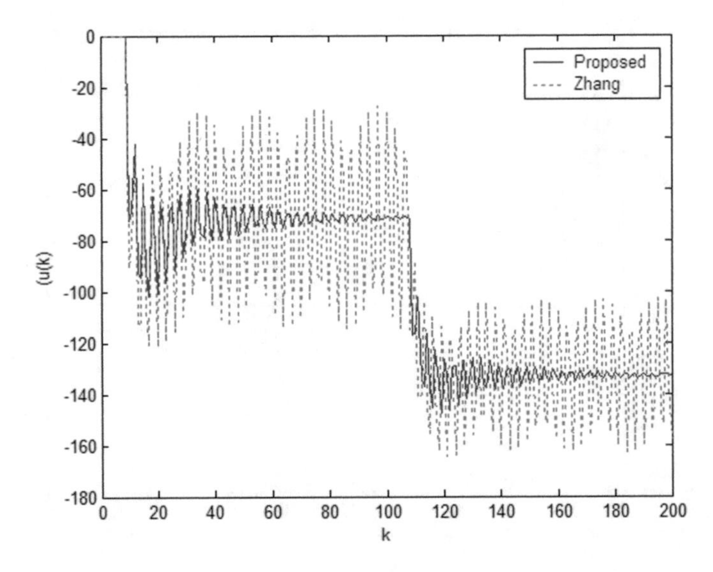

Fig. 9.10 Closed-loop control signals under output disturbance for case 3

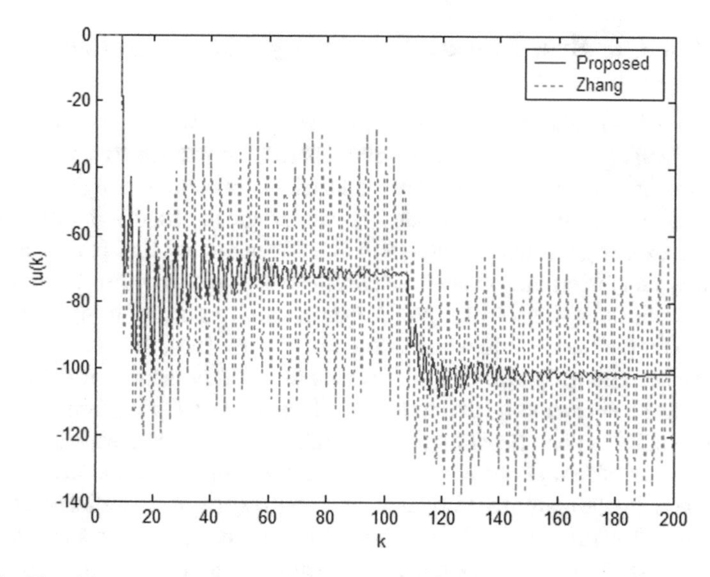

Fig. 9.11 Closed-loop control signals under input disturbance for case 3

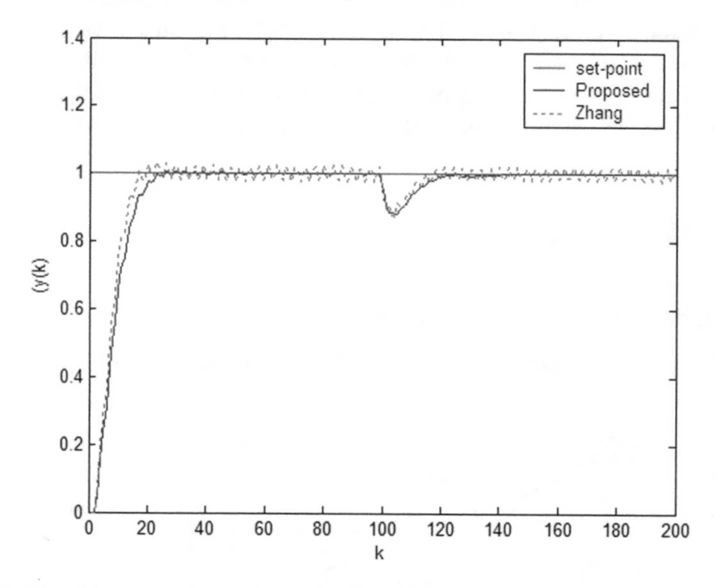

Fig. 9.12 Closed-loop output responses under input disturbance for case 3

9.3 RNA-GA-Based Fuzzy Neuron Hybrid Controller

With the precise mathematic model of the process, many advanced model-based control strategies can achieve satisfying control performances in simulations. However, most of them fail in practical industrial applications. Single-neuron non-model control strategies are more appropriate when the model is difficult to be established or has great uncertainty [34]. But for the controlled plant with severe nonlinearity and uncertainty characteristics, the basic neuron controller is also difficult to achieve satisfying control performance. Considering a strong complementarity between the fuzzy system and the neuron control [35], they can be combined to improve the control performance. However, the determination of the fuzzy rule requires a deep understanding of the controlled plant. The adaptive formula proposed in [36] is actually an analytical description of fuzzy rules, which greatly simplifies the implementation of fuzzy controllers. Therefore, the simple formula fuzzy controller is combined with the neuron controller to obtain a fuzzy neuron parallel control structure. Since the neuron gain has a great impact on the control performance, the fuzzy controller is used for online self-tuning of the neuron gain. Although the fuzzy neuron control system can improve the controller's performance, it sacrifices the simplicity of the traditional PID controller actually. In the design of the fuzzy neuron controller, many controller parameters are introduced to be tuned in advance. Then RNA-GA is used to optimize the controller parameters, and finally is applied to the liquid level control of the continuous casting process with uncertainty and nonlinearity.

9.3.1 Neuron Controller

The basic neuron controller in [28] is illustrated in Fig. 9.13.

where E is the surroundings of the neuron. The neuron output $u(t)$ can be marked as

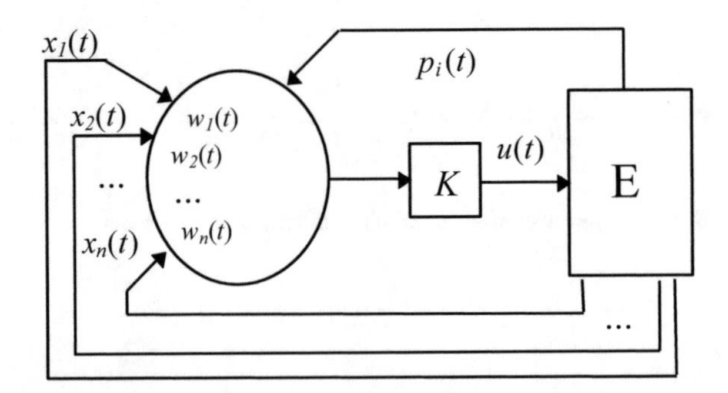

Fig. 9.13 The neuron model for control

$$u(t) = K \sum_{i=1}^{n} w_i(t) x_i(t) \tag{9.23}$$

where K is the neuron proportional coefficient and $K > 0$; $x_i(t)$ $(i = 1, 2, \ldots, n)$ represent the neuron inputs; $w_i(t)$ are the connection weights of $x_i(t)$, which can be determined by some learning rule. It is widely believed that a neuron can self-organize by modifying its synaptic weights. According to the well-known hypothesis proposed by D. O. Hebb, the learning rule of a neuron is formulated as

$$w_i(t+1) = w_i(t) + dp_i(t) \tag{9.24}$$

where d is the learning rate and $d > 0$; $p_i(t)$ denotes learning strategy, which is suggested for control purposes as follows [28]:

$$p_i(t) = z(t)u(t)x_i(t) \tag{9.25}$$

It expresses that an adaptive neuron using the learning way that integrated Hebbian learning and supervised learning, even it makes actions and reflections to the unknown outsides with the associative search. That means that the neuron self-organizes the surrounding information under supervising of the teacher's signal $z(t)$, emits the control signal, and implies a critic on the neuron actions.

According to the neuron model and its learning strategy described above, the neuron model-free control method is derived as follows:

$$\begin{cases} u(t) = \frac{K \sum_{i=1}^{n} w_i(t) x_i(t)}{\sum_{i=1}^{n} w_i(t)} \\ w_i(t+1) = w_i(t) + de(t)u(t)x_i(t) \end{cases} \tag{9.26}$$

$$e(t) = r(t) - y(t) \tag{9.27}$$

where $y(t)$ is the output of the plant, respectively, $u(t)$ is the control signal produced by the neuron, $r(t)$ is the set point, and the neuron inputs, and $x_i(t)$ can be selected by the demands of the control system designs.

9.3.2 Simple Fuzzy PI Control

When a basic fuzzy system is designed, the following three problems should be solved.

(1) Fuzzification of the input variables.
(2) Design of a fuzzy rule base for the inference engine.
(3) Defuzzification of the output U of the inference engine.

The key to having good performance of a fuzzy control system is the regulation of the fuzzy rule base according to the controlled plant. A regulating method for fuzzy control rule base was presented in [34], which can express the fuzzy control inference process by a simple formula as follows:

$$U = < \lambda E + (1 - \lambda) EC > \tag{9.28}$$

where E, EC are the fuzzy input variables of a control system, i.e., error $e(t)$ and its change $\Delta e(t)$, respectively,$\lambda \in (0,\ 1)$ is the factor regulating the fuzzy rule base. $\langle x \rangle$ denotes the inference engine to have the nearest integer of x. By changing the factor λ, the rule base can be regulated, and the performance of fuzzy controller can be changed conveniently, U is the output of the inference engine. It has been proved that Eq. (9.28) has the same function as a conventional Mamdani-type fuzzy inference engine with symmetrical triangle membership functions. Thus, the fuzzy system can be written as

Fuzzifier:

$$E = < k_e e(t) >, \quad EC = < k_{ec} \Delta e(t) > \tag{9.29}$$

Fuzzy inference:

$$U = < \lambda E + (1 - \lambda) EC > \tag{9.30}$$

Defuzzifier:

$$u_f(t) = k_u U \tag{9.31}$$

where k_e, k_{ec} are fuzzification factors of inputs $e(t)$ and $\Delta e(t)$, respectively, $\lambda \in (0,\ 1)$. k_u is the defuzzification coefficient of the fuzzy inference engine output, $u_f(t)$ is the fuzzy system output. The above fuzzy controller is obviously a PD-type controller in terms of Eqs. (9.29)–(9.31). To obtain the PI controller, the integration is introduced and fuzzy PI-type controller is constructed as follows:

$$u_{FC}(t) = \sum_{i=0}^{t} u_f(i) \tag{9.32}$$

Substituting Eqs. (9.29)–(9.31) into Eq. 9.32 can lead to

$$u_{FC}(t) = k_u \sum_{i=0}^{t} < \lambda < k_e e(i) > +(1 - \lambda) < k_{ec} \Delta e(i) \gg \tag{9.33}$$

Thus, the simple fuzzy PD-type controller becomes a fuzzy PI-type controller.

9.3.3 Fuzzy Neuron Hybrid Control (FNHC)

The fuzzy neuron hybrid control system is then set up in Fig. 9.14. In the hybrid structure, the gain of the neuron controller is tuned by the fuzzy algorithm, the sum of the outputs of the fuzzy PI controller and neuron controller is the output of the hybrid controller. $u(t)$ is the control signal produced by the hybrid controller, $x_i(t)$ are the inputs of the neuron, $u_{NC}(t)$ is the control action produced by the neuron, $u_f(t)$ is the output of the fuzzy PD controller, which is used to update the neuron gain $K(t)$. $u_{FC}(t)$ is the control action of the fuzzy PI controller.

The output of the fuzzy neuron hybrid controller is obtained as follows:

$$u(t) = u_{NC}(t) + u_{FC}(t) \tag{9.34}$$

where $u_{NC}(t)$ is the output of the neuron controller:

$$\begin{cases} u_{NC}(t) = K(t)\sum_{i=1}^{3} w_i(t)x_i(t) \Big/ \sum_{i=1}^{3} |w_i(t)| \\ w_i(t+1) = w_i(t) + d_i e(t) u_{NC}(t) x_i(t) \end{cases} \tag{9.35}$$

$K(t)$ is the gain of the neuron model-free controller, which is regulated by

$$K(t) = K(t-1) + u_f(t) \tag{9.36}$$

where $u_f(t)$ is obtained by Eq. (9.31). $K(t) \geq k_1 e^{-u_1}$, where k_1 is a constant to be chosen. Considering the demands of different set points, the neuron inputs are chosen as follows:

$$x_1(t) = u_1 r(t), \quad x_2(t) = u_1 e(t), \quad x_3(t) = \Delta e(t) \tag{9.37}$$

where u_1 can be a parameter of the controlled plant.

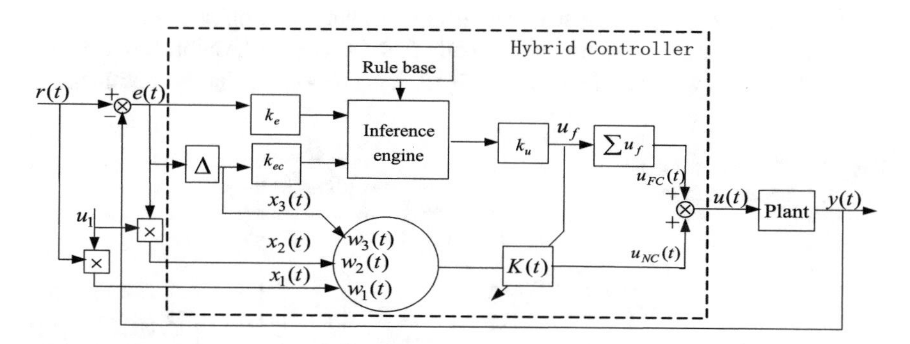

Fig. 9.14 Fuzzy neuron hybrid control system

In Eq. (9.34), $u_{FC}(t)$ is the output of the Fuzzy PI controller given by Eq. (9.33). Hence, the whole fuzzy neuron hybrid control method is as follows:

$$\begin{cases} u_{NC}(t) = K(t)\sum_{i=1}^{3} w_i(t)x_i(t) \Big/ \sum_{i=1}^{3} |w_i(t)| \\ u_{FC}(t) = \sum_{j=0}^{t} u_f(j) \\ u(t) = u_{NC}(t) + u_{FC}(t) \\ E = < k_e e(t) >, \quad EC = < k_{ec}\Delta e(t) > \\ u_f(t) = k_u < \lambda E + (1-\lambda)EC > \\ K(t) = K(t-1) + k_u U, \ K(0) = k_1 e^{-u_1} \\ w_i(t+1) = w_i(t) + d_i(r(t) - y(t))u_{NC}(t)x_i(t) \\ x_1(t) = u_1 r(t), \quad x_2(t) = u_1 e(t), \quad x_3(t) = \Delta e(t) \end{cases} \tag{9.38}$$

where $k_e, k_{ec}, k_u, \lambda, k_1, d_i$ are the parameters of the hybrid controller to be optimized.

9.3.4 Parameters Optimization of RNA-GA

Since the parameters $(k_e, k_{ec}, k_u, \lambda, k_1, d_i)$ of the hybrid controller will seriously affect the control performance, it is crucial to optimize these parameters appropriately. The parameters range should be set before RNA-GA optimization. Moreover, the commonly accepted performance criterion, such as ITAE, is selected, which takes into account the rising time, the overshoot, and the output of controller:

$$J_{\min} = \sum_{i=1}^{\mathrm{T}} (|e(i)| + u(i)) \tag{9.39}$$

where T is the control time when the parameters are given at one generation of RNA-GA. $e(i)$ is the tracking error, and $u(i)$ is the controller output. However, as a computationally intensive optimization method, RNA-GA for the hybrid controller is executed off-line, which is shown in Fig. 9.15. It is noteworthy that the optimization

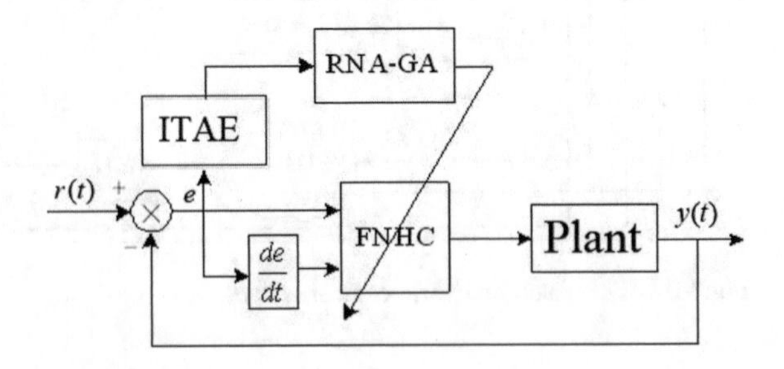

Fig. 9.15 The framework of RNA-GA-based controller optimization

process is based on the mathematic model, however, the constructed model is always different from the actual plant.

For a class of ITAE linear integral performance indices, Wang et al. gave the following sufficient conditions that the performance of the model can reflect the actual plant [36]:

$$|\Delta J - \Delta J_m| \leq 2\varepsilon T \tag{9.40}$$

where ε is set as $\max_k |y(k) - \hat{y}(k)|$, ΔJ_m indicates the performance index variation by applying two groups of controller parameters to the constructed model, ΔJ indicates the performance index variation by applying the same two groups of controller parameters to the actual plant. If Eq. (9.40) is satisfied, the obtained optimal controller parameters can be used to the actual process. Though the model is identified based on the specific case, the self-learning capability of FNHC can overcome the uncertainty and time-variance characteristics of the controlled plant.

9.3.5 Continuous Steel Casting Description

The continuous steel casting is the process of molding molten metal into solid blooms. A schematic diagram of this process is illustrated in Fig. 9.16 [37], where the ladle [a] is acting as a reservoir for the molten metal and the valve [c] regulating its flow into the mold [d]. The tundish [b] acts as an intermediate reservoir that retains a constant supply to mold when an emptied ladle is being replaced by a full one. The cast metal undergoes two cooling processes. Primary cooling occurs in the mold and

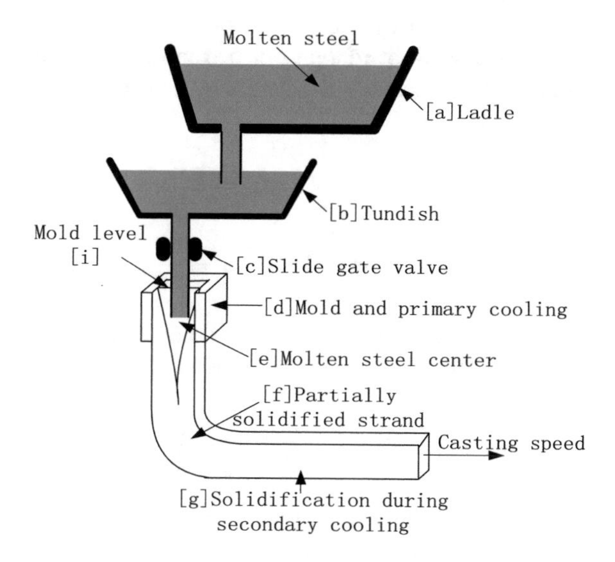

Fig. 9.16 Simplified continuous steel caster

Fig. 9.17 Slide gate valve

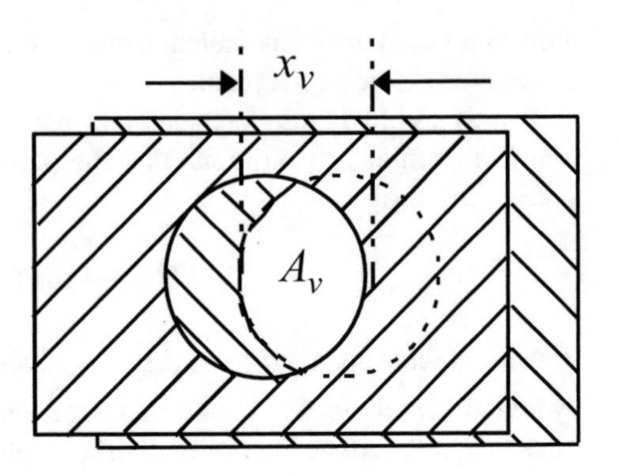

produces a supporting shell around the still liquid center [e]. This semielastic strand is then continuously withdrawn from the mold through a series of supporting rolls containing the secondary cooling stage [f, g], after which the new strand is cut into blooms by torch cutters. [i] is the mold level, it can be measured by a mold-level sensor.

Being complex, with great uncertainty, nonlinearity, and running under a high-temperature condition, it is hard to model the continuous steel casting process. This caster mainly consists of hydraulic actuators, a slide gate valve, and a mold.

The hydraulic actuators are prone to have non-smooth nonlinearity such as slip-stick friction and backlash. The non-smooth nonlinearity can be compensated by a high bandwidth controller [28]. Thus, the approximate transfer function of the valve position loop can be gained as $k_v/(\beta s+1)$, where k_v, β are the parameters of the loop.

The slide gate valve consists of three identical plates, with the outer two fixed and the center one sliding between two outer parts in Fig. 9.17. All plates contain an orifice of radius R, so that the effective area of matter flowing into the mold is determined by the overlapping orifice areas. When the center plate is at the specified location x_v, elementary trigonometric considerations show that the effective flow area can be given by

$$A_v = R^2(\alpha - \sin(\alpha)) \tag{9.41}$$

$$\alpha = 2\cos^{-1}(1 - \frac{x_v}{2R}), \qquad 0 \le x_v \le 2R \tag{9.42}$$

The process also has non-smooth nonlinearity due to the flow dynamics and valve geometry. The nonlinear model is given by

$$\frac{dy(t)}{dt} = \frac{A_v}{A_m} c_v c_c \sqrt{2gh} - u_1 \tag{9.43}$$

where A_m is the casting cross-sectional area, c_v is a velocity coefficient dependent on the viscosity of the steel grade being cast, c_c is a coefficient of contraction with

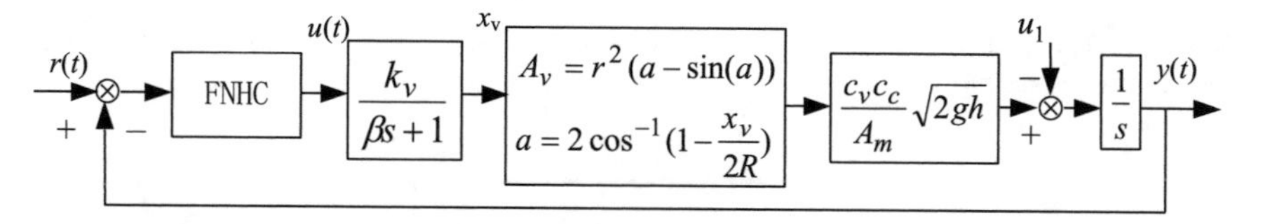

Fig. 9.18 Mold-level control system for the continuous steel casting

value 0.6 for a new valve with sharp edges and 0.95 for a worn valve with rounded edges, and h is the height of matter in the tundish. It can be seen that this plant is time-varying.

By analyzing the continuous steel casting, the mold-level control system is illustrated in Fig. 9.18, where, $r(t)$ is the set point, $y(t)$ is the mold level, $u(t)$ is the control signal, FNHC is the fuzzy neuron hybrid controller, u_1 is the casting speed.

9.3.6 FNHC Controller Performance Analysis

To verify the effectiveness of the proposed control method, the simulation tests of mold-level control for the continuous steel casting process are executed. The parameters of the plant are given as: $g = 9.8$, $h = 0.9$, $A_m = 1$, $c_v = 0.24$, $c_c = 0.74$, $R = 0.8$, $\beta = 1$, $k_v = 1$. All of those experiments are carried out to track the reference trajectory using the same controller parameters under the case of $u_1 = 1$, $c_c = 0.74$, $k_v = 1$. The parameters of the fuzzy neuron hybrid controller optimized by RNA-GA are given as: $k_e = 8.1$,$k_{ec} = 8.3$, $k_u = 0.007$, $\lambda = 0.8$, $k_1 = 3.2$,$d_1 = 100$, $d_2 = 25$, $d_3 = 80$. The sampling period is set to be $T = 0.6$ s. In order to inspect the performance of the proposed model-free control method, the robustness tests are also made under the conditions of different casting speeds, new slide gate valve and old slide gate valve, and the changing of the valve position loop gain.

At the case of $u_1 = 0.6$, 1000 input data are generated randomly, and the corresponding outputs can be obtained, with the meantime, $\pm 3\%$ noise is added to simulate the actual condition. Since most of parameters of the continuous casting process are known in advance, only k_v, β, and $c_c c_v$ are the unknown parameters, the model structure of the nonlinear controlled plant adopts the structure shown in Fig. 9.18. The model parameters of the system are identified by the RNA-GA, the identification process can refer to Sect. 5.2. The obtained parameters are: $k_v = 3.3644$, $\beta = 4.9545$, $c_c c_v = 0.1318$. The model outputs and its errors by applying the optimal parameters are illustrated in Figs. 9.19 and 9.20. It is obvious that the optimal model has achieved good modeling accuracy under the given conditions, and the maximum modeling error is only 0.1106. Therefore, the identified model can be used as the controlled plant to optimize the controller parameters, and the ITAE is selected as the optimization index. RNA-GA is also applied to optimize the parameters of the FNHC controller. The range of values of each controller parameter is set by trial

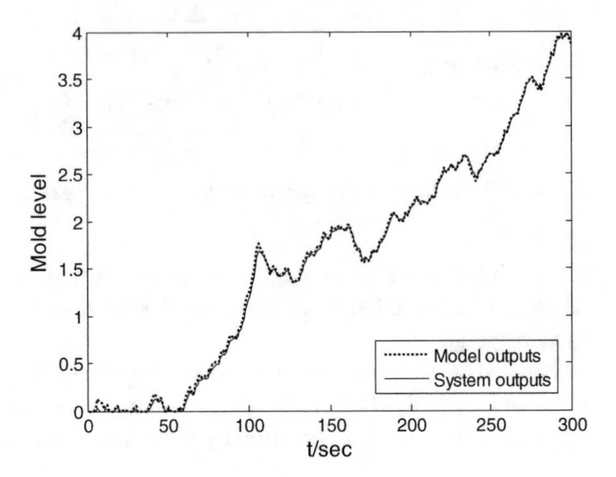

Fig. 9.19 The modeling of the continuous casting process

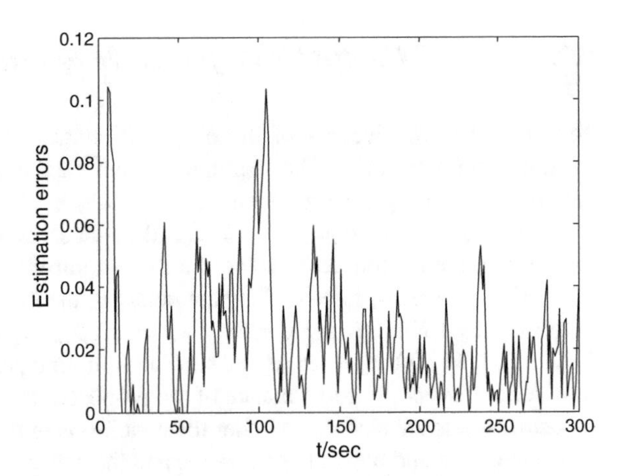

Fig. 9.20 The modeling errors of the continuous casting process

and error: $k_e \in [1, 50]$, $k_{ec} \in [1, 50]$, $k_u \in [0.0001, 0.01]$, $k_u \in [0.0001, 0.01]$, $K(0) \in [0.1, 1]$, $\lambda \in [0, 1]$, $d_i \in [1, 150]$, $w_i \in [0.1, 20]$ (Fig. 9.20).

To further illustrate the efficiency of the constructed model, two groups of controller parameters for FNHC optimized by RNA-GA, denoted as P_1 and P_2, are selected and applied to the constructed model and the actual plant, respectively. Accordingly, the ITAE indexes can be obtained as: $J_{m1} = 38.0649$, $J_{m2} = 29.2784$, $J_1 = 41.7868$, $J_2 = 31.9412$. Thus, $|\Delta J - \Delta J_m| = 1.0591$, $2\varepsilon T = 39.82$, satisfying Eq. (9.40). RNA-GA runs 10 times, and the best results are listed as follows: $k_e = 9.8738$, $k_{ec} = 9.9934$, $k_u = 0.0063$, $K(0) = 0.7455$, $\lambda = 0.8748$, $d_1 = 133.36$, $d_2 = 30.47$, $d_3 = 40.22$, $w_1(0) = 12.5732$, $w_2(0) = 5.8283$, $w_3(0) = 4.6209$. The simulation results obtained by the constructed model are illustrated in Fig. 9.21. It shows that the controller optimized by RNA-GA under the identified model can obtain good control tracking performance. Applying the same FNHC to the controlled plant in Fig. 9.18, the simulation results under the same

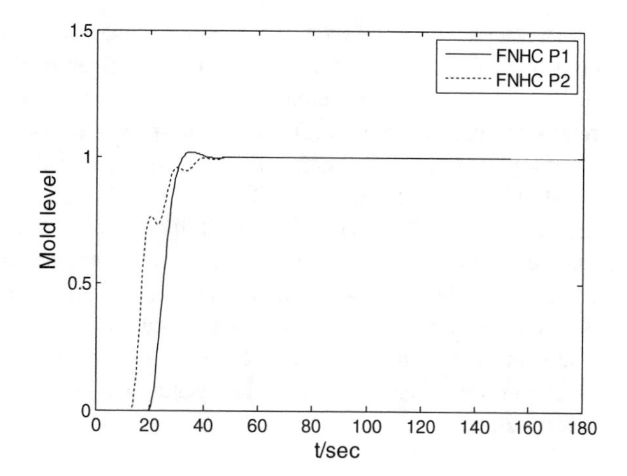

Fig. 9.21 The control results applied on the constructed model

conditions are illustrated in Fig. 9.22. It tells us that the controller is effective for the actual plant.

For the purpose of verifying the effectiveness of the proposed FNHC controller in self-learning capability, various robustness test experiments are carried out at different continuous casting rates and valve ages. The PID controller and single-neuron controller are selected to compare with the FNHC, where the incremental PID controller $\Delta u(k) = k_p[e(k) - e(k-1)] + k_i e(k) + k_d[e(k) - 2e(k-1) + e(k-2)]$ is used. Both the PID and single-neuron controller are optimized by RNA-GA. The optimal parameters of PID controller are: $k_p = 1.3405$, $k_i = 0.4548$, $k_d = 15.5937$, and the parameters of single-neuron controller are: $K = 1.7416$, $d_1 = 149.22$, $d_2 = 37.40$, $d_3 = 51.47$, $w_1(0) = 5.9273$,$w_2(0) = 5.3694$, $w_3(0) = 12.1955$. In all tests, the controller parameters and the molt steel level set to be the same.

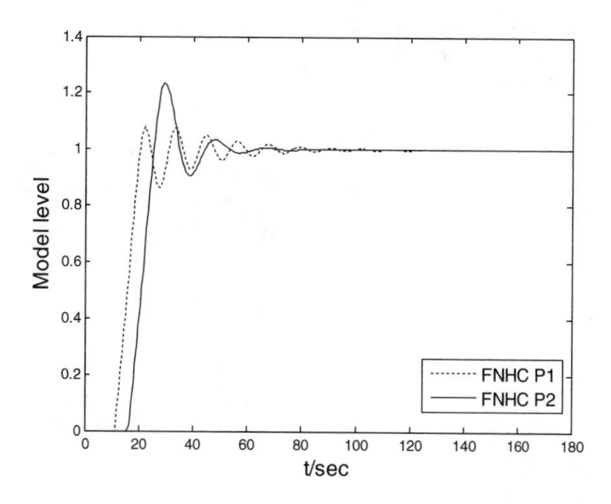

Fig. 9.22 The control results applied on the actual plant

The step response of the FNHC controller is shown in Figs. 9.23 and 9.24, and the control performance of the PID controller is given in Figs. 9.25 and 9.26. Because the plant has nonlinear characteristics, the tracking performance of the FNHC's is relatively superior to the PID controller. When the work conditions change, the PID control performance is greatly deteriorated due to lack of adaptive and self-learning capability of the PID controller. The step response of single-neuron control is shown in Figs. 9.27 and 9.28. Due to the lack of online self-adjustment of gain K, although the single neuron has a similar control performance to FNHC's under the initial conditions, when the work conditions change, single-neuron control performance is also deteriorated, a large overshoot occurs and the set points cannot be tracked. The gain self-tuning process of the FNHC controller is shown in Fig. 9.29, which illustrates the effectiveness of the fuzzy unit to improve the performance of the neuron controller.

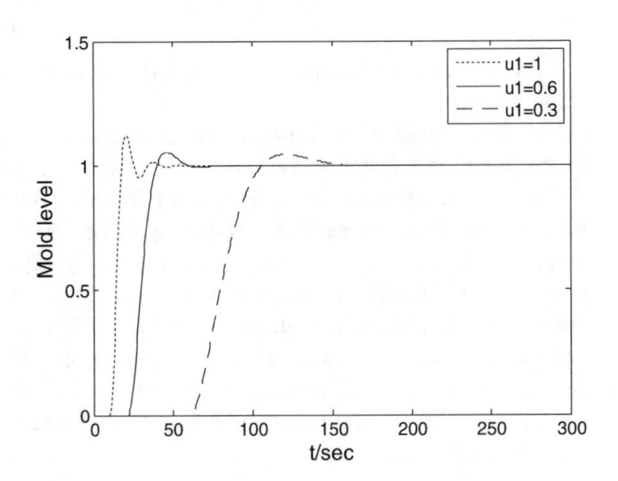

Fig. 9.23 Step response of FNHC under different speeds

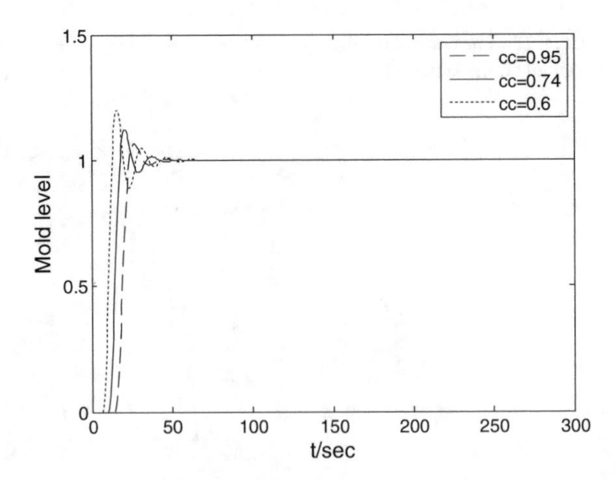

Fig. 9.24 Step response of FNHC under valve with different ages

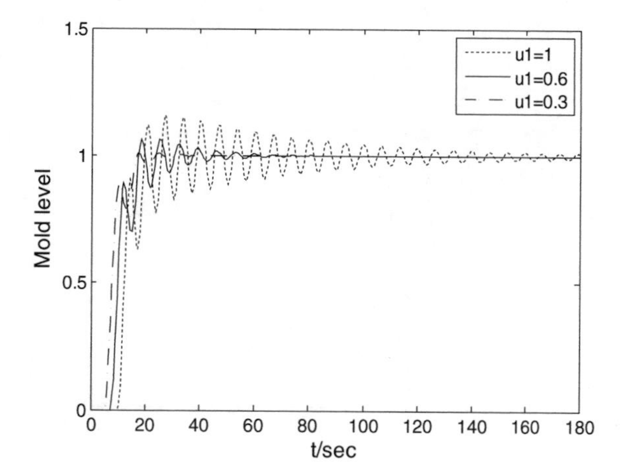

Fig. 9.25 Step response of PID under different speeds

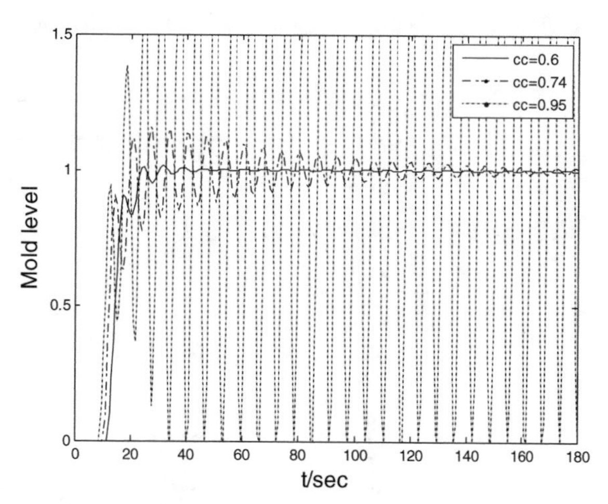

Fig. 9.26 Step response of PID under valve with different ages

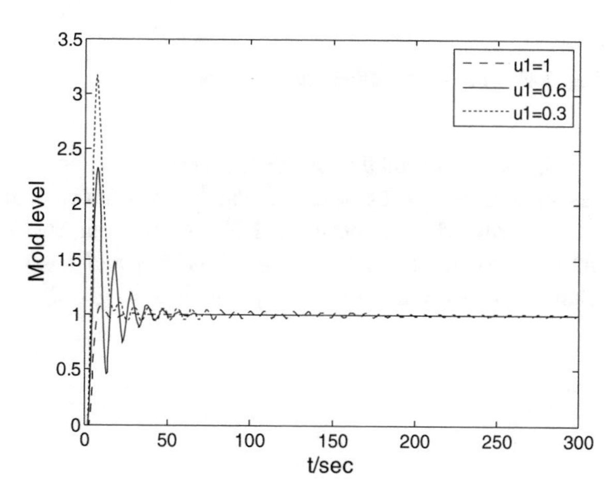

Fig. 9.27 Step response of SNC under different speeds

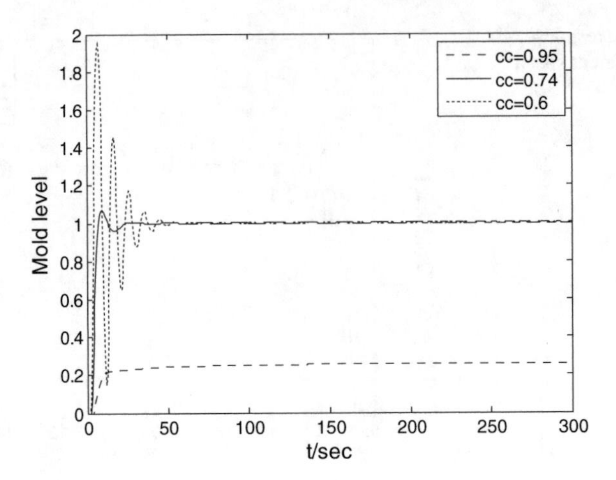

Fig. 9.28 Step response of SNC under valve with different ages

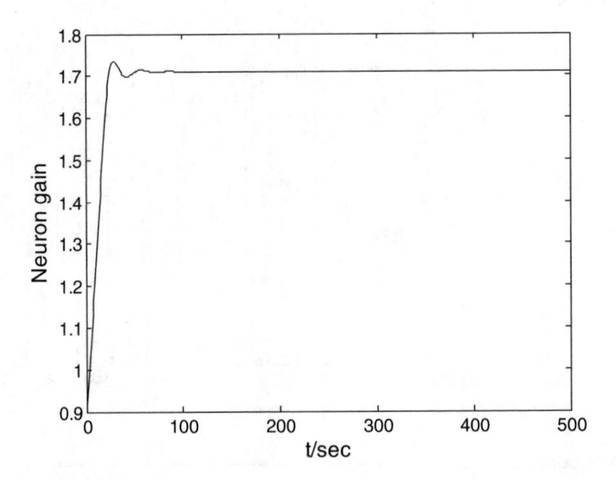

Fig. 9.29 The gain adjusting process of FNHC

The above simulation results illustrate that good performance is obtained in all test cases and the set points of the mold level can be reached. The control system can respond quickly, smoothly, and almost without overshoot. The proposed model-free controller has very strong robustness and adaptability, even if the dynamic characteristics of the nonlinear plant changes greatly.

9.4 Stabilization Subspaces Based MOGA for PID Controller Optimization

There are many processes that can be expressed as unstable first-order time-delay (FOTD) plants. When the lag time is large, the performance of PID controllers will deteriorate greatly. Since the controller parameter optimization can greatly improve the performance of the system control, the optimization problem of the PID controller design for unstable first-order time-delay plant is tried to be solved by using a genetic algorithm. However, the difficulty in the PID controller design is to accurately obtain the stable region of the PID controller parameters [38]. The generalized Hermite-Biehler theorem (abbreviated as H-B theorem) provides sufficient and necessary conditions for obtaining the parameter stabilization space of the PID controller [39, 40]. However, for unstable FOTD plants, the generalized H-B theorem cannot guarantee that the stable closed space can always be obtained. Therefore, two-level control structure in [41] is adopted, the unstable plant is first stabilized by the inner loop, then the generalized stable plant is obtained, and the stabilization search space can be obtained, which will be utilized by GA.

9.4.1 *Generalized Hermite-Biehler Theorem*

In this section, the generalized Hermite-Biehler theorem is introduced. Let $\delta(s) = \delta_0 + \delta_1 s + \cdots \delta_n s^n$ be a given real polynomial of degree n. Rewrite $\delta(s) = \delta_e(s^2) + s\delta_o(s^2)$, where $\delta_e(s^2)$, $\delta_o(s^2)$ are the components of $\delta(s)$ made up of even-order part and odd-order part of s, respectively. For each frequency $\omega \in R$, denote $\delta(j\omega) = p(\omega) + jq(\omega)$, where $p(\omega) = \delta_e(-\omega^2)$, $q(\omega) = \omega\delta_o(-\omega^2)$, and define the feature sign of the polynomial $\delta(s)$ by $\sigma(\delta(s))$, where $\sigma(\delta(s))$ can be derived by subtracting number of roots in the right half plane of $\delta(s)$, n_δ^R from the number of roots in the left half plane of $\delta(s)$, n_δ^L, i.e., $\sigma(\delta(s)) = n_\delta^L - n_\delta^R$. Thus, the generalized Hermite-Biehler theorem can be described as follows.

Generalized Hermite-Biehler Theorem [40]: Let $\delta(s)$ be a given real polynomial of degree n without $j\omega$ axis roots, except for the root at the origin. Let $0 = \omega_0 < \omega_1 < \omega_2 < \ldots < \omega_{m-1} < w_l = \infty$ be the real, nonnegative, distinct finite zeros of $q(\omega)$ with odd multiplicities. Then

$$\sigma(\delta(s)) = \begin{cases} \{\mathrm{sgn}[p(\omega_0)] - 2\mathrm{sgn}[(p(\omega_1)] + 2\mathrm{sgn}[(p(\omega_2)] + \cdots \\ +(-1)^{m-1}2\mathrm{sgn}[(p(\omega_{m-1})] + (-1)^m 2\mathrm{sgn}[(p(\omega_m)]\} \cdot \gamma \\ \quad \text{if}\, n \text{ is even} \\ \{\mathrm{sgn}[p(\omega_0)] - 2\mathrm{sgn}[(p(\omega_1)] + 2\mathrm{sgn}[(p(\omega_2)] + \cdots \\ \quad +(-1)^{m-1}2\mathrm{sgn}[(p(\omega_{m-1})] \cdot \gamma \\ \quad \text{if}\, n \text{ is odd} \end{cases} \tag{9.44}$$

where $\gamma = (-1)^{l-1}\mathrm{sgn}[q(k_p, \infty)]$, and

$$\operatorname{sgn}(x) = \begin{cases} x/|x| & x \neq 0 \\ 0 & x = 0 \end{cases} \tag{9.45}$$

If the polynomial $\delta(s)$ is Hurwitz, the interlacing property can be immediately implied in the generalized Hermite-Biehler theorem.

9.4.2 Hermite-Biehler Theorem Based PID Controller Stabilizing

In [40], the generalized Hermite-Biehler theorem is used to provide a complete analytical solution to the PID controller, the control system is illustrated in Fig. 9.30.

where $r(t)$ is the set point, and $y(t)$ is the output of a controlled plant, $C(s)$ is the PID controller, $C(s) = k_p + \frac{k_i}{s} + k_d s$, $G(s)$ is the controlled plant, $G(s) = \frac{N(s)}{D(s)}$.

The closed-loop characteristic polynomial becomes

$$\delta(s) = sD(s) + (k_i + s^2 k_d)N(s) + sk_p N(s) \tag{9.46}$$

The stabilization problem of using a PID controller is to determine the values of k_p, k_i, and k_d for which the closed-loop characteristic polynomial $\delta(s)$ is Hurwitz.

According to the generalized Hermite-Biehler theorem and its synthesis of stabilizing PID controller, $p(\omega)$ and $q(\omega)$ can be obtained.

$$p(\omega) = p_1(\omega) + (k_i - k_d\omega^2)p_2(\omega) \tag{9.47}$$

$$q(\omega) = q_1(\omega) + k_p q_2(\omega) \tag{9.48}$$

where $p_1(\omega) = -\omega^2[D_o(-\omega^2)N_e(-\omega^2) - D_e(-\omega^2)N_o(-\omega^2)]$, $p_2(\omega) = N_e^2(-\omega^2) + \omega^2 N_o^2(-\omega^2)$, $q_2(\omega) = \omega p_2(\omega)$, $q_1(\omega) = \omega[N_e(-\omega^2)D_e(-\omega^2) + \omega^2 N_o(-\omega^2)D_o(-\omega^2)]$, $D_e(s^2)$, $D_o(s^2)$, and $N_e(s^2)$, $N_o(s^2)$ are the components of $D(s)$ and $N(s)$ made up of even and odd orders of s, respectively, $s = j\omega$.

In Eq. (9.48), for every fixed k_p, the zeros of $q(\omega, k_p)$ do not depend on k_i or k_d, so we can first calculate the stable range of k_p, then use generalized Hermite-Biehler theorem to determine the stabilizing sets of k_i and k_d.

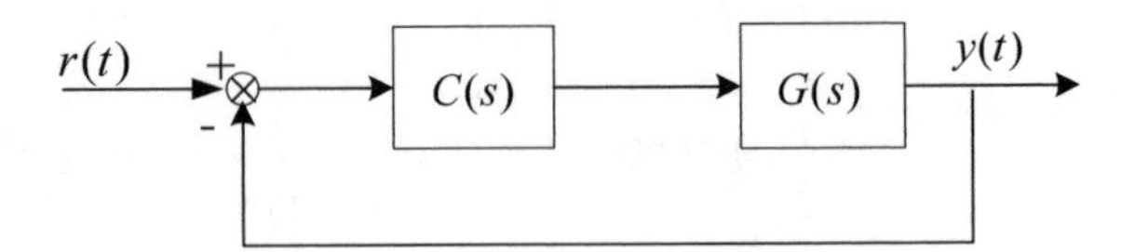

Fig. 9.30 Unity feedback control system

9.4.2.1 Stable Region of K_p for Unstable FOPDT Processes

The FOPDT process is described by the transfer function $G(s) = \frac{ke^{-\theta s}}{Ts-1}$, which is usually reduced to Eq. (9.49) by normalizing the dead time ($L = \theta/T$) and absorbing the gain into the controller:

$$G(s) = \frac{e^{-Ls}}{s-1} \tag{9.49}$$

Assume $C(s) = k_p$ in Fig. 9.30, the closed-loop characteristic polynomial $\delta(s)$ can be derived:

$$\delta(s) = D(s) + k_p N(s) \tag{9.50}$$

There are several classic methods that can be used to determine the value of k_p satisfying the condition that $\delta(s)$ is Hurwitz, such as the root locus technique, the Nyquist stability criterion, and the Routh-Hurwitz criterion and so on. Since the former two methods are graphical in nature and fail to provide us an analytical characterization of all stabilizing values of k_p. Hence, The Routh-Hurwitz criterion can be adopted to obtain an analytical solution. However, there will not always exist k_p to stabilize the unstable FOPDT process when using Routh-Hurwitz criterion. To obtain the stable range of k_p, the two-controller structure is adopted as illustrated in Fig. 9.31 [39].

In Fig. 9.31, $C_1(s)$ is the P or PD controller in order to stabilize the controlled plant, while $C(s)$ is a PID controller to improve the control performance of the inner loop. According to the time delay (L), the controller structure is selected as follows [39]:

$$C_1(s) = \begin{cases} k_{p1} & L \leq 1 \\ k_{p1}(1 + T_{d1})s & 1 < L < 2 \end{cases} \tag{9.51}$$

If the controlled plant is known, the controller structure can be fixed. If $C_1(s)$ is PD controller, the range of T_{d1} is given as

$$L - 1 < T_{d1} < 1 \tag{9.52}$$

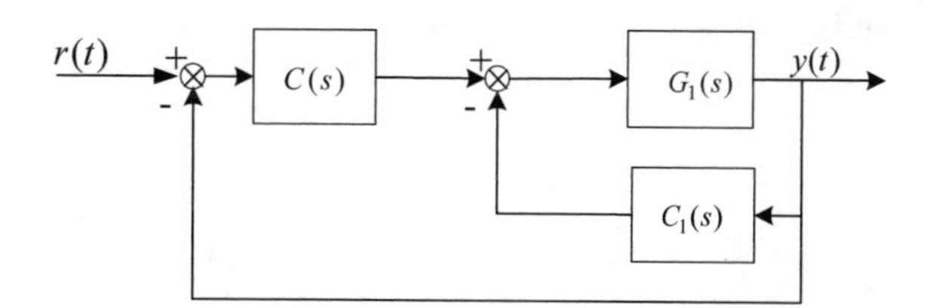

Fig. 9.31 Two-loop controller structure

By selecting T_{d1} among the given range, the stable range of k_{p1} can be derived according to Routh-Hurwitz criterion. The exponential transfer function e^{-Ls} can be approximated by using the Pade approximation formula. Thus, the controller $C_1(s)$ is designed, and the generalized controlled plant $G(s)$ in Fig. 9.30 is also obtained:

$$G(s) = \frac{G_1(s)}{1 + G_1(s)C_1(s)} \tag{9.53}$$

The stable range of k_p can be derived by the same method used in $C_1(s)$. Because $G(s)$ is stable, the stable interval of k_p can always be guaranteed.

9.4.2.2 Stable Domains of K_i and K_d

Since zeros of $q(\omega, k_p)$ in Eq. (9.48) is independent on k_i or k_d, we can obtain a stable interval of k_i and k_d corresponding to each k_p. The solution can be implemented using the following procedure [40].

Step 1: For a fixed k_p, determine $(0 = \omega_0 < \omega_1 < \omega_2 < \ldots < \omega_{l-1} < \omega_l = \infty)$, the real, non-negative, distinct finite zeros of $q(\omega, k_p)$, also define n as the degree of $\delta(s)$ and m' as the degree of $N'(s)$,where $N'(s) = N(-s)$.

Step 2: Choose i_t such that

$$n - \sigma(N') = \begin{cases} \{i_0 - 2i_1 + 2i_2 + \cdots + (-1)^{l-1}2i^{l-1} + (-1)^l i_l\} \cdot \gamma \\ \text{for } m' + n \text{ even} \\ \{i_0 - 2i_1 + 2i_2 + \cdots + (-1)^{l-1}2i^{l-1}\} \cdot \gamma \\ \text{for } m' + n \text{ odd} \end{cases}$$

, where γ is defined in Eq. (9.44).

Step 3: Determine the stable intervals of k_i and k_d by solving the following linear inequalities:

$$\begin{cases} p_1(\omega_t) + (k_i - k_d\omega_t^2)p_2(\omega_t) > 0 & \text{if } i_t = 1 \\ p_1(\omega_t) + (k_i - k_d\omega_t^2)p_2(\omega_t) < 0 & \text{if } i_t = -1 \end{cases} \tag{9.54}$$

for $t = 0, 1, 2, \ldots, l$ and $p_2(\omega_t) \neq 0$.

By solving the above inequalities, the stable intervals of (k_i, k_d) can be obtained for which $\delta(s)$ is Hurwitz. As to the stable controlled plant, the stable interval of k_i and k_d can always be obtained.

9.4.3 Optimizing PID Controller Parameters Based on Stabilization Subspaces

9.4.3.1 The Objective Functions

Since the stable regions of (k_p, k_i, k_d) are obtained based on Routh-Hurwitz criterion and generalized Hermite-Biehler theorem in Sect. 9.4.2, the initial separation of search space can be accomplished. Assuming that the tracking error and energy consumption is to be minimized, meanwhile, the system overshoot and steady-state error are required to meet certain constraints. Thus, two objective functions with constraints are given as follows:

$$\begin{aligned} \min \quad f_1 = \sum_{k=1}^{T} |e(k)|, \quad & f_2 = \sum_{k=1}^{T} u^2(k) \\ \text{s.t.} \quad & O_{\max} \leq \xi_1 r(k) \\ & |e_{ss}| \leq \xi_2 r(k) \end{aligned} \tag{9.55}$$

where $e(k) = r(k) - y(k)$ is the system tracking error, $u(k)$ is the controller output $O_{\max} = \max\limits_{k} y(k) - r(k)$ is the maximal overshoot, e_{ss} is the steady-state error. The objective functions in Eq. (9.55) is to minimize the accumulated error and the energy consumption satisfying the maximum overshoot and steady-state error inequality constraints. Generally, ξ_1 and ξ_2 are set as: $0 < \xi_1 \leq 1, 0 \leq \xi_2 \leq 0.05$.

9.4.3.2 Dominating with Constraint Handling

Since there are two constraints in Eq. (9.55), the constraints should be combined in performing non-dominated sorting algorithm. When the ith and jth solutions satisfy one of the following conditions:

1. The ith solution is feasible, and the j th solution is not feasible;
2. Neither solutions are feasible, the constraint offset of the ith solution is smaller than the constraint offset of the jth solution;
3. Both solutions are feasible, at least one objective function of the ith solution is better than the jth solution.

Thus, the ith solution dominates the jth solution. According to the above Pareto dominance relationship with constraint handling, the Pareto sorting can then be implemented.

9.4.3.3 The Procedure of PID Controller Optimization

The main procedure of implementing PID controller parameters optimization by MOGA based on stabilization space is as follows:

1. Compute the stabilization spaces of the parameters for PID controller

Step 1 Assume $C(s) = k_p$, obtain the stable region of k_p based on the Routh-Hurwitz criterion, $k_p \in (k_{p\min}, k_{p\max})$.

Step 2 Choose the number of search subspaces: N_z, and gain the scan step of k_p: $\eta = (k_{p\max} - k_{p\min})/N_z$.

Step 3 Select k_p among $k_{p\min}, k_{p\min} + \eta, \ldots, k_{p\min} + N_z \cdot \eta$ in turns, and calculate the corresponding stable regions of k_i and k_d based on Sec. 9.4.2.

Step 4 Derive N_Z search subspaces, where the search domains of k_p is as follows: $[k_{p\min}, k_{p\min} + \eta], \ldots, [k_{p\min} + (N_z - 1)\eta, k_{p\min} + N_z\eta]$. Because each k_p has the corresponding stable regions of k_i and k_d in step 3, there exist a cluster of 2-dimensional regions of (k_i, k_d), choose the biggest region of (k_i, k_d) as the search subspace.

2. Execute MOGA in Ch.4 to solve the constrained bi-objective optimization problem. The steps are as follows:

Step 1 Initialize the population size N, the maximum evolution generations G, and the number of grids K_i in the stable space.

Step 2 Adopt decimal encoding, perform the constraint non-dominated sorting of the Pareto frontier based on f_1 and f_2.

Step 3 Use the tournament selection to produce the parents, and the elitists with rank 1 are directly kept as the parents of the crossover and mutation operators.

Step 4 Perform an analog binary crossover operator and a polynomial mutation operator.

Step 5 Repeat steps 2 to 4 until the termination condition is met. The termination condition is the maximum evolution generation.

3. Select one of the PID controllers according to the application requirement.

9.4.4 *Simulation for Optimization of PID Controllers*

In order to indicate the efficiency of the proposed method, two first-order with dead time (FOPDT) unstable processes are given as follows.

$$G_1(s) = \frac{4e^{-2s}}{4s - 1} \tag{9.56}$$

$$G_2(s) = \frac{e^{-1.5s}}{s - 1} \tag{9.57}$$

According to Eq. (9.49), the values of L are 0.5 and 1.5 for plants $G_1(s)$ and $G_2(s)$, respectively, then, the structure of $C_1(s)$ in Fig. 9.31 can be selected in terms of Eq. (9.51). The controlled plants $G_1(s)$ and $G_2(s)$ are stabilized first by the inner loop. In terms of Eq. (9.52), the range of T_{d1} for $G_2(s)$ can be gained as $0.5 < T_{d1} < 1$. For convenience of comparing with the method in [41], denoted as X&N method, k_{p1} and T_{d1} are set the same values as X&N method's, and the results of the inner loop controller $C_1(s)$ are listed in Table 9.2. Once the inner loop is chosen, the stable generalized controlled plant can be obtained. Thus, the stable region of k_p can be calculated based on Routh-Hurwitz criterion. Suppose there are three subspaces, that is, $N_Z = 3$, the search subspaces based on Hermite-Biehler theorem are given in Table 9.3, where the maximal region of the subspace is selected as the final search space. Obviously, the stability spaces of the outer-loop controller for $G_1(s)$ are: $0 < k_p < 0.314$, $0 < k_i < 31.949$, $-0.9879 < k_d < 1$, while the $G_2(s)$'s are: $0 < k_p < 0.0341$, $0 < k_i < 53.8427$, $-1.5868 < k_d < 0.41$.

When using MOGA to optimize the parameters of the PID controller, the parameters of MOGA are set as: $\xi_1 = 50\%$, $\xi_2 = 1\%$, $G = 1000$, $K_i = 30$, $N = 60$. After being optimized, the Pareto frontier of PID controllers and their control performances are shown in Figs. 9.32 and 9.33.

From the Pareto frontier distribution in Figs. 9.32a and 9.33a, it can be seen that f_1 and f_2 are contradictory. The smaller the error accumulation, the larger the required energy consumption. From Figa. 9.32b and 9.33b, it can also be seen that the optimal PID controllers satisfy the constraint conditions with small steady error

Table 9.2 The results of inner loop of $G_1(s)$ and $G_2(s)$

Plants	Inner loop	Stable range of k_{pl}	The results
$G_1(s)$	k_{p1}	$0.25 < k_{p1} < 0.634$	$k_{p1} = 0.41$
$G_2(s)$	$k_{p1}(1 + T_{d1}s)$	$T_{d1} = 0.59$,$-0.05 < k_{p1} < 1.8$	$k_{p1} = 1.019$,$T_{d1} = 0.59$

Table 9.3 The search subspaces of generalized control plant

Plants	Stable range of k_p	Subspaces of PID controller parameters		
$G_1 = (S)$	$0 < k_p < 0.314$ $N_Z = 3$	$0.001 < k_p <$ 0.1047	$0.97864 < k_i <$ 31.3168	$-0.9306 < k_d < 1$
		$0.1047 < k_p <$ 0.2093	$0.001 < k_i <$ 31.6365	$-0.9905 < k_d < 1$
		$0.2093 < k_p <$ 0.314	$0.001 < k_i <$ 31.9492	$-0.9879 < k_d < 1$
$G_2 = (S)$	$0 < k_p < 0.0341$ $N_Z =$ 3	$0.001 < k_p <$ 0.0113	$0.001 < k_i <$ 53.7929	$-1.5868 < k_d <$ 0.41
		$0.0113 < k_p <$ 0.0227	$0.001 < k_i <$ 53.8171	$-1.5866 < k_d <$ 0.41
		$0.0227 < k_p <$ 0.0341	$0.001 < k_i <$ 53.8427	$-1.5864 < k_d <$ 0.41

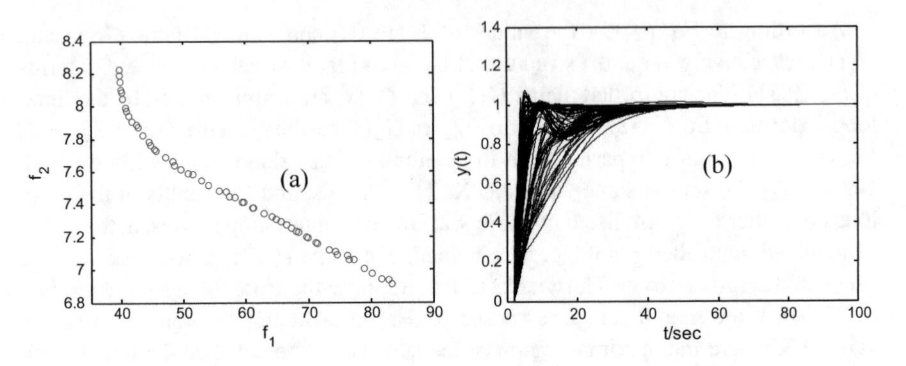

Fig. 9.32 **a** Pareto frontier of PID controllers for $G_1(s)$ and **b** their control performances

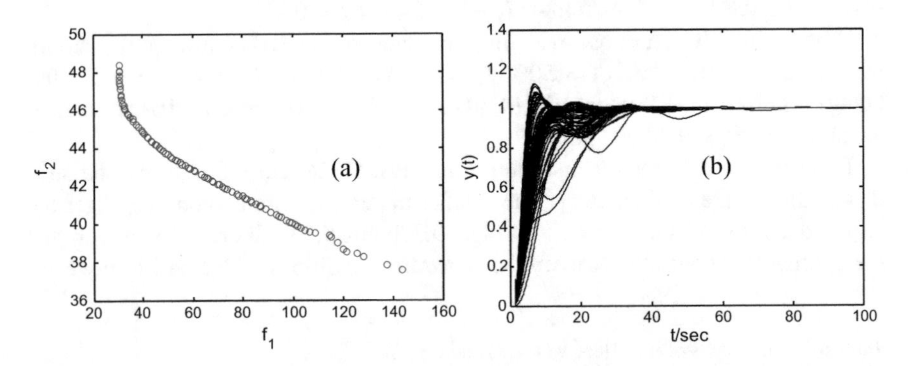

Fig. 9.33 **a** Pareto frontier of PID controllers for $G_2(s)$ and **b** their control performances

and overshoot. How to choose the PID controller should be determined according to the application requirement. Assuming that the overshoot is required to be less than 1% and the rising time to be the shortest, the final PID controller can then be found, which is listed in Table 9.4.

In order to indicate the effectiveness of the proposed method, the simulation results are compared with the X&N method, which also adopts a two-level control structure. The parameters of the outer-loop PID controller are set according to the specified phase-amplitude characteristics. The tuning parameters are also listed in Table 9.4. The simulation results are shown in Figs. 9.34 and 9.35. It is obvious that the control performance obtained by MOGA is better than that of the X&N method.

Table 9.4 The parameters of PID controller using two methods

Plants	X&N Method	DNA-MOGA
	$[k_p, K_i, k_d]$	$[k_p, K_i, k_d]$
G_1(s)	[0.0681,0.0421,0.2133]	[0.1095, 0.0192 0.2353]
G_2(s)	[0.0080, 0.0054, 0.0729]	[0.0116, 0.0042, 0.0547]

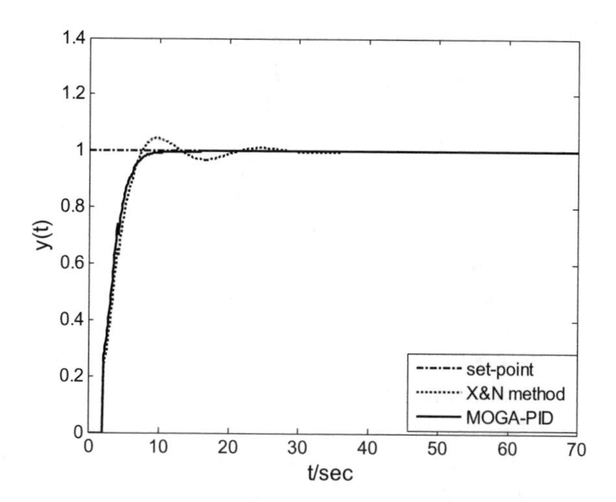

Fig. 9.34 Comparison results for $G_1(s)$

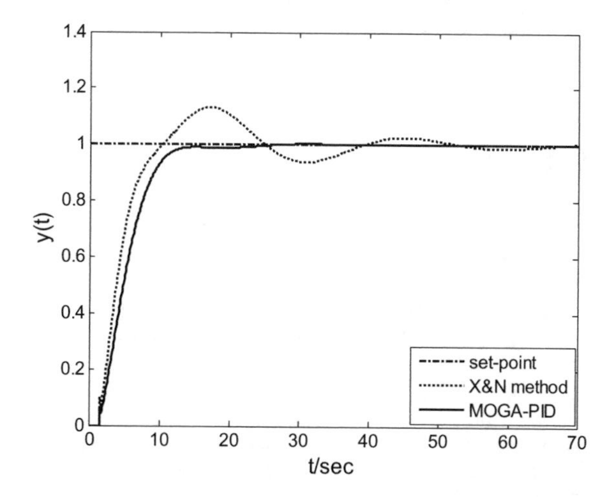

Fig. 9.35 Comparison results for $G_2(s)$

9.5 Summary

In this chapter, GA is used to optimize three types of control systems:

(1) A new PID controller based on ENMSSPFC and GA optimization is proposed and tested on the chamber pressure in a coke furnace. Since the PID controller structure is based on the prediction model and the parameters are optimized in the controller design, the control performance can be further improved, and at the same time, facilitates with a simple PID structure.

(2) Though the PID controller has achieved good control performance through the optimization of the control system structure and parameters, the limitation of the PID controller is also demonstrated. Therefore, the improved neuron

controller is designed and optimized by RNA-GA. The hybrid controller is constructed by the fuzzy PI controller and the neuron controller. The gain of the neuron controller is tuned online by a fuzzy algorithm and the parameters of the proposed controller are optimized by RNA-GA. The simulation tests under various conditions are made. The results illustrate that high precision mold-level control is reached and the proposed control method can efficiently control the mold level for the plant with big uncertainties and grave nonlinearities. This model-free controller has good performance, very strong robustness, and adaptability.

(3) An MOGA based on stabilization subspaces to optimize PID controller parameters for unstable FOPDT processes is developed. Because generalized Hermite-Biehler theorem cannot always obtain stable intervals of PID controller parameters for an unstable FOPDT process, two-level PID controller structure is utilized, where the plant is first stabilized by the inner loop and the outer loop of PID controller is optimized by MOGA based on stabilization subspaces to improve the control performance. Simulation results of several unstable FOPDT plants show the efficiency of the proposed methods.

References

1. Zhang, R., S. Wu, and J. Tao. 2018. A new design of predictive functional control strategy for batch processes in the two-dimensional framework. *IEEE Transactions on Industrial Informatics* 15 (5): 2905–2914.
2. Dorf, R.C. and R.H. Bishop. 2011. *Modern control systems*. Pearson.
3. Jacquot, R.G. 2019. *Modern digital control systems*. Routledge.
4. Nocedal, J. and S. Wright. 2006. *Numerical optimization*. Springer Science & Business Media.
5. Bryson, A.E., Y.C. Ho, and G.M. Siouris. 2007. Applied optimal control: optimzation, estimation, and control. *IEEE Transactions on Systems Man Cybernetics* 9 (6): 366–367.
6. Skogestad, S. 2004. Simple analytic rules for model reduction and PID controller tuning. *Modeling Identification Control Engineering Practice* 13 (4): 291–309.
7. Wang, Q.G., C.C. Hang, and X.P. Yang. 2001. Single-loop controller design via IMC principles. *Automatica* 37 (12): 2041–2048.
8. Marchetti, G., C. Scali, and D.R. Lewin. 2001. Identification and control of open-loop unstable processes by relay methods. *Automatica* 37 (12): 2049–2055.
9. Tyreus, B.D., and W.L. Luyben. 1992. Tuning PI controllers for integrator/dead time processes. *Industrial and Engineering Chemistry Research* 31 (11): 2625–2628.
10. Luyben, W.L. 1996. Tuning proportional—integral—derivative controllers for integrator/deadtime processes. *Industrial Engineering Chemistry Research* 35 (10): 3480–3483.
11. Ziegler, J.G., and N.B. Nichols. 1993. Optimum settings for automatic controllers. *Asme Trans* 64 (2B): 759–768.
12. Ramasamy, M., and S. Sundaramoorthy. 2008. PID controller tuning for desired closed-loop responses for SISO systems using impulse response. *Computers Chemical Engineering* 32 (8): 1773–1788.
13. Rico, J.E.N., and J.L. Guzmán. 2012. Unified PID tuning approach for stable, integrative and unstable dead-time processes. *IfAC Proceedings* 45 (3): 35–40.
14. Zhang, R., A. Xue, and S. Wang. 2011. Modeling and nonlinear predictive functional control of liquid level in a coke fractionation tower. *Chemical Engineering Science* 66 (23): 6002–6013.

15. Zhang, R., et al. 2014. Real-Time implementation of improved state-space MPC for air supply in a coke furnace. *IEEE Transactions on Industrial Electronics* 61 (7): 3532–3539.
16. Zhang, R., et al. 2012. An improved state-space model structure and a corresponding predictive functional control design with improved control performance. *International Journal of Control* 85 (8): 1146–1161.
17. Zhang, R., and F. Gao. 2012. State space model predictive control using partial decoupling and output weighting for improved model/plant mismatch performance. *Industrial Engineering Chemistry Research* 52 (2): 817–829.
18. Exadaktylos, V., and C.J. Taylor. 2010. Multi-objective performance optimisation for model predictive control by goal attainment. *International Journal of Control* 83 (7): 1374–1386.
19. Zhang, R., A. Xue, and S. Wang. 2011. Dynamic modeling and nonlinear predictive control based on partitioned model and nonlinear optimization. *Industrial Engineering Chemistry Research* 50 (13): 8110–8121.
20. Tao, J., Z. Yu, and Z. Yong. 2014. PFC based PID design using genetic algorithm for chamber pressure in a coke furnace. *Chemometrics Intelligent Laboratory Systems* 137 (20): 155–161.
21. Zhang, R., et al. 2014. New PID controller design using extended nonminimal state space model based predictive functional control structure. *Industrial Engineering Chemistry Research* 53 (8): 3283–3292.
22. Tang, K.S., et al. 2001. An optimal fuzzy PID controller. *IEEE Transactions on Industrial Electronics* 48 (4): 757–765.
23. Grum, J. 2008. Book review: Fuzzy controller design, theory and applications by Z. Kovacic and S. Bogdan. *International Journal of Microstructure Materials Properties, 3*(2/3): 465–466.
24. Zeng, G.Q., et al. 2019. Adaptive population extremal optimization-based PID neural network for multivariable nonlinear control systems. *Swarm Evolutionary Computation* 44: 320–334.
25. Chen, J., and T.C. Huang. 2004. Applying neural networks to on-line updated PID controllers for nonlinear process control. *Journal of Process Control* 14 (2): 211–230.
26. Chen, M., and D.A. Linkens. 1998. A hybrid neuro-fuzzy PID controller. *Fuzzy Sets and Systems* 99 (1): 27–36.
27. Kim, S.M., and W.Y. Han. 2006. Induction motor servo drive using robust PID-like neuro-fuzzy controller. *Control Engineering Practice* 14 (5): 481–487.
28. Wang, N. 1993. Neuron intelligent control for electroslag remelting process. *Acta Automatica Sinica* 38 (3): 178–180.
29. Muyeen, S., et al. 2009. A variable speed wind turbine control strategy to meet wind farm grid code requirements. *IEEE Transactions on Power Systems* 25 (1): 331–340.
30. Gaing, Z.L. 2004. A particle swarm optimization approach for optimum design of PID controller in AVR system. *IEEE Transactions on Energy Conversion* 19 (2): 384–391.
31. Coelho, L.D.S., and M.W. Pessôa. 2011. A tuning strategy for multivariable PI and PID controllers using differential evolution combined with chaotic Zaslavskii map. *Expert Systems with Applications* 38 (11): 13694–13701.
32. Hung, M.H., et al. 2008. A Novel intelligent multiobjective simulated annealing algorithm for designing robust PID controllers. *IEEE Transactions on Systems Man Cybernetics Part A Systems Humans* 38 (2): 319–330.
33. Zhang, J., et al. 2009. Self-organizing genetic algorithm based tuning of PID controllers. *Information Sciences* 179 (7): 1007–1018.
34. Tao, J., and N. Wang. 2005. Fuzzy neuron hybrid control for continuous steel casting. *IFAC Proceedings* 38 (1): 121–126.
35. Kikuchi and Pursula. 1998. Treatment of uncertainty in study of transportation: Fuzzy set theory and evidence theory. *Journal of Transportation Engineering* 124 (1): 1–8.
36. Wang Y.N. 2006. *Intelligent control system.* Hunan University Press.
37. Graebe, S.F., G.C. Goodwin, and G. Elsley. 1995. Control design and implementation in continuous steel casting. *IEEE Control Systems* 15 (4): 64–71.
38. Datta, A, M.T. Ho., and S.P. Bhattacharyya. 2013. Structure and synthesis of PID controllers. Springer Science & Business Media.

39. Roy, A, and K. Iqba. 2005. Synthesis of stabilizing PID controllers for biomechanical models. in *Proceedings of 2005 IFAC World Congress*. Praha.
40. Ho, M.T., A. Datta, and S.P. Bhattacharyya. 2000. Generalizations of the Hermite-Biehler theorem: the complex case. *Linear Algebra and its Applications* 320 (1): 23–36.
41. Xiang, C, and L.A. Nguyen. 2005. Control of unstable processes with dead time by PID controllers. *International Conference on Control and Automation. IEEE*, 2: 703–708.

Chapter 10
Further Idea on Optimal Q-Learning Fuzzy Energy Controller for FC/SC HEV

With the development of intelligent algorithms, the learning-based algorithm has been considered as viable solutions to various optimization and control problems. GA can also be efficient to optimize the new emerging intelligent algorithm. Here, an adaptive fuzzy energy management control strategy (EMS) based on Q-Learning algorithm is presented for the real-time power split between the fuel cell and supercapacitor in the hybrid electric vehicle (HEV) in order to adapt the dynamic driving pattern and decrease the fuel consumption. Different from the driving pattern recognition based method, Q-Learning controller observes the driving states, takes actions, and obtains the effects of these actions. By processing the accumulated experience, the Q-Learning controller progressively learns an appropriate fuzzy EMS output tuning policy that associates suitable actions to the different driving patterns. The environment adaptation capability of fuzzy EMS is then improved needless of driving pattern recognition. To enhance the learning capability and decrease the effect on the initial values of Q-table, GA can also be utilized to optimize the initial values of Q-Learning based fuzzy energy management.

10.1 Introduction

With the energy crisis and environment pollution, new fuel cell (FC) energy vehicles (EVs) are drawing more and more attention because of their high reliability and low pollutant emission [1]. However, with slow dynamic response and limited load following capability, the vehicle equipped with fuel cell is often acted as the main power [2], and the energy storage devices, such as supercapacitor, are usually selected as a power buffer during climbing, acceleration, and braking [3, 4]. For fuel cell hybrid energy vehicle (HEV), an efficient energy management control strategy (EMS) is critical to improve the fuel economy and prolong the lifetime of the fuel cell.

J. Tao et al., *DNA Computing Based Genetic Algorithm*,
https://doi.org/10.1007/978-981-15-5403-2_10

Generally, there are two types of energy management strategies: rule-based and optimization-based [5, 6]. The former strategies can be subdivided into the deterministic rule and fuzzy rule based methods, while the latter is conducive to the combination of advanced control theory, such as the dynamic programming(DP) [7], GA [8], or predictive control [9], to achieve optimal control of energy management. The deterministic rule based methods are used widely because of its easy implementation [10–12]. However, the rule-based strategy is often sub-optimal and highly dependent on the expert experience, so efforts have focused on improving the optimization-based strategy.

Most of DP and genetic algorithm based energy management strategy could determine the best fuel economy once the driving cycle was given [13, 14]. Because they required a high computation load, the above optimal algorithm cannot be used in a real-time scene.

To combine the advantages of rule-based and optimization-based methods, some rule-based control strategies are optimized by minimizing a loss function that generally represents the control objectives under a fixed driving cycle [15]. However, such a strategy is no longer optimal at various driving patterns. Since driving patterns have an important impact on the energy economy of HEVs, Dayeni et al. obtained better control performances using a prior knowledge of the driving cycle [16]. However, it is impossible in practice to know the driving cycles in advance except for global positioning system (GPS), geographic information system (GIS), intelligent transport system (ITS) [17–19]. Comparing with external traffic information, the driving information derived by on-vehicle sensors is much more reliable. Several methods, such as k-nearest neighbor [20], fuzzy logic classifier [16], neural networks [21], support vector machine [22], have been utilized to recognize the driving patterns. There are usually four typical driving patterns in the literature, but the actual traffic is much complex and the EMS should be adaptive to the varying environment. How to improve the adaptability of the energy management controller without complicated structure and heavy computation? The learning-based energy management method may provide a viable solution to solve electric power system decision and control problems [23]. A learning-based energy management system can learn to take actions directly from the states without any prediction or predefined rules, and converge to an optimal policy. Additionally, the learning-based energy management system has shown its self-learning capability on the adaption of different driving conditions [24–26]. However, the deep reinforcement learning algorithm is not stable and its convergence is relatively slow [27].

In this chapter, a Q-Learning based fuzzy energy management controller is considered for real-time HEV energy split that satisfies the driver' demand and achieves optimization of energy consumption and load fluctuation. In particular, we focus on improving the Q-Learning (QL) driven agent's adaption to different driving cycles. And the Q-Learning strategy is utilized to tune fuzzy output to be adaptive to different driving patterns. To avoid the initialization effect of Q-table, GA can be utilized to optimize its initial value. Moreover, in order to enhance the learning capability, the elitist is maintained, and the converging process is speeded up by getting rid of the greedy process in the late stage.

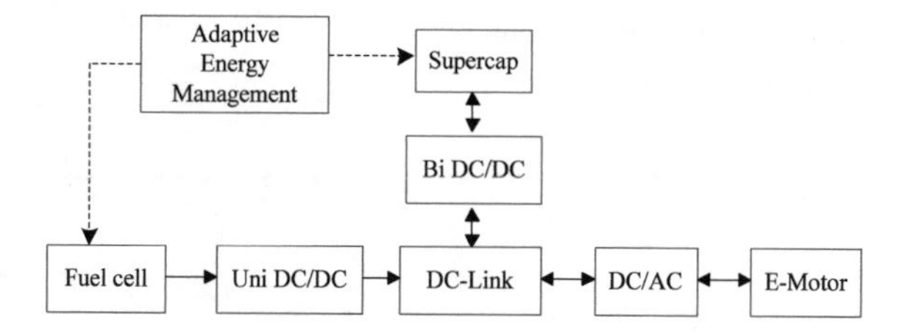

Fig. 10.1 The structure and main components in the powertrain

10.2 FC/SC HEV System Description

The series architecture of the powertrain for FC/SC HEV is shown in Fig. 10.1. The primary power is applied by the fuel cell and the supercapacitor acts as the power buffer, which provides the peak power during cold start, hard acceleration and absorbs regenerative braking energy. The load is a 49 kW alternating current (AC) permanent magnet motor. A unidirectional DC/DC converter is connected to FC and a bidirectional DC/DC converter to SC, while a DC/AC converter is connected to the AC motor. The Q-Learning based adaptive fuzzy energy management controller is proposed to split the demand power of HEV between fuel cell and supercapacitor. The target vehicle is a VW Jetta modified hybrid vehicle and its main parameters can be found in [28].

10.3 Q-Learning Based Fuzzy Energy Management Controller

Since Q-Learning alone has the limitation in that it is too hard to visit all the state–action pairs and fall into local optimum [29]. In addition, there are many hidden states that cannot be visited due to the discretization process. More seriously, the algorithm may be difficult to converge and become unstable [27]. Hence, we extend the learning algorithm by using a fuzzy EMS to generalize Q-Learning over the continuous state space. In addition, GA is used as an optimization tool to tune the initial values of the Q-table. The framework of the proposed strategy is shown in Fig. 10.2. Fuzzy logic controller (FLC) is the main energy management controller, and a Q-Learning strategy is designed to compensate the adaptation of FLC for different driving conditions. Moreover, GA is introduced to initialize the Q-table, and its learning algorithm is improved to speed up the convergence process, which will compensate for the limitation of the Q-Learning. The instantaneous management of the power flow between the FC and the SC is to meet the power demand of the HEV and minimize the fuel consumption. Meanwhile, the current fluctuation of the FC is to be reduced to prolong its cycling life.

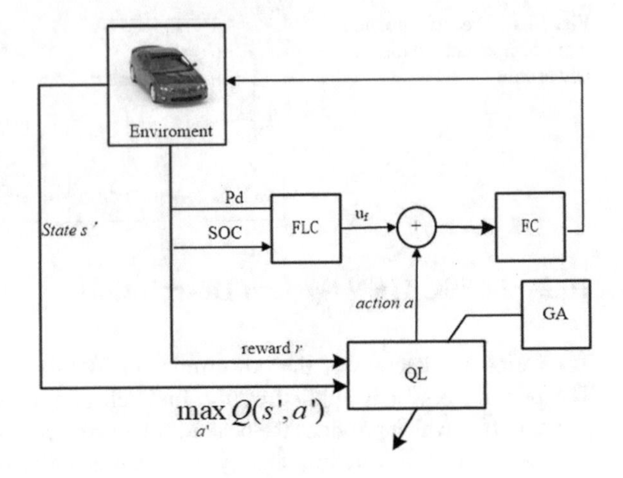

Fig. 10.2 The framework of the proposed method

10.3.1 Fuzzy Energy Management Controller

A block diagram of fuzzy energy management system is shown in Fig. 10.3. It has two inputs: the positive demand power P_{dem} required by the vehicle and the SoC of supercapacitor, and the output u_f is the ratio of P_{dem} assigning to the fuel cell. To guarantee the safety of supercapacitor, the energy management controller is executed when the demand power is positive and the SoC is larger than 0.45. If the SoC of supercapacitor is less than 0.45, the fuel cell provides all the required power in its power capability.

I_1, I_2 are the inputs by fuzzifying P_{dem} and SoC into the fuzzy domain [0, 1], [0, 1], respectively.

$$I_1 = \frac{P_{\mathrm{dem}}}{P_{\max}},\ I_2 = \frac{\mathrm{SoC} - \mathrm{SoC}_{\min}}{\mathrm{SoC}_{\max} - \mathrm{SoC}_{\min}} \tag{10.1}$$

where $P_{\max}$ is the maximal demand power, and $\mathrm{SoC}_{\max}$ $\mathrm{SoC}_{\min}$ are the maximal and minimal safe values of SoC.

In terms of expert knowledge about the energy management system, the fuel cell delivers as much as possible the required power when the demand power is high and the SoC of supercapacitor is low. When the demand power is low and the SoC of

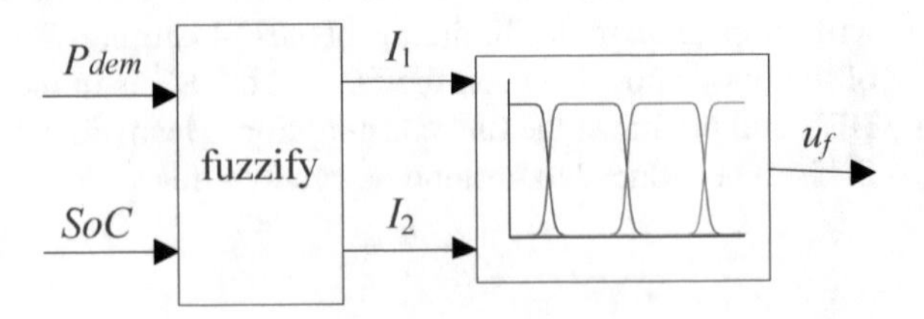

Fig. 10.3 The block diagram of fuzzy energy management controller

Table 10.1 The rules of fuzzy logic controller

SoC	u_f	P_{dem}	VL L M H
L			L M MH H
M			ML L M MH
H			VL ML L M

supercapacitor is high, the fuel cell stack delivers relatively low power. The whole rule base is listed in Table 10.1.

Each linguistic value is assigned by a membership function (MF). Here, a Gaussian MF has been selected: $\text{Gaussion}(x; \sigma, c) = \exp(-\|x - c\|^2 / 2\sigma^2)$, where c represents MF's center and σ determines MF's width. The fuzzy membership functions for SoC, P_{dem}, and u_f are shown in Fig. 10.4.

Takagi–Sugeno–Kang fuzzy inference system (TSKFIS) is adopted, and ith rule, for example, is described as follows: If P_{dem} is H and SoC is L, then u_f is H. Using centroid defuzzification, the fuzzy output can be formulated as follows:

$$u_f(k) = \frac{\sum_{j=1}^{r_1} \sum_{i=1}^{r_2} \mu_{I_1}^{j}(k) \mu_{I2}^{i}(k) \Delta\mu_u^{ij}(k)}{\sum_{j=1}^{r_1} \sum_{i=1}^{r_2} \mu_{I_1}^{j}(k) \mu_{I2}^{i}(k)} \tag{10.2}$$

where r_1 is 4 and r_2 is 3, as given from Table 10.1. $\mu_x(k)$ is the degree of the membership function of SoC and P_{dem}, respectively, and $\Delta\mu_u^{ij}(k)$ is obtained by using the Mamdani product and maximization fuzzy inference scheme.

The fuzzy EMS output u_f cooperating with the Q-Learning controller is used as the coefficient to assign the power of fuel cell, which is derived as follows:

$$P_{FC} = (u_f + a) P_{dem} \tag{10.3}$$

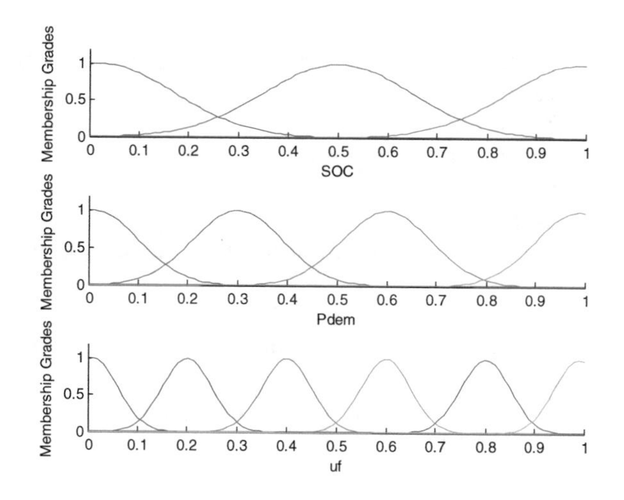

Fig. 10.4 Membership functions for inputs and output of fuzzy EMS

where a is the action of Q-Learning controller and the total split coefficient should be not more than 1.

MATLAB code for the calculation of fuzzy controller output u_f is shown as follows:

```
function out = fuzzyEMSGauss(input x, input y, parameters)
% input x, input y are the inputs of fuzzy controller, parameters are the widths of
fuzzy MFs
c1 = [0.02, 0.5, 0.99];
c2 = [0.01, 0.3, 0.6, 0.99];
c3 = [0.01, 0.2, 0.4, 0.6, 0.8, 0.99];
% The centers of MFs for inputs and output
point_n = 100;
x = linspace(0, 1, point_n);
a = parameters(1:3); % obtain the width of 3 MFs

ante_param_x = [];
temp = [a(1) a(1) a(1)]; % the width for the input1 MF
ante_param_x = [ante_param_x; temp; c1(1:3);]';
 temp = [];
temp = [a(2) a(2) a(2) a(2)]; % the width for the input2 MF
ante_param_y = [];
ante_param_y = [ante_param_y; temp; c2(1:4);]';
temp = [];
temp = [a(3) a(3) a(3) a(3) a(3) a(3)]; % the width for the output MF
cons_param = [];
cons_param = [cons_param; temp; c3(1:6);]';
for i = 1:3,
    ante_mf_x(i, :) = gauss_mf(x, ante_param_x(i, :));
end
for i = 1:4,
    ante_mf_y(i, :) = gauss_mf(x, ante_param_y(i, :));
end
for i = 1:6,
    cons_mf(i, :) = gauss_mf(x, cons_param(i, :));
end
for k = 1:3
    mfx(k) = gauss_mf(inputx, ante_param_x(k, :));
end
for k = 1:4
    mfy(k) = gauss_mf(inputy, ante_param_y(k, :));
end
  rules = [3 4 5 6; % Table of fuzzy rules
      2 3 4 5;
      1 2 3 4;];
```

```
kk = 1; qualified_cons_mf = [];
     for k = 1:3
       for j = 1:4
          w(k, j) = min(mfx(k), mfy(j));
          qualified_cons_mf(kk,:) = w(k, j)*cons_mf(rules(k, j), :);
          kk = kk + 1;
       end
end
 overall_out_mf = max(qualified_cons_mf);
 out = defuzzy(x, overall_out_mf, 1);
```

10.3.2 Q-Learning in HEV Energy Control

Given an episode under the defined driving cycle, Q-Learning controller is utilized to compensate the fuzzy EMS output to adapt to the varying driving conditions. During the learning process, the Q-Learning driven controller observes the state of the driving condition, then performs the action and calculates the reward value. The value function accumulating the total rewards over time is then updated. When the value function converges, the learning process ends, and the control policy is obtained. The key concepts applied in the QL are formulated as follows:

Policy: The policy is generated from a lookup Q-table filled with value functions. The Q-table is a multiple dimension array that contains the state space and action space. When the agent is in one of these states, the action can be derived by checking the maximal value function in the Q-table.

State space: The instantaneous demand power P_{dem} and SoC of the SC are selected as the system state denoted as s. In order to discretize the continuous state variables, they are discretized by Eqs. 10.4 and 10.5:

$$P_d = \frac{P_{\mathrm{dem}}}{d_1} + 1, d_1 = \frac{P_{\max}}{n_1 - 1} \tag{10.4}$$

$$\mathrm{SoC}_d = \frac{\mathrm{SoC} - \mathrm{SoC}_{\min}}{d_2} + 1, d_2 = \frac{\mathrm{SoC}_{\max} - \mathrm{SoC}_{\min}}{n_2 - 1} \tag{10.5}$$

where d_1, d_2 represent the discretization degree of P_{dem}, SoC, n_1, n_2 represent the number of the states, respectively. After discretization, P_d and SoC_d delegate the searching index in the state space.

Action space: The compensating output for different driving patterns is chosen as the control action. The same discretization technique is applied to the action output as shown in Eq. 10.6:

$$u_d = kd_3 + u_{d\min}, d_3 = \frac{u_{d\max} - u_{d\min}}{n_3 - 1} \tag{10.6}$$

where d_3 is the discretization degree, n_3 represents the number of action, $u_{d\max}$ and $u_{d\min}$ is the maximal and minimal u_d for compensation, and k is the one-based index to the action obtained in terms of Q-table.

Reward function: Since the control objectives of the HEV is to satisfy the demanding power, minimize the fuel consumption and the fluctuation of FC's current, leave the SoC of the SC to the safe range, the reward function is then defined as follows:

$$r_t = \begin{cases} -1000 & \text{SoC} > \text{SoC}_{\max} \text{ or SoC} < \text{SoC}_{\min} \\ \Delta I^2 + \Delta P^2 & \text{otherwise} \end{cases} \tag{10.7}$$

where r_t is the immediate reward at time t, $\Delta P = P_{\text{dem}} - P_{\text{FC}} - P_{\text{SC}}$. P_{SC} is the power provided by SC.

Value Function: An estimation of future total rewards at state s, which is formulated as the expectation of the sum of future immediate rewards as described in Eq. 10.8:

$$Q(s_t, a_t) = \text{E}(r_{t+1} + \gamma r_{t+2} + \gamma^2 r_{t+3} + \cdots |s_t, a_t) \tag{10.8}$$

where γ is the discount factor that assures the infinite sum of rewards to converge. Define Q^* as the optimal value function representing the maximum accumulative rewards; Q^* can be expressed by the Bellman equation:

$$Q^*(s_t, a_t) = \text{E}(r_{t+1} + \gamma \max_{a_{t+1}} Q^*(r_{t+1}, a_{t+1})|r_t, a_t) \tag{10.9}$$

where the first part is the immediate reward r_{t+1}, and the second part is the discounted value of successor state $Q(s_{t+1}, a_{t+1})$. To obtain Q^*, the Bellman equation is to iterate the value function as follows:

$$Q_{t+1}(s_t, a_t) = Q_t(s_t, a_t) + \eta(r_{t+1} + \gamma \max_{a_{t+1}} Q_t(s_{t+1}, a_{t+1}) - Q_t(s_t, a_t)) \tag{10.10}$$

where $\eta \in (0, 1)$ is the learning rate, by using Eq. 10.10, Q_t will converge to the optimal action value function Q^* as $t \to \infty$. After obtaining Q^*, the action can be obtained in terms of the current input states.

The procedure of Q-Learning strategy is executed off-line and shown as follows:

Step 1: Initialize value function $Q(s_t, a_t)$ randomly, set the maximal learning epochs N_p, the driving cycles T.

Step 2: In the former 10% driving cycles, with probability ξ select a random action a_t, otherwise select $a_t = \arg\max_{a_t} Q(s_t, a_t)$. Take action a_t and add to the fuzzy EMS output in terms of Eq. 10.12, calculate s_{t+1}, and reward r_{t+1}. Update Q_{t+1} according to Eq. 10.19.

Step 3: If the reward is −1000, the learning process during the driving cycles T will break immediately, otherwise repeat step 2 until T is terminal, thus, $Q_{t+1}(s_t, a_t) = Q_t(s_t, a_t) + \eta(r_{t+1} - Q_t(s_t, a_t))$ is the learned Q-table.

Step 4: Repeat steps 1–3 until the maximal epochs N_p is satisfied.

MATLAB code for the action calculation is given below as follows:

```
Function action = QL(P, SOC)
% P is the demand power, SOC is the state of charge of SC, action is the action
value.
p_dem_index = round((P/300 + 1);
soc_index = round((SOC-0.4)/0.01 + 1);
greedy = rand(1); %choose action by greedy search
  if(greedy <=epsilon)
    action_index = randperm(num_action,1);
    action = (action_index-1)/(num_action-1)*0.5 + 0.4;
 else
    temp = Q(p_dem_index, soc_index,:); %Q*
    [value, action_index] = max(temp);
    action = (action_index-1)/(num_action-1)*0.5 + 0.4;
end
```

10.3.3 GA Optimal Q-Learning Algorithm

As shown in 10.10, Q-Learning algorithm is a gradient descent optimal algorithm, and its results will be affected by the initial value of Q-table. Moreover, it is easy to trap into the local optima. GA is then introduced to improve the performance of the learning algorithm.

10.3.4 Initial Value Optimization of Q-Table

(1) Objective function

For Q-Learning based fuzzy EMS, the initialization optimization of Q-table is to minimize fuel consumption to save the energy and the current fluctuation of the fuel cell to prolong the cycle period of the fuel cell. In addition, some constraints have to be satisfied to guarantee the safety of EMS. For example, in order to avoid reactant starvation, the maximal current of the fuel cell is limited to 150 A; the power change rate of the fuel cell is restricted to 10 kW/s with the chemical response lag of the reactant supply system. Once the stack voltage falls below 60 V, the fuel cell will be shut down. For SC, its transient power is limited to 30 kW and the current is less than 150 A considering the power limit of bidirectional DC/DC converter. Moreover, the

SoC of the SC is kept in [0.45 0.95] in order to absorb the regenerative braking power and provide the transient power. The objective function is formulated by using the weighted-sum method:

$$
\begin{aligned}
\text{Min}\quad & J = \omega \sum_{k=1}^{K} \Delta I(k)^2 + \sum_{k=1}^{K} m_{H_2}(k) \\
\text{s.t.}\quad & P_{FC} + P_{SC} = P_{dem} \\
& 0 < P_{FC} \leq 40 \\
& 0 < i_{FC} \leq 150 \\
& -5 \leq \Delta P_{FC} \leq 5 \\
& -30 \leq P_{SC} \leq 30 \\
& -150 \leq i_{SC} \leq 150 \\
& v_{FC} \geq 60
\end{aligned}
\tag{10.11}
$$

where K is the number of samples in the whole driving trip, i_{FC} and i_{SC} are the currents of fuel cell and supercapacitor, respectively. $\Delta I(t) = i_{FC}(t) - i_{FC}(t-1)$ is the current variation of the fuel cell at time t, $\Delta P_{FC}(t) = P_{FC}(t) - P_{FC}(t-1)$ is the power variation of the fuel cell, and ω is the weight coefficient of the two objectives. The inequality and equality constraints are handled as a penalty factor, which is added to J. The constraint handling is similar to that in Chap. 3.

(2) Genetic encoding and operators

As described in Sect. 10.3.2, Q is an $n_1 \times n_2 \times n_3$ matrix. Provided there are N chromosomes, the elements in the ith chromosome (C_i) using decimal encoding are randomly initialized between (0, 1). The selection, crossover, and mutation operators have been adopted to make the objective evolve to the optimal one.

Roulette wheel selection is used, and its probability distribution is computed in terms of the objective function:

$$
\mathbf{p} = [p_1, \cdots, p_N] = \left[\frac{f_1}{\sum_{i=1}^{N} f_i}, \frac{f_1 + f_2}{\sum_{i=1}^{N} f_i}, \ldots, \frac{\sum_{i=1}^{N-1} f_i}{\sum_{i=1}^{N} f_i}, 1 \right] \tag{10.12}
$$

where $f_i = 1/J_i$, J_i is computed according to 10.11 for the ith individual. $\xi \in (0, 1)$ is randomly generated, and the individuals satisfying $\xi < p_i$ will be found; however, only one individual at the first index is selected as the parent. Totally, $N - 1$ Roulette wheel selections are executed, and the elitism is maintained in the parents. After parent selection, the crossover operator is executed with probability p_c between individuals $\mathbf{C}_i$ and $\mathbf{C}_{i+1}$. The offspring $\mathbf{C}_i^{'}$, $\mathbf{C}_{i+1}^{'}$ are then generated:

$$
\begin{aligned}
\mathbf{C}_{\mathbf{i}}^{'} &= \beta \mathbf{C}_{\mathbf{i}} + (1-\beta)\mathbf{C}_{\mathbf{i+1}} \\
\mathbf{C}_{\mathbf{i+1}}^{'} &= \beta \mathbf{C}_{\mathbf{i+1}} + (1-\beta)\mathbf{C}_{\mathbf{i}}
\end{aligned} \tag{10.13}
$$

where β is gained randomly between (0, 1).

For a better exploration, the mutation operator is carried out among N offspring with probability p_m. Once the element in $\mathbf{C}_i^{'}$ is mutated, its value will be reproduced randomly between (0, 1).

10.3.4.1 Application of Improved Q-Learning Algorithm

The initial value of Q-table is obtained by GA to decrease the effect of different initializations. Moreover, to avoid trapping into the local optima too quickly, the action index is randomly generated with probability ξ in the first 10% learning steps. Else, the action index is obtained by maximizing the Q value function. The Q-Learning process is iterated according to the best Q-table with the maximal reward in Eq. 10.7. When a training trip is finished, the whole learning process will be re-evaluated:

$$R_{\text{total}}(i) = \sum_{t=1}^{T} m_{H_2}(t), i = 1 : N_p \tag{10.14}$$

The Q-table with minimal R_{total} in the learning process is denoted as Q_{bt}, and Eq. 10.10 is then rewritten as follows:

$$Q_{b,t+1}(s_t, a_t) = Q_{bt}(s_t, a_t) + \eta(r_{t+1} + \gamma \max_{a_{t+1}} Q_{bt}(s_{t+1}, a_{t+1}) - Q_{bt}(s_t, a_t)) \tag{10.15}$$

The whole evaluation of Q-Learning based EMS is beneficial to keep the elitism of Q-table and speed up the convergence of the learning process.

10.3.5 Procedure of Improved Q-Learning Fuzzy EMS

The GA and Q-Learning optimization processes are done off-line, which is shown as follows:

Step 1: Initialize the maximal generation G, population size N, crossover and mutation operator probabilities p_c, p_m, and its weight coefficient ω. Initialize the chromosomes Qs randomly.

Step 2: For each chromosome, implement Q-Learning fuzzy EMS for one time and calculate the performance J.

Step 3: Produce the offspring using tournament selection and elitism maintaining strategy. Execute the crossover and mutation operator with probability p_c and p_m, respectively.

Step 4: Repeat steps 2–3 until the maximal evolution generation G is met and obtain the optimal Q initialization.

Step 5: Set the number of training epochs N_p to update the value function and the length of the driving episode K. The energy management policy is performed at each step within the episode duration with GA initialized Q-table.

Step 6: The action is selected and performed for the corresponding state with probability ξ in the first 10% learning steps. Otherwise, the action index is obtained by maximizing the Q value function. The SoC and reward function are then obtained.

Step 7: The value function is updated by the Bellman Eq. 10.10. Steps 6–7 repeated until the episode ends.

Step 8: If the sum of fuel consumption in Eq. 10.14 is better in the latter training epochs, the Q_b is gained and the learning process is updated according to Eq. 10.15.

After off-line optimization, the Q-Learning fuzzy EMS is obtained which can be applied online to adapt to various driving conditions.

10.3.6 Real-Time Energy Management

The GA optimal Q-Learning fuzzy EMS for HEV energy management in the above sections is implemented off-line, which means the agent is trained under the specific driving cycle and expected to adapt to different driving cycles for real-time energy management. Unlike the traditional algorithms based on driving cycle recognition, the HEV energy management algorithm trains the values of Q-table and does not depend on the driving pattern recognition. The framework in power system decision and control is described in Fig. 10.5. It can be seen that the learned agent in the learning module is directly applied in the execution module. In the simulation environment, the agent tries to explore more information by the action generated from greedy strategy in the early learning stage. In this way, the agent can enlarge

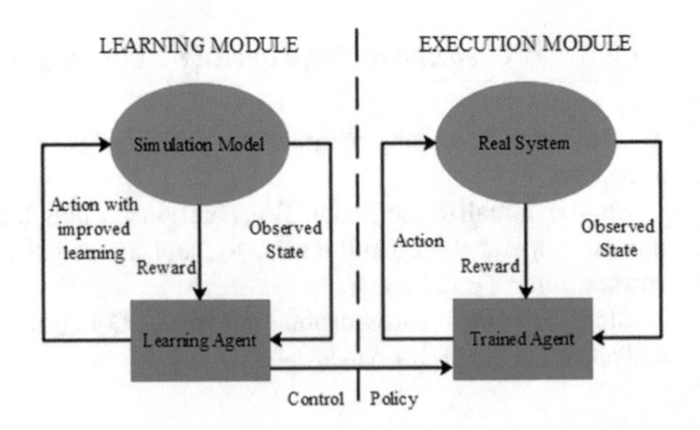

Fig. 10.5 The learning and real-time agents for driving conditions

the scope of the cognition about the environment through the Q-table as complete as possible and keep the best Q-table. However, in practice, the agent no longer takes risks to obtain more information by the greedy algorithm, but still receives reward from the environment to help in adapting to different driving conditions. That is,a_t is obtained according to $\arg\max_{a_t} Q(s_t, a_t)$ without the greedy process.

10.4 Summary

The Q-Learning based adaptive fuzzy EMS has a potential to adapt to different driving conditions and obtain minimal fuel consumption of the fuel cell. The characteristics of the quick charge and discharge of supercapacitor can be utilized adequately and the slow response and hydrogen starvation of the fuel cell can be compensated by SC bank during the transient variation of the required power. The perturbation minimization of the output current and voltage of the fuel cell may be helpful to prolong the lifetime of the fuel cell. Little expert knowledge is required to define the fuzzy rules carefully, and GA can also be utilized to automatically initialize the value of Q-table to speed up the convergence of Q-Learning controller. Thus, the driving pattern is not required in advance, and the environment information can be learned automatically.

References

1. Ralph, T.R. 2006. Principles of fuel cells. *Platinum Metals Review* 50 (4): 200–201.
2. Meacham, et al. 2006. Analysis of stationary fuel cell dynamic ramping capabilities and ultra capacitor energy storage using high resolution demand data. *Journal of Power Sources 156*(2): 472–479.
3. Khaligh, A., and Z. Li. 2010. Battery, ultracapacitor, fuel cell, and hybrid energy storage systems for electric, hybrid electric, fuel cell, and plug-in hybrid electric vehicles: state of the art. *IEEE Transactions on Vehicular Technology* 59 (6): 2806–2814.
4. Liu, C., et al. 2010. Graphene-based supercapacitor with an ultrahigh energy density. *Nano Letters* 10 (12): 4863–4868.
5. Hofman, T., et al. 2007. A Rule-based energy management strategies for hybrid vehicles. *International Journal of Electric Hybrid Vehicles* 1 (1): 71–94.
6. Trovão, J.P., et al. 2013. A multi-level energy management system for multi-source electric vehicles—an integrated rule-based meta-heuristic approach. *Applied Energy* 105 (2): 304–318.
7. Chen, B.C., Y.Y. Wu, and H.C. Tsai. 2014. Design and analysis of power management strategy for range extended electric vehicle using dynamic programming. *Applied Energy* 113 (1): 1764–1774.
8. Golchoubian, P., and N.L. Azad. 2017. Real-time nonlinear model predictive control of a battery-supercapacitor hybrid energy storage system in electric vehicles. *IEEE Transactions on Vehicular Technology* 66 (11): 9678–9688.
9. Panday, A., and H.O. Bansal. 2016. Energy management strategy for hybrid electric vehicles using genetic algorithm. *Journal of Renewable Sustainable Energy* 8 (1): 646–741.
10. Jalil, N., N.A. Kheir, and M. Salman. 1997. Rule-based energy management strategy for a series hybrid vehicle. In *American Control Conference*.

11. Hemi, H., J. Ghouili, and A. Cheriti. 2014. A real time fuzzy logic power management strategy for a fuel cell vehicle. *Energy Conversion Management* 80 (4): 63–70.
12. Zhang, R., J. Tao, and H. Zhou. 2019. Fuzzy optimal energy management for fuel cell and supercapacitor systems using neural network based driving pattern recognition. *IEEE Transactions on Fuzzy Systems* 26 (4): 1833–1843.
13. Wu, L., et al. 2011. Multiobjective optimization of HEV fuel economy and emissions using the self-adaptive differential evolution algorithm. *IEEE Transactions on Vehicular Technology* 60 (6): 2458–2470.
14. Opila, D.F., et al. 2012. An energy management controller to optimally trade off fuel economy and drivability for hybrid vehicles. *IEEE Transactions on Control Systems Technology* 20 (6): 1490–1505.
15. Glavic, M., R. Fonteneau, and D. Ernst. 2017. Reinforcement learning for electric power system decision and control: past considerations and perspectives. *IFAC-PapersOnLine* 50 (1): 6918–6927.
16. Dayeni, M.K., and M. Soleymani. 2016. Intelligent energy management of a fuel cell vehicle based on traffic condition recognition. *Clean Technologies Environmental Policy* 18 (6): 1–16.
17. Johnson, D.A. and M.M. Trivedi. 2011. Trivedi. Driving style recognition using a smartphone as a sensor platform. In *International IEEE Conference on Intelligent Transportation Systems.* 2011.
18. Stenneth, L., et al. 2011. Transportation mode detection using mobile phones and GIS information. In *Acm Sigspatial International Symposium on Advances in Geographic Information Systems.*
19. Gong, Q., Y. Li, and Z.R. Peng. 2007. Optimal power management of plug-in HEV with intelligent transportation system. In *IEEE/ASME International Conference on Advanced Intelligent Mechatronics.*
20. Liaw, B.Y. 2004. Fuzzy logic based driving pattern recognition for driving cycle analysis. *Journal of Asian Electric Vehicles* 2 (1): 551–556.
21. Wang, J., et al. 2015. Driving cycle recognition neural network algorithm based on the sliding time window for hybrid electric vehicles. *International Journal of Automotive Technology* 16 (4): 685–695.
22. Xing, Z., et al. 2015. Embedded feature-selection support vector machine for driving pattern recognition. *Journal of the Franklin Institute* 352 (2): 669–685.
23. Yuan, Z., et al. 2016. Reinforcement learning-based real-time energy management for a hybrid tracked vehicle. *Applied Energy* 171: 372–382.
24. Qi, X., et al. 2016. Data-driven reinforcement learning-based real-time energy management system for plug-in hybrid electric vehicles. *Journal of the Transportation Research Board* 2572 (1): 1–8.
25. Teng, L., et al. 2017. Reinforcement learning optimized look-ahead energy management of a parallel hybrid electric vehicle. *IEEE/ASME Transactions on Mechatronics PP*(99): 1497–1507.
26. Yue, H., et al. 2018. Energy management strategy for a hybrid electric vehicle based on deep reinforcement learning. *Applied Sciences* 8 (2): 187–198.
27. Volodymyr, M., et al. 2015. Human-level control through deep reinforcement learning. *Nature* 518 (7540): 529.
28. Caux, S., et al. 2010. On-line fuzzy energy management for hybrid fuel cell systems. *International Journal of Hydrogen Energy* 35 (5): 2134–2143.
29. Sutton, R.S., and A.G. Barto. 1998. Reinforcement learning: an introduction. *IEEE Transactions on Neural Networks* 9 (5): 1054.

Printed in the United States
by Baker & Taylor Publisher Services